NUTRITION AND HEALTH SERIES

Adrianne Bendich, PhD, FACN, SERIES EDITOR

For further volumes:
http://www.springer.com/series/7659

Nathan S. Bryan, Ph.D.
Joseph Loscalzo, M.D., Ph.D.
Editors

Nitrite and Nitrate in Human Health and Disease

Humana Press

Editors
Nathan S. Bryan, Ph.D.
Brown Foundation Institute
of Molecular Medicine
The University of Texas
Health Science Center at Houston
Houston, TX
USA
Nathan.Bryan@uth.tmc.edu

Joseph Loscalzo, M.D., Ph.D.
Department of Medicine
Brigham and Women's Hospital
and
Harvard Medical School
Boston, MA
USA
jloscalzo@partners.org

Series Editor
Adrianne Bendich, Ph.D., FACN
President, Consultants in Consumer
Healthcare, LLC
Morristown, NJ
USA
abconsult@verizon.net

ISBN 978-1-60761-615-3 e-ISBN 978-1-60761-616-0
DOI 10.1007/978-1-60761-616-0
Springer New York Dordrecht Heidelberg London

Library of Congress Control Number: 2011925549

Printed on acid-free paper

Humana Press is part of Springer Science+Business Media (www.springer.com)

Series Editor Page

The great success of the Nutrition and Health Series is the result of the consistent overriding mission of providing health professionals with texts that are essential because each includes: 1) a synthesis of the state of the science, 2) timely, in-depth reviews by the leading researchers in their respective fields, 3) extensive, up-to-date fully annotated reference lists, 4) a detailed index, 5) relevant tables and figures, 6) identification of paradigm shifts and the consequences, 7) virtually no overlap of information between chapters, but targeted, inter-chapter referrals, 8) suggestions of areas for future research, and 9) balanced, data-driven answers to patient as well as health professionals questions which are based upon the totality of evidence rather than the findings of any single study.

The Series volumes are not the outcome of a symposium. Rather, each editor has the potential to examine a chosen area with a broad perspective, both in subject matter as well as in the choice of chapter authors. The editor(s), whose training(s) is (are) both research and practice oriented, have the opportunity to develop a primary objective for their book, define the scope and focus, and then invite the leading authorities to be part of their initiative. The authors are encouraged to provide an overview of the field, discuss their own research and relate the research findings to potential human health consequences. Because each book is developed de novo, the chapters are coordinated so that the resulting volume imparts greater knowledge than the sum of the information contained in the individual chapters.

"Nitrite and Nitrate in Human Health and Disease", edited by Nathan S. Bryan, Ph.D. and Joseph Loscalzo, M.D., Ph.D., clearly exemplifies the goals of the Nutrition and Health Series. For many of us who learned about the potential adverse effects of nitric oxide as an environmental air pollutant and the potentially carcinogenic effects of nitrosamines, seeing a volume on nitrate and nitrite may appear at first glance as a toxicology volume. This is not the case. Nitric oxide (NO) still remains an air pollutant; however, within human blood vessels, this short lived gas acts to dilate vessels and lower blood pressure and also maintains the integrity of platelets. This first comprehensive review of the science behind active nitrogen-containing molecules and their effects in humans is of great importance to the nutrition community as well as for health professionals who have to answer client questions about this new area of clinical research.

The two editors are highly qualified in nitrate and nitrite investigations. Dr. Bryan is currently Assistant Professor at the University of Texas Health Sciences Center's Institute of Molecular Medicine in Houston TX. He has been a pioneer in recognizing the salubrious effects of dietary nitrite and nitrate and elucidating the novel metabolism of nitrate and nitrite in humans. He has published extensively in this area. Dr. Bryan is a member of the Society for Free Radical Biology and Medicine, American Physiological Society, Nitric Oxide Society, American Heart Association and the American Association for the Advancement of Science. Joseph Loscalzo, M.D., Ph.D is the Hersey Professor of the Theory and Practice of Medicine at Harvard Medical School and serves as Chairman of the Department of Medicine and Physician-in-Chief at Brigham and Women's Hospital in Boston, Massachusetts. Dr. Loscalzo is recognized as an outstanding cardiovascular scientist, clinician, and teacher. Dr. Loscalzo was one of the early pioneers in the nitric oxide field with many of his seminal publications and discoveries dating from the early 1980s. He was the first to demonstrate the formation of S-nitrosoproteins and to elucidate the antithrombotic and antiplatelet mechanisms of NO, as well as many other contributions to this field. He has received many awards, including the Clinician-Scientist Award, the Distinguished Scientist Award, the Research Achievement Award, and the Paul Dudley White Award from the American Heart Association; a Research Career Development Award, a Specialized Center of Research in Ischemic Heart Disease Award, and a MERIT Award from the National Institutes of Health; the George W. Thorn Award for Excellence in Teaching at Brigham and Women's Hospital, and Educator of the Year Award in Clinical Medicine from Boston University; the Glaxo Cardiovascular Research Award, and the Outstanding Investigator Prize from the International Society for Heart Research; and election to the American Society for Clinical Investigation, the Association of American Physicians, and the Institute of Medicine of the National Academy of Sciences.

"Nitrite and Nitrate in Human Health and Disease" represents the first comprehensive compilation of the newest science of inorganic nitrite and nitrate in human health, and it is to the credit of Drs. Bryan and Loscalzo that they have organized this volume so that it provides an in-depth overview of the natural occurrence and biochemistry of relevant nitrogen-containing molecules; human exposure to nitrate and nitrite; and the latest research on the role of nitrate and nitrite in normal health, cardiovascular disease, and related conditions, as well as their use as therapeutic agents. Of importance, this volume includes an in-depth review in the final section that re-examines the role of nitrosamines (which can be formed from nitrite and nitrate) in human cancer risks.

The volume is organized into three comprehensive sections. The first section includes four chapters that define the biochemistry, metabolism, and physiology of the nitrogen-containing compounds. The first chapters review the major processes involved in the biochemical steps from the capturing of nitrogen from the atmosphere through the generation of nitric oxide (NO), nitrite, and nitrate within the human body. The chapters provide the reader with a basic understanding of the interactions between the inert gas, nitrogen, that comprises 78% of our atmosphere and natural physical events, such as lightning strikes and volcanic eruptions, that result in the formation of active compounds. The most abundant source of active nitrogen compounds result from plant fixation of nitrogen with the help of anerobic microorganisms that are capable of conversion of atmospheric nitrogen into inorganic nitrogen-containing compounds that can be incorporated into essential amino acids and proteins, nucleic acids, and enzymes. In addition to a full description of the traditional nitrogen cycle, there is also a complete and detailed discussion of the novel nitrogen cycle that was first elucidated in the 1980s and resulted in naming NO as the Molecule of the Year in 1992 and the awarding of the Nobel Prize in Medicine to Dr. Louis J. Ignarro, Robert Furchgott, and Ferid Murad. Dr. Ignarro has generously provided a most insightful Foreword for this important volume.

The two final chapters of the first section describe the formation of NO within the human body and its role in delivering oxygen to tissues under very low oxygen pressures. Nitrite binds to hemoglobin, and when it is delivered to ischemic or hypoxic tissue beds, it dilates the arterioles permitting more of the oxygen-carrying hemoglobin to reach these tissues. These highly complex reactions are excellently illustrated in the figures in the chapter. The genetic polymorphisms that are associated with alterations in the enzymes required for conversion of active nitrogen compounds and the effects on delivery of oxygen to tissues are also reviewed.

The next chapter, that follows logically from the previous chapter, discusses the sources of nitrate and nitrite, and reminds us that naturally occurring nitrogen-containing compounds have been used throughout human history to reduce heart pain (angina) as well as to reduce the spoiling of foods (salt peter, as an example). Ground water is a major source of nitrate, as are fruits and vegetables. This chapter also includes an examination of the potential sources of nitrosamines and reviews the environmental conditions that can result in their formation within the body. The take-home message from the introductory chapters and throughout the volume is that the level and duration of exposure to nitrate and nitrite is the keys to whether these molecules are of benefit or may be harmful to health.

The second section contains six chapters on food and environmental exposures to nitrite and nitrate. The chapter on the history of use of nitrate in foods postulates that one potential reason that many of the randomized controlled clinical trials of antioxidants have not shown significant decreases in cardiovascular disease is that there was no inclusion of nitrate or nitrite sources that may have helped optimize vascular function and health. There is an important review of the Mediterranean and DASH diets as these diets contain higher than average intakes of fruits and vegetables that are major sources of nitrate and nitrite. Dr. Willett, in the next chapter, outlines the critical issues involved in capturing the level of exposure to nitrate and nitrite in the human diet and examines the methodologies used in collecting epidemiological data on intakes and biochemical indicators of status. At present there is no validated assessment of nitrate or nitrite content in foods and the lack of this critical tool has resulted in many differing values for exposures. Also, there is no acceptable biomarker of long term intake. The suggestion is to utilize currently validated food frequency questionnaires, estimate nitrate and nitrite levels based upon published studies and measure intakes at multiple time points in prospective cohort studies as this model is the least compromised in terms of recall and other potential biases.

The next chapter reviews the sources of nitrate and nitrite in our food supply and environment. The novel discovery of the oral cavity's role in nitrate balance within the human body is explained with great depth. This chapter provides a clear understanding of the importance of anerobic bacteria that live within the tongue and in our mouth and describes the metabolism of nitrate, its functions within the blood stream, and its elimination. Salivary nitrate and nitrite levels are much higher than those in the blood and represent a recycling of both exogenous and endogenous sources. The potential effects of nitrate and nitrite are modulated by other constituents, such as antioxidant vitamins and flavanols, found in the same food sources. The next chapter reviews the important role of flavanols and polyphenols in affecting the activity of the key enzyme involved in NO generation within blood vessels, endothelial nitric oxide synthase, that results in vascular dilation. The chapter includes an extensive discussion of methodologies used to measure vascular health as well as the biochemistry of NO synthesis. Two unique chapters complete this section of the volume. The first examines the critical role of nitrate and nitrite found in human breast milk. Colostrum contains higher concentrations than later milk and may be involved in protecting the neonate from infections when the immune system is not yet fully developed. The concentrations of nitrate and nitrite in human milk are compared to those found in infant formula, cow's milk and soy milk formulas, all of which have lower concentrations than seen in human breast milk. FDA regulations of infant formula levels are reviewed and this information complements the next chapter that describes the numerous regulations in the US and globally of nitrate/nitrite levels in ground water and processed meats. Regulatory bodies include the EPA, FDA, USDA, and the WHO. The regulations include the addition of antioxidants to processed meats to reduce potential nitrosamine formation. Often, the concentrations that represent the upper levels of intake by the regulatory bodies are significantly lower than the concentrations found naturally in certain foods such as spinach. Several authors suggest that regulations be re-examined based upon the new data on the importance of nitrate and nitrite for cardiovascular health and other health benefits. Authors and the editors agree that using a process such as that used by the US Institute of Medicine to determine intake levels for essential nutrients, may be of value in determining safe and effective levels of intake for nitrate and nitrite.

The third section reviews the newest data on nitrate and nitrite use in therapeutics and disease conditions and includes seven chapters. In order to develop therapies that utilize the synthesis of NO, there must be a clear understanding of the enzymatic reactions that naturally result in NO formation, the dietary sources of precursor molecules, and the endogenous and environmental factors that modulate the formation as well as destruction of these biologically active nitrogen-containing species. The next chapters review these reactions and provide relevance to the studies of L-arginine as a precursor and the development of NO forming drugs used to treat angina and erectile dysfunction by dilating relevant blood vessels. Inhaled NO is currently being used in serious therapeutic cases to increase pulmonary vasodilation. Owing to the potential for NO to combine with oxygen and create highly reactive molecules that can damage the lung and other tissues, the control of the concentration of NO that is utilized is critical. This is especially true when NO is used in the treatment of premature infants who lack maturity in their lungs and may develop bronchopulmonary dysplasia. Inhaled NO is also used to treat acute respiratory distress syndrome and acute lung injury. There are detailed descriptions of the animal models that have been developed to better understand the value and limits of NO inhalation in severe lung diseases. The medical conditions associated with the use of nitrovasodilators are described in detail in the next chapter. Nitrovasodilators are used in the treatment of stable and unstable angina, congestive heart failure and myocardial infarction. The complex pharmacology and metabolism of the drugs are reviewed. As an example, cyanide radical is released when sodium nitroprusside is metabolized; control of exposure to this toxic molecule is obviously essential.

The next three chapters examine the potential for nitrate and nitrite to serve as therapeutic agents under different disease conditions. The chapter on ischemia-reperfusion (IR) injury describes the effects of IR injury on the heart, lungs, liver, kidney and brain. Active nitrogen-containing molecules can blunt the formation of reactive oxygen radicals known to be the major causes of IR injury. Other conditions that are also prone to IR cellular damage include sickle cell anemia and peripheral artery disease and low dose nitrite therapies are being explored as therapies in these and other relevant conditions. Traditional Chinese Medicines (TCM) that have been used for thousands of years for cardiovascular indications are often composed of the roots, stems and leaves of plants. The plants have been recently analyzed for their nitrate, nitrite contents, the values are tabulated in the next chapter, and a full discussion of the combinations of herbs and other plants are described in detail. As indicated earlier, the question of the association of nitrate and nitrite with cancer risk is examined in great detail in the next chapter. It appears that there are inconsistent data from human epidemiological studies. The literature consistently reports a decreased risk of many cancers, including lung, breast, and colorectal in populations with the highest intakes of fruits and vegetables, especially green and yellow vegetables that have high concentrations of nitrate and nitrite. The literature also reports a higher risk of stomach cancer in individuals with the highest intakes of pickled, processed, and cured meats that are also concentrated sources of nitrate and nitrite. The laboratory animal studies may not be relevant to human intake levels as animals are often exposed to very high intakes of nitrate and nitrite. One argument that is raised is that stomach cancer has significantly decreased over the past decades and this may be linked to the addition of antioxidants to cured meats. The critical issue, discussed in the final chapter written by the volume's editors, is the perception of the negative effects of nitrate and nitrite in the minds of many health professionals based upon the early findings of increased cancers in animals exposed to very high intakes of nitrate and nitrite and the well recognized fact that nitrosamines are mutagenic and may be carcinogenic under the appropriate conditions. This volume presents compelling data to support a paradigm shift in the understanding that nitrate and nitrite can be of great benefit to health if consumed at the right levels and if used therapeutically with the proper precautions.

The logical sequence of chapters enhances the understanding of the latest information on the current standards of practice for clinicians, related health professionals including the dietician, nurse, pharmacist, physical therapist, behaviorist, psychologist, and others involved in the team effort required for successful treatment of cardiovascular and other relevant diseases, as well as conditions that adversely affect normal metabolic processes. This comprehensive volume also has great value for academicians involved in the education of graduate students and post-doctoral fellows, medical students, and allied health professionals who plan to interact with patients with disorders that can be beneficially affected by the addition of nitrate and nitrite to the diet or by use of pharmacological agents that utilize active nitrogen-containing drugs.

Cutting edge discussions of the roles of signaling molecules, growth factors, hormones, cellular and nuclear receptors, and all of the cells directly involved in NO metabolism are included in well-organized chapters that put the molecular aspects into clinical perspective. Of great importance, the editors have provided chapters that balance the most technical information with discussions of its importance for clients and patients as well as graduate and medical students, health professionals and academicians.

The volume contains over 40 detailed tables and figures that assist the reader in comprehending the complexities of the nitrogen cycles and the metabolism of nitrate and nitrite within the body. The over-riding goal of this volume is to provide the health professional with balanced documentation and awareness of the newest research and therapeutic approaches including an appreciation of the complexity of this relatively new field of investigation. Hallmarks of the 17 chapters include, key words, and bulleted key points at the beginning of each chapter, complete definitions of terms with the abbreviations fully defined for the reader and consistent use of terms between chapters. There are over 1,400 up-to-date references; all chapters include a conclusion to highlight major findings. The volume also contains a highly annotated index.

This unique text provides practical, data-driven resources based upon the totality of the evidence to help the reader understand the basics, treatments, and preventive strategies that are involved in the understanding of the role of a critical gas, NO, in the control of blood flow through the body as well as the modulation of platelet function. The overarching goal of the editors is to provide fully referenced information to health professionals so they may have a balanced perspective on the value of various preventive and treatment options that are available today as well as in the foreseeable future.

In conclusion, "Nitrite and Nitrate in Human Health and Disease," edited by Nathan S. Bryan, Ph.D., and Joseph Loscalzo, M.D., Ph.D., provides health professionals in many areas of research and practice with the most up-to-date, well referenced and comprehensive volume on the current state of the science and medical uses of active nitrogen-containing molecules. This volume will serve the reader as the most authoritative resource in the field to date and is a very welcome addition to the Nutrition and Health Series.

Adrianne Bendich, Ph.D, FACN
Series Editor

Foreword

The short-lived, free radical molecule nitric oxide (NO) has emerged as one of the most versatile cell signaling transmitters produced by mammalian biological systems. NO, identified as 'endothelium-derived relaxing factor' and proclaimed 'Molecule of the Year' in 1992, functions critically in physiology, neuroscience, and immunology. The vascular effects of NO alone include vasodilatation, inhibition of platelet aggregation and leukocyte adhesion to the endothelium, scavenging of superoxide anions, and inhibition of smooth muscle cell hyperplasia. Early studies on NO stemmed from work with nitroglycerin in an attempt to elucidate the mechanism through which it relieved pain due to angina pectoris. It was discovered that the formation of NO from nitroglycerin accounts for its therapeutic efficacy for angina by dilating constricted and diseased blood vessels in the heart. Not surprisingly, some of the most prevalent diseases result, at least in part, from decreased NO availability, for example, hypertension, atherosclerosis, diabetes mellitus, and hypercholesterolemia.

The discovery of the formation of NO from the semi-essential amino acid L-arginine through one of three isoforms of nitric oxide synthase provided a key therapeutic target, which is still the focus of much research today. Dietary supplementation of L-arginine has been shown to enhance NO production in healthy individuals (despite already saturated extracellular concentrations), and this may provide both cardiovascular protective effects and enhance athletic performance. Indeed, endothelial dysfunction, an early sign of cardiovascular disease, has been reversed through enhanced NO production. This observation leads us to believe that intervention through the NO-pathway is a viable route for treatment and prevention of vascular dysfunction.

Recently, the oxidative 'waste' products of nitric oxide, nitrite and nitrate, have been evaluated in a new context, due to their ability to form NO independent of nitric oxide synthase enzymes, through reductive electron exchanges. Since nitrate (as well as nitrite) are primarily ingested in the form of fruits and vegetables, which have been known for some time to protect against diseases from atherosclerosis to cancer, a new paradigm has emerged regarding the role of these once feared nitrogen oxides. Both public and scientific perception of nitrite and nitrate still revolve around fears of nitrosamine formation and carcinogenesis. What has not been considered, however, is the fact that consumption of antioxidants with nitrite and nitrite (both significant components of fruits and vegetables) inhibits the formation of nitrosamines in the gastric milieu. Furthermore, a human nitrogen cycle consisting of commensal bacteria in the oral cavity, which serve a reductive role in the conversion of approximately 20% of ingested nitrate to nitrite, now appears to provide a significant NOS-independent source of NO generation.

This body of work may have revolutionary implications in terms of developing strategies to combat heart disease and many other contemporary diseases associated with a NO deficiency. Furthermore we may finally have an explanation for the many known and undisputed benefits of the Mediterranean diet. Perhaps now we should consider nitrite and nitrate as the bioactive food components that account for the protective benefits of certain foods and diets. Numerous clinical trials of supplementation with various antioxidants borrowed from heart-healthy diets, such as those typical of Mediterranean countries, have consistently failed to replicate the protective effects of the foods themselves. Consistently absent, but the primary human source, is dietary nitrate and nitrite. Recent work has shown various cardioprotective effects from modest supplementation of nitrite and nitrate. Nitrite, in particular, has been shown to prevent hypercholesterolemic microvascular inflammation and protect against injury from ischemic events.

The broader context of research regarding nitrate, nitrite, and nitric oxide suggests these simple nitrogen oxides serve as a critical dietary component for protection against various chronic diseases. Currently, heart disease and cancer lead the nation in cause of deaths. Concurrently, the dietary patterns of the West have transitioned towards heavily processed foods and lack significant quantities of fruits and vegetables. The explanations have been varied but overlook simple molecules known to play critical roles in multiple organ systems through the chemical messenger NO. The dietary contributions to normal NO homeostasis would not only help explain significantly lower rates of cardiovascular disease in those who regularly consume fruits and vegetables, but also arm scientists and physicians with a relatively simple and inexpensive therapeutic intervention.

This text effectively overviews the important role nitrite and nitrate play in biological systems and NO homeostasis. A risk benefit analysis has shown nitrite and nitrate present no danger when consumed in modest quantities and preferably with antioxidants. In fact, research appears to suggest nitrite acts as a redundant NO reservoir when NOS activity is insufficient or stress requires a secondary source. The future use of nitrite/nitrate in dietary considerations will likely have a significant impact on current public health policy. This book brings the NO-story full circle and presents novel thought on the future treatment for many of the country's most pressing health issues. This is a relatively new area of nitric oxide research but a very exciting one. The L-arginine pathway for NO synthesis may turn out to be only part of the story. The symbiosis between humans and the bacteria that reside in and on our body may be just as important in terms of utilizing nitrate and nitrite to make NO under conditions when the oxidation of L-arginine is dysfunctional. Drs. Nathan S. Bryan and Joseph Loscalzo have assembled the world's experts to present a first of its kind, comprehensive work on nitrite and nitrate in human health and disease, carefully examining the context for a risk benefit assessment.

Louis J. Ignarro, Ph.D.
1998 Nobel Laureate in Medicine

Preface

Our major objective and driving force in developing *Nitrite and Nitrate in Human Health and Disease* is to consolidate all the key research and knowledge in one volume in order to establish a framework based on the totality of evidence for nitrite and nitrate and the effects of these two anions on human health and disease. Although the biomedical science community is excited and optimistic about the potential for developing new therapeutics and perhaps regimens of disease prevention based on their ability to generate nitric oxide under appropriate conditions, epidemiologists, nutritionists, and cancer biologists have cause for concern due to the inherent nitrosative chemistry of nitrite and NO that could form potentially carcinogenic N-nitrosamines. This is the first book in the Springer/Humana Nutrition Series dedicated to understanding the nutritional aspects of nitrite and nitrate for human health. It is our intent to deliver a comprehensive review of nitrite and nitrate, from basic biochemistry to the complex physiology and metabolism of these two naturally occurring molecules in the human body.

Overall, the book contains well-organized and well referenced chapters by respected scientists and physicians that covers the rich history of nitrite and nitrate, sources of exposure and physiological effects when consumed through foods containing nitrite and nitrate. The first portion of the book describes the biochemistry, metabolism, and physiology of nitrite and nitrate, how these molecules get incorporated into the foods we eat, and how they are systematically metabolized to bioactive nitric oxide. This involves the environmental processes of nitrogen fixation and the presence of a human nitrogen cycle involving symbiotic bacteria that reside in and on the human body. The book then shifts focus to the sources of exposure to nitrite and nitrate, both environmental and dietary, as a means to quantify exposure estimates and what this may mean for human health. We discuss the epidemiology and dietary effects on the nitric oxide pathway. This portion of the book also examines systems in nature in which this pathway is exploited, including the breast milk of nursing mothers. Finally, the last section of the book discusses nitric oxide-based therapeutics and how nitrite and nitrate biochemistry can be harnessed safely and efficaciously to improve human health through the production of nitric oxide. We end with a summary of the collective body of knowledge presented in the book and what we might expect going forward. Each chapter begins with Key Points that outline the concepts presented to assist the reader in understanding the fundamental principles presented.

It is undisputed in the biomedical community that NO is one of the most important molecules the human body produces. If NO is such an important molecule in practically every organ system in our body, why, then, is there only a singular pathway for its production, i.e., the complex oxidation of L-arginine? Phosphorylation is another fundamental cellular process that is just as important in cell signaling, with over 500 recognized kinases and many phosphatases to regulate this biochemical process; by contrast, there is only one class of enzymes, the nitric oxide synthases, to produce NO. Most physiological systems are rich in redundancy, allowing backup systems to support the primary system. The provision of nitrate and nitrite as sources of NO may then be viewed as a system of redundancy. After all, a one-electron reduction is energetically and kinetically favorable to a five-electron oxidation.

According to the World Health Organization, cardiovascular disease is the number one killer of both men and women in the United States. These deaths represent a staggering 40% of all deaths. Close to 1 million people die each year and more than 6 million are hospitalized due to cardiovascular disease. Therefore, developing new strategies to correct NO insufficiency and replete NO availability is of paramount importance and could potentially save millions of lives worldwide and lessen the burden on the health care system. We now appreciate that reduced or insufficient NO production or activity is a hallmark of a number of disorders, including many complex, chronic cardiovascular diseases and even Alzheimer's diseases. Therefore, developing new strategies to restore and replete bioactive NO is of paramount importance and could potentially lessen the burden of disease for society. Thus, understanding the biological activity of nitrite and nitrate may not only lead to novel treatments for disease but may lead to strategies to prevent disease development or progression and even the physiological basis for the benefits of certain diets such as the Mediterranean diet. To achieve this laudable goal, we must first establish the context for potential benefit while preventing unwanted risks or harm. We hope the information provided in this text will begin to help define that context, be a source of valuable information, and be useful for anyone who wants the most important and updated information about nitrate and nitrite.

We have invited the world's leading experts to share their research and perspectives which we hope will help define the context for benefits vs. any potential risks associated with nitrite and nitrate, either through dietary ingestion or therapeutic dosing. This diverse collection of authors includes muscle biologists, physiologists, physicians, epidemiologist, cancer biologist, registered dietician, chemist, and public health experts from five countries around the world in both academia and government. This approach provides a fair and balanced view of nitric oxide biochemistry, nitrite and nitrate biochemistry in physiology and in the food sciences. As a result, we are indebted to these many individuals. We realize the time and dedication it takes to compose a book chapter on the latest body of knowledge, and, thus, appreciate these authors' taking time to help up develop this volume. We also sincerely thank Springer-Humana Press for including this body of work in the Nutrition and Health Series with Dr. Adrianne Bendich as Series Editor. This is an exciting time in NO and nitrite-based research. This has been – and we predict will continue to be – an area of intense research in the future. It is our hope that the information contained herein will educate and inform scientists, physicians, health care professionals, nutritionists, dieticians, and even the general public on the effects of nitrite and nitrate in human health and disease.

Houston, TX Nathan S. Bryan Ph.D.
Boston, MA Joseph Loscalzo M.D., Ph.D.

Contents

Contributors

Nigel Benjamin, M.D.
Peninsula Medical School, St. Luke's Campus, Exeter, UK
Nigel.Benjamin@phnt.swest.nhs.uk

Pamela D. Berens, M.D.
Department of Obstetrics, Gynecology and Reproductive Sciences,
The University at Houston, Houston, TX, USA

Kenneth D. Bloch, M.D.
Anesthesia Center for Critical Care Research, Department of Anesthesia, Critical Care, and Pain Medicine, Cardiovascular Research Center, Massachusetts General Hospital, Harvard Medical School, Boston, MA 02114, USA
kdbloch@partners.org

Nathan S. Bryan, Ph.D.
Brown Foundation Institute of Molecular Medicine, The University of Texas Health science Center at Houston,
Houston, TX, USA
Nathan.Bryan@uth.tmc.edu

John P. Cooke, M.D., Ph.D.
Department of Medicine, Stanford School of Medicine,
Cardiovascular Medicine Clinic,
Stanford, CA, USA
John.Cooke@stanford.edu

Ho-Leung Fung, Ph.D.
Department of Pharmaceutical Sciences, School of Pharmacy and Pharmaceutical Sciences, The University of Buffalo,
Buffalo, NY, USA
hlfung@buffalo.edu

Yong-Jian Geng, M.D., Ph.D.
Cardiovascular Medicine, The University of Texas School of Medicine at Houston, Houston, TX, USA
Yong-Jian.Geng@uth.tmc.edu

Mark Gilchrist, Ph.D.
Peninsula Medical School, St. Luke's Campus, Exeter, UK

Mark T. Gladwin, M.D.
Department of Medicine, University of Pittsburgh School of Medicine, Pittsburgh, PA, USA
gladwinmt@upmc.edu

Susheel Gundewar, M.D.
Department of Surgery, Emory University School of Medicine, Atlanta, GA, USA

Norman G. Hord, Ph.D., M.P.H., R.D.
Department of Food Science and Human Nutrition, Michigan State University, East Lansing, MI, USA
hord@msu.edu

Louis J. Ignarro, Ph.D.
Department of Molecular and Medical Pharmacology, University of California Los Angeles, Los Angeles, CA, USA

Jimmy T. Keeton, Ph.D.
Department of Nutrition and Food Science, Texas A&M University, *Office:* Kleberg Center
jkeeton@tamu.edu

Malte Kelm, M.D., Ph.D.
Department of Medicine, The University of Duesseldorf, Düesseldorf, Germany;
Klinik für Kardiologie, Pneumologie, Angiologie, Moorenstrasse 5, Düesseldorf, Germany

David M. Klurfeld, Ph.D.
United States Department of Agriculture, Agricultural Research Service, Beltsville, MD, USA
david.klurfeld@ars.usda.gov

Martin Lajous, M.D.
Department of Epidemiology, Harvard School of Public Health, 665 Huntington Avenue, Boston, MA 02115, USA and
Centro de Investigación en Salud Poblacional, Instituto Nacional de Salud Pública, Av. Universidad # 655, Col. Santa Maria Ahuacatitlán, Cuernavaca Morelos 62508, México
mlajous@hsph.harvard.edu; mlajous@insp.mx

Jack R. Lancaster Jr. Ph.D.
Department of Anesthesiology, The University of Alabama at Birmingham, Birmingham, AL, USA
doctorno@uab.edu

Madhav Lavu, M.D.
Department of Surgery, Emory University School of Medicine,
Atlanta, GA, USA

David J. Lefer, Ph.D.
Division of Cardiothoracic Surgery, Department of Surgery,
Emory University School of Medicine, Atlanta, GA, USA
dlefer@emory.edu

Lisa Lohmeyer
Department of Medicine, Harvard Medical School, Boston, MA, USA

Joseph Loscalzo, M.D., Ph.D.
Department of Medicine, Brigham and Women's Hospital, and Harvard Medical School Boston, MA, USA
jloscalzo@partners.org

Jon O. Lundberg, M.D., Ph.D.
Institute of Molecular Medicine, University of Texas Health Science Center at Houston, Pressler Street 1825, 77030, Houston, TX, USA
jon.lundberg@ki.se

Rajeev Malhotra, M.D.
Department of Medicine, Massachusetts General Hospital,
and
Harvard Medical School, Boston, MA, USA
rmalhotra@partners.org

Andrew L. Milkowski, Ph.D.
Department of Animal Science, University of Wisconsin,
Madison, WI, USA
amilkowski@ansci.wisc.edu

Nathaniel Page, Pharm D.
Department of Pharmaceutical Sciences, School of Pharmacy
and Pharmaceutical Sciences, The University of Buffalo,
Buffalo, NY, USA

Tienush Rassaf, M.D., Ph.D.
Department of Medicine, The University of Duesseldorf,
Duesseldorf, Germany
Klinik für Kardiologie, Pneumologie, Angiologie, Moorenstrasse 5,
Düsseldorf, Germany
Tienush.Rassaf@med.uni-duesseldorf.de

Sruti Shiva, Ph.D.
Department of Medicine, University of Pittsburgh School of Medicine,
Pittsburgh, PA, USA

Andrea Steinbicker, M.D.
Department of Medicine, Massachusetts General Hospital, and Harvard Medical School, Boston, MA, USA
ASteinbicker@partners.org

Matthias Totzeck, M.D., Ph.D.
Department of Medicine, The University of Duesseldorf, Duesseldorf, Germany;
Klinik für Kardiologie, Pneumologie, Angiologie, Moorenstrasse 5, Düesseldorf, Germany

Eddie Weitzberg, M.D.
Department of Physiology and Pharmacology, Karolinska Institute, Stockholm, Sweden
eddie.weitzberg@ki.se

Walter Willett, M.D., M.P.H.
Department of Epidemiology, Harvard School of Public Health, Boston, MA, USA;
Department of Nutrition, Fredrick John Stare Professor of Epidemiology and Nutrition, Boston, MA, USA
wwillett@hsph.harvard.edu

Wing Tak Wong, Ph.D.
Department of Medicine, Stanford School of Medicine, Cardiovascular Medicine Clinic, Stanford, CA, USA

Part I
Basic Biochemistry, Metabolism, and Physiology

Chapter 1
Introduction

Nathan S. Bryan and Joseph Loscalzo

The discovery of the nitric oxide (NO) pathway in the 1980s represented a critical advance in the understanding of cell signaling and subsequently into major new advancements in many clinical areas including, but not limited to cardiovascular medicine. This seminal finding was viewed as so fundamentally important that the Nobel Prize in Physiology or Medicine was awarded to its discoverers, Drs. Louis J. Ignarro, Robert Furchgott, and Ferid Murad in 1998, a short 11 years after NO was identified. The Swedish Nobel Assembly sagely noted, "The signal transmission by a gas that is produced by one cell, penetrates through membranes and regulates the function of another cell, represents an entirely new principle for signaling in biological systems." It was shocking to realize that NO, a colorless, odorless gas, was able to perform such important biochemical functions. Dr. Valentin Fuster, then president of the American Heart Association, noted in a 1998 interview that "the discovery of NO and its function is one of the most important in the history of cardiovascular medicine." More than a decade after the Nobel Prize was awarded for the discovery of NO and after more than 100,000 scientific papers have been published on it, we still don't have a firm grasp on its production and regulation or understand all of its biological functions.

Although one of the simplest biological molecules in nature, NO has found its way into nearly every phase of biology and medicine, ranging from its role as a critical endogenous regulator of blood flow and thrombosis to a principal neurotransmitter-mediating penile erectile function, as well as a major pathophysiological mediator of inflammation and host defense. Continuous generation of NO is essential for the integrity of the cardiovascular system, and decreased production and/or bioavailability of NO is central to the development of many cardiovascular disorders. Research has since shown that NO is just as important in other organ systems. NO is important for communication in the nervous system,

N.S. Bryan (✉)
Brown Foundation Institute of Molecular Medicine,
The University of Texas, Health Science Center at Houston,
1825 Pressler Street 530C, Houston, TX 77030, USA
e-mail: Nathan.Bryan@uth.tmc.edu

N.S. Bryan and J. Loscalzo (eds.), *Nitrite and Nitrate in Human Health and Disease*,
Nutrition and Health, DOI 10.1007/978-1-60761-616-0_1,

and is a critical molecule used by the immune system to kill invading pathogens, including bacteria and cancer cells. The production of NO from L-arginine is a complex and complicated biochemical process involving a 5-electron oxidation with many cofactors and prosthetic groups. As a result, there are many steps and/or factors that may be altered and affect ultimate NO production. Once produced, NO can be quickly scavenged before it has a chance to perform its actions. It is, therefore, the generation of NO reflects the results of a war of attrition. It is truly a remarkable feat that this short-lived gas is responsible for so many essential cellular activities.

The major metabolic pathway for NO is the stepwise oxidation to nitrite and nitrate. In plasma NO is oxidized almost completely to nitrite or nitrate. The half-life of nitrite in human blood is about 110 s. Nitrate on the other hand has a circulating half life of 5–8 h. During fasting conditions with a low previous intake of nitrite/nitrate, enzymatic NO formation accounts for the majority of steady-state blood levels of nitrite. On the basis of these observations, it was believed that bioactive NO was acutely terminated by oxidation to nitrite and nitrate. Consistent with this view, for the past 20 years, we have used plasma nitrite as a biomarker to reflect acute changes in endogenous NO production. Nitrate on the other hand is somewhat sensitive to chronic changes in endogenous NO production; however, dietary factors often confound the use of nitrate as a biomarker of NO activity. Research over the past 10–15 years has now questioned this NO inactivation cascade. A new paradigm has emerged whereby nitrate and nitrite, instead of being inert oxidation production of NO, can be recycled to reform NO under certain conditions. In 1994, two groups independently presented evidence for the generation of NO in the stomach resulting from the acidic reduction of inorganic nitrite. Benjamin and colleagues demonstrated that the antibacterial effects of acid alone were markedly enhanced by addition of nitrite, which is present in saliva, whereas Lundberg and colleagues measured high levels of NO in expelled air from the stomach in humans. These levels were abolished after pretreatment with a proton pump inhibitor and markedly increased after ingestion of nitrate, showing the importance of both luminal pH and the conversion of nitrate to nitrite for stomach NO generation. These were the first reports of NO synthase-independent formation of NO in vivo. In the classical NO synthase pathway NO is formed by oxidation of the guanidino nitrogen of L-arginine with molecular oxygen as the electron acceptor. This complex reaction is catalyzed by specific heme-containing enzymes, the NO synthases, and the reaction requires several cofactors. A very simplified schematic of the L-arginine-NO pathway is illustrated in Fig. 1.1. The alternative pathway was fundamentally different; instead of L-arginine, it used the simple inorganic anions nitrate (NO_3^-) and nitrite (NO_2^-) as substrates in a stepwise reduction process that did not require NO synthase or cofactors.

The biochemical pathway and biological effects of nitrate reduction to nitrite and further on to NO in the gastrointestinal tract is now better understood. Oral commensal bacteria are essential for the first step in the nitrate–nitrite–NO pathway since they are responsible for the enzymatic reduction of the higher nitrogen oxide nitrate to form nitrite. It was known from the literature that the salivary glands extract nitrate from plasma, but the reason for this active process was not explained.

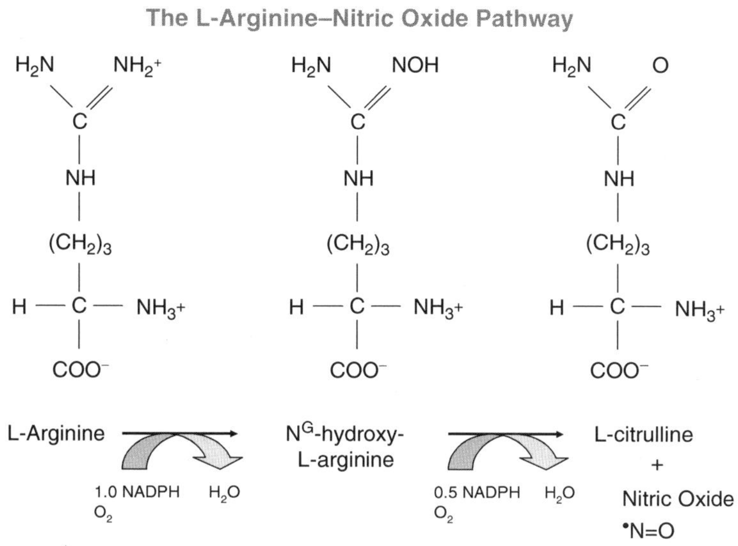

Fig. 1.1 The L-arginine-nitric oxide pathway

This active process leads to levels of salivary nitrate that are 10–20-fold higher than in plasma. Oral facultative anaerobic bacteria, residing mainly in the crypts of the tongue, then reduce nitrate to nitrite by the action of nitrate reductase enzymes. This relatively effective bacterial nitrate reduction results in salivary levels of nitrite that are 1,000-fold higher than those found in plasma. When nitrite-rich saliva meets the acidic gastric juice, nitrite is protonated to form nitrous acid (HNO_2), which then decomposes to NO and other nitrogen oxides. It is now established that oral commensal bacteria are pivotal in gastric NO formation. The presence of antioxidants can facilitate the reduction of nitrite to NO. This pathway becomes extremely important when you consider that nitrite and nitrate are part of our diet, which means that there is now a recognized pathway to affect endogenous NO production based on which foods we eat. The nitrate–nitrite–NO pathway is illustrated in Fig. 1.2.

This emerging paradigm of the benefits of dietary sources of nitrite and nitrate may come as a surprise to many since historically nitrite and nitrate have been viewed as adversely affecting our food supply and drinking water. The early implications of dietary nitrite and nitrate causing methemoglobinemia and their propensity to form potentially carcinogenic N-nitrosamines have led to government efforts to quantify exposure rates in humans and establish regulations on the amounts in our food supply and drinking water. Since the early 1970s there have been both government and academic research efforts to understand the toxicology of nitrite, nitrate, and N-nitrosamines. That collective body of science is presented in several of the chapters with the hope that the reader can distill the burden of evidence on the risks and benefits of exposure to sources of nitrite and nitrate.

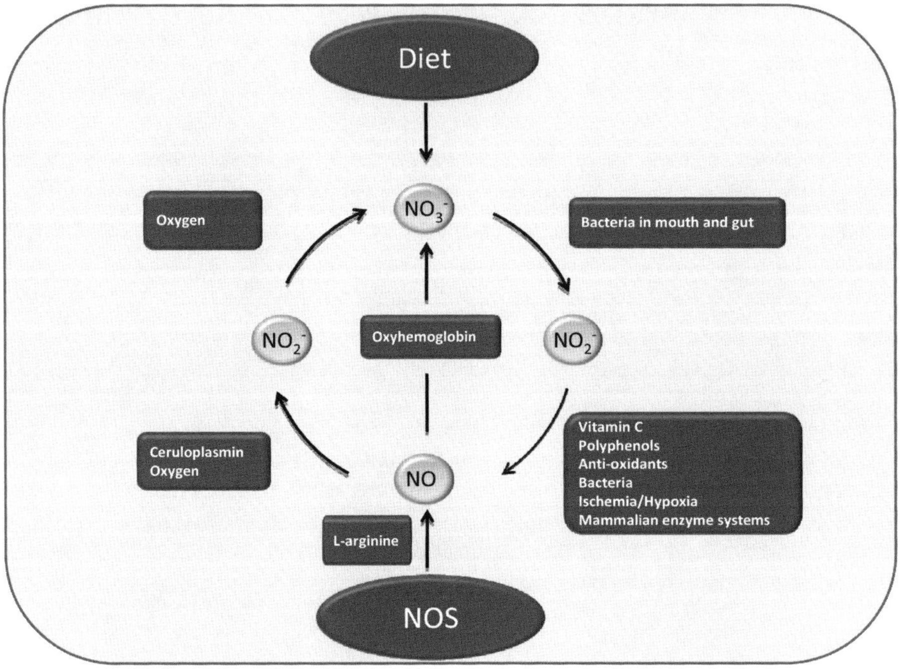

Fig. 1.2 The nitrite–nitrate–NO pathway

As we move forward with paradigm shifts and medical discoveries, the scientific community's main objective is to understand mechanisms of disease development to the extent needed to design rational therapies but with the ultimate goal of developing strategies for the *prevention* of human diseases. One could make a strong argument that diet should be a first target for disease prevention. Very little can affect our health more than what we choose to eat and our daily lifestyle habits. The realization of a nitrate–nitrite–NO pathway suggests that NO can be modulated by the diet independent of its enzymatic synthesis from L-arginine, e.g., the consumption of nitrite- and nitrate-rich foods, such as fruits, leafy vegetables, and some meats along with antioxidants. Antioxidants, such as vitamin C and polyphenols, can positively affect NO production from both pathways. First, they help protect the essential cofactors for the NOS pathway, such as tetrahydrobiopterin, from becoming oxidized and, therefore, promote L-arginine conversion to NO. Second, the presence of antioxidants protects NO from being scavenged once it is produced. Third, vitamin C and polyphenol can facilitate the reduction of nitrite to NO in the presence of an electron acceptor, thereby providing a recycling pathway. Lastly, vitamin C and polyphenol are very potent inhibitors of nitrosation reactions. Collectively, antioxidants promote beneficial effects of nitrate and nitrite reduction to NO and prevent any unwanted nitrosation reactions.

Regular intake of nitrate- and nitrite-containing foods may ensure that blood and tissue levels of nitrite and NO pools are maintained at a level sufficient to compensate for any disturbances in endogenous NO synthesis. Since low levels of supplemental nitrite and nitrate have been shown to enhance blood flow, dietary sources of NO metabolites can, therefore, improve blood flow and oxygen delivery, and protect against various cardiovascular disease states or any condition associated with NO insufficiency. This view has led to the current belief that a healthy diet focus not only on reducing fat and caloric intake, but on adding foodstuffs promoting NO bioactivity, which can include foods enriched in L-arginine and antioxidants to promote NO production and availability. This pathway is also the current focus of a number of biotechnology and pharmaceutical companies in their attempts to develop NO and nitrite-based therapeutics.

In 2006, total healthcare expenditures in the U.S. exceeded $2 trillion, or $6,700 per person. This trend is expected to increase over the coming years, reaching $4 trillion in 2015. Currently, costs associated with chronic diseases such as obesity, diabetes, hypertension, and coronary artery disease account for 75% of the nation's annual healthcare costs. According to the American Heart Association, an estimated 81 million people had one or more forms of cardiovascular disease in the U.S. in 2006, including hypertension, coronary artery disease, myocardial infarction, angina pectoris, stroke, and heart failure. Most, if not all, of the chronic conditions mentioned above are the result of a dysfunctional endothelium and inability to produce NO and/or maintain NO homeostasis and signaling. Understanding and developing new strategies to restore NO homeostasis will have a profound impact on public health and on the healthcare system. Defining the context for the role of nitrite and nitrate in human health and disease is an essential first step in the process.

Chapter 2
From Atmospheric Nitrogen to Bioactive Nitrogen Oxides

Mark Gilchrist and Nigel Benjamin

Key Points

- Nitrogen is the most abundant element in the atmosphere.
- N_2 chemically inert.
- Conversion to biologically active, "fixed," nitrogen requires considerable energy input.
- Nitrogen, inorganic nitrogen oxides, and organic nitrogen participate in complex biological cycle.
- The individual steps in the cycle may be used for respiration, anabolic processes, detoxification, or host defense.
- Plants and animals often have symbiotic relationships with specific micro-organisms which catalyze the parts of the cycle which they are unable to do.
- Mammals, including humans, have a complex nitrogen cycle which is, as yet, only partly understood.
- An understanding of the handling of nitrogen by micro-organisms may broaden our understanding of the role of nitrate, nitrite, and nitric oxide in health and disease.

Keywords Nitrogen fixation • Bacteria • Symbiosis • Nitrate reductase • Fertilizers • Denitrification

---

N. Benjamin (✉)
Peninsula Medical School, St. Luke's Campus, Exeter, UK
e-mail: Nigel.Benjamin@phnt.swest.nhs.uk

N.S. Bryan and J. Loscalzo (eds.), *Nitrite and Nitrate in Human Health and Disease*, Nutrition and Health, DOI 10.1007/978-1-60761-616-0_2,

Introduction

Nitrogen atoms are a constituent part of a vast array of biologically important chemicals from the complex, such as proteins and nucleic acids, to the apparently more simple dimolecular nitric oxide. Nitrogen (N_2) gas makes up approximately 78% of the atmosphere. It is highly inert, being bound together by a triple bond that requires considerable energy to break (bond energy 940 kJ mol^{-1}). Higher organisms lack the apparatus to do so. The conversion of nitrogen from inert N_2 to fixed, and, thus, biologically available nitrogen is essential to life on earth and requires dedicated multi-step enzymatic pathways; vast amounts of energy in the case of lightning; or a combination of heat, pressure, and metallic catalysts in the Haber–Bosch process. The overview of the biological nitrogen cycle is illustrated in Fig. 2.1.

Theories exist that life on earth in its earliest and most primitive stages, when the gaseous constitution of the atmosphere was very different from that found today, used synthesized NO to detoxify ozone or oxygen [1] or, indeed, developed NO reductases to remove NO from the cell. Whatever the reason, the chemical properties of NO, NO_2^-, and NO_3^- played a key role in the lives of the most primitive organisms. It is striking that as evolution progressed, higher organisms became utterly dependent on the nitrogen fixing and cycling abilities of their more primitive antecedents. Indeed, the ability to process nitrogen is no less important to familiar plants such as clover and their symbionts to extremophile bacteria.

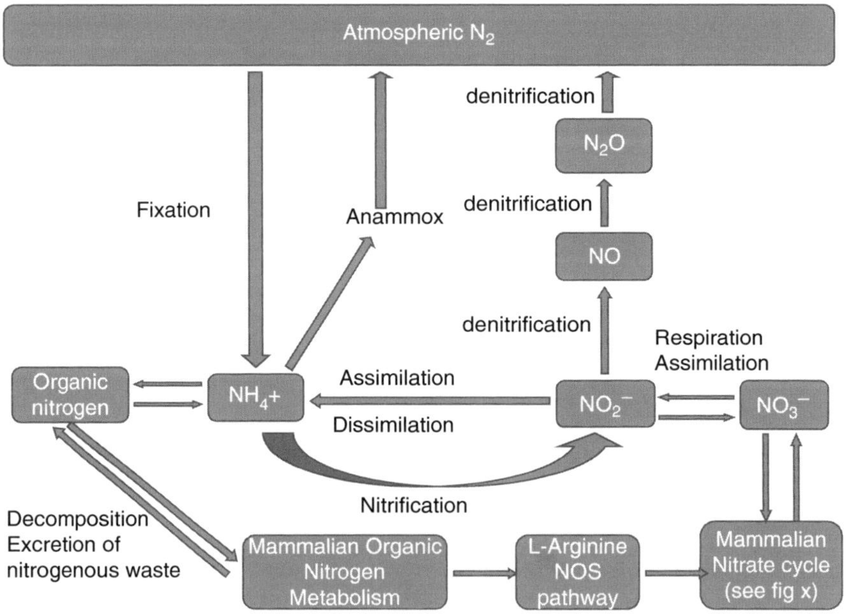

Fig. 2.1 Overview of the biological nitrogen cycle. For details see text

The surge in knowledge and interest in the microbial/environmental nitrogen cycle is mirrored by the burgeoning interest in the mammalian nitrate cycle and the role of its multifarious reactive nitrogen oxide intermediates in health and disease. The two are, however, rarely considered together.

Nitrate, Agriculture, and Economics

The importance of nitrogen for plant growth was realized in the nineteenth century, first by Boussingault and later by Gilbert and Lawes' Rothamsted experiments [2]. Thus, readily available sources of fixed nitrogen were of considerable economic value. Guano, the nitrogen rich desiccated droppings of sea birds, proved to be a very valuable export commodity for Peru in the early and mid 1800s. Indeed, such was the economic value of fixed nitrogen that Chile contested a war with Bolivia and Peru from 1879 to 1883 [3]. Bolivia increased taxation on the export of nitrate mined by Chilean companies. In retaliation, Chile seized the Bolivian port of Antofogasta. Bolivia declared war on Chile with Peru joining the Bolivian side in line with a secret treaty. Chile won the conflict which ensued; consequently, the nitrate rich zones of Bolivia and Peru were ceded to Chile. The redrawn national boundaries also left Bolivia landlocked.

While these sources were economically important at the time, they did not represent a sustainable source of fixed nitrogen. Sustained plant growth requires continuing availability of fixed nitrogen. This can be derived from any of the following sources.

Nitrogen Fixation

Atmospheric nitrogen can be fixed via the following processes

1. Microbial dinitrogen fixation
2. Industrially by the Haber–Bosch process
3. Fossil fuel combustion
4. Lightning

Microbial Dinitrogen Fixation

The conversion of atmospheric nitrogen to a fixed, biologically available nitrogen species is the exclusive preserve of prokaryotes, eubacteria, and archea. Eukaryotes are not capable of this action. Collectively these organisms are known as diazotrophs. They fall into three distinct groups: free-living, symbiotic (within plant roots), and those with a loose association with another organism, typically plant roots.

Nitrogen fixation can only occur in anaerobic conditions. Nitrogenase, the enzyme responsible for fixing atmospheric nitrogen, is exquisitely sensitive to oxygen.

To maintain anaerobic conditions, a variety of approaches may be adopted: a very high respiration rate to maintain a low internal oxygen tension and/or binding of O_2 to leghemoglobin, a hemoglobin homologue, in the roots of leguminous plants [4].

The symbiotic relationship between rhizobia and leguminous plants is worthy of special consideration. The plant provides rhizobia, which are obligate anaerobes, with an anaerobic environment and sucrose in return for fixed nitrogen. The rhizobia infect plant roots of compatible species. A signaling cascade takes place between the plant and rhizobia resulting in changes in gene transcription in the plant driving root nodule formation and progressive infection of the nodule.

Fixed nitrogen in the form of NH_3 is transferred from an infected root cell to an adjacent uninfected cell for incorporation into amides, typically glutamate or asparagine, or ureides, such as allantoin for transport to the upper part of the plant.

Haber–Bosch

Such was the economic importance of fixed nitrogen for agriculture, the development of an industrial process became a prime concern for the chemical industry. In 1910 Fritz Haber began to develop what would ultimately become known as the Haber–Bosch process. This process, still widely used today, involves reacting N_2 and H_2 gases at approximately 500°C and a pressure of 300 atm over an iron catalyst to yield ammonia.

$$N_2 + 3H_2 \xrightarrow{\text{Fe}} 2NH_3$$

Some estimates suggest that 40% of the plant derived protein consumed by humans globally is derived from fertilizers utilizing nitrogen fixed by the Haber–Bosch process [5].

Fossil Fuel Combustion

The burning of fossil fuel for energy is responsible for less than 10% of the fixed biologically available nitrogen deposited on the terrestrial surface [6]. The majority of the nitrogen made biologically available via this route is already fixed. Burning fossil fuels simply releases it from long term sequestration in geological stores [7]. At sufficiently high temperatures some de-novo nitrogen fixation occurs.

Atmospheric Nitrogen Fixation by Lightning

While nitrogen fixation by lightning is a relatively minor player in the overall turnover of nitrogen within the global cycle, it is worth mentioning as it serves as a potent reminder of the magnitude of the energy investment required to fix nitrogen.

The conversion of N_2 to NO and rapid oxidation to NO_2 (nitrogen dioxide) in the upper atmosphere by lightning is notoriously difficult to study and quantify. The primary interest in this process pertains to its importance to stratospheric and tropospheric chemistry and its impact on climate [8], as comparatively little of the nitrogen fixed in this way finds its way into biological systems.

Once fixed the nitrogen will enter anyone of a variety of biological pathways.

Nitrate Assimilation in Plants and Bacteria

Most plants do not have a symbiont partner to provide them with fixed nitrogen. For those plants and bacteria unable to fix nitrogen independently, nitrate, found in soil, is the preferred source of fixed nitrogen. Nitrate is initially reduced to nitrite and ultimately ammonium before being acted upon by glutamate synthase and incorporated into carbon skeletons as described above for transport elsewhere in the plant. Certain plant species reduce nitrate rapidly on entry into the root system. Others transport unmetabolized nitrate to the upper part of the plant.

Nitrification

Decomposition of organic nitrogen-containing compounds such as proteins by a wide variety of bacteria and fungi produces ammonium. This ammonium is oxidized to nitrite and ultimately nitrate in reactions catalyzed by the chemolithoautotrophs nitrosomonas and nitrobacter which use the energy released to assimilate carbon from CO_2.

$$2NH_4^+ + 3O_2 \rightarrow 2NO_2^- + 4H^+ + 2H_2O + \text{energy}$$

$$2NO_2^- + O_2 \rightarrow 2NO_3^- + \text{energy}$$

It is worth noting that under certain circumstances, nitrifiers are capable of denitrification.

Denitrification

Fixed nitrogen is converted back to inert nitrogen gas by micro-organisms, which utilize this multi-step reductive process as an alternative respiratory pathway to oxygen-dependent respiration in anaerobic or near anaerobic conditions. Oxygen respiration is more energetically favorable and is, thus, preferred. It is only when oxygen tension falls to levels that limit the microbe's ability to respire that the enzymes of denitrification are expressed. Conversely, in conditions of rising oxygen tension, denitrification is inhibited.

Some organisms are only capable of catalyzing part of the denitrification pathway. This may represent a defense mechanism in some important human pathogens, such as *Neisseria gonorrhoea* [9], which may use it as a device for surviving NO attack from the host.

Denitrification is a four stage process with each step catalyzed by complex multi-site metallo-enzymes.

$$\underset{\text{Nitrite}}{NO_3^-} \rightarrow \underset{\text{Nitrite}}{NO_2^-} \rightarrow \underset{\text{Nitric oxide}}{NO} \rightarrow \underset{\text{Dinitrogen monoxide}}{N_2O} \rightarrow \underset{\text{Nitrogen}}{N_2}$$

Typically, nitrate reduction by membrane-bound nitrate reductase enzyme (NAR) occurs on the cytoplasmic side. Some organisms express a periplasmic nitrate reductase, NAP. Subsequent steps occur in the periplasm or on the periplasmic side of membrane-bound proteins. Thus denitrifying bacteria have developed specialized nitrate and nitrite transport mechanisms [10]. The nitrite produced by this process is further reduced to nitric oxide by either a heme or copper containing nitrite reductase, NIR [11]. The NO generated must be rapidly reduced to N_2O as accumulation of NO would be fatal to the organism. Bacterial nitric oxide reductases (NOR) contain the typical components of the main catalytic subunit of heme/copper cytochrome oxidases [12]. The final step in the denitrification pathway is the reduction of N_2O to N_2 by another copper containing enzyme, N_2O reductase; while it represents the terminal step in a complete denitrification pathway, it also represents a separate and independent respiratory process [13].

It is worth noting that NO, N_2O, or N_2 may be released in the denitrification process. The stoichiometry of the gas mix will be affected by the relative activities of the pertinent reductases.

The denitrification process is usually thought to occur in aerobic soil or sediment environments. Recently, it has emerged that a complete denitrification pathway exists in human dental plaque [14], a finding that may have important implications in the mammalian nitrate cycle.

Dissimilatory Reduction of Nitrate to Ammonia

A third route for nitrate reduction exists that provides neither energy nor components for anabolic processes within the cell. Dissimilatory nitrate reduction appears to exist as a method for detoxifying excess nitrite following respiratory reduction of nitrate or balancing an excessive quantity of reductants [15]. It is exclusively an anaerobic process. Under certain conditions such as may be found during photoheterotrophic growth [16], anaerobic marine sediments, and for some organisms in the human gastrointestinal tract [17] this may be the principal nitrate reduction pathway. A reductant-rich environment has a paucity of electron acceptors. Reduction of NO_2^- to NH_4^+ consumes six electrons compared with the two to three utilized in denitrification [18]. In these circumstances, dissimilatory nitrate reduction to ammonia is a key to maintaining the overall redox balance of the cell.

Anammox

The biological nitrogen cycle was widely held to be complete until the discovery of a pathway for anaerobic ammonium oxidation in 1990 [19]. When a wastewater plant reported higher than expected generation of dinitrogen gas, investigations revealed that certain bacteria were able to oxidize ammonium using nitrite as the electron acceptor to generate energy for growth in anoxic conditions. This process was thought to provide sufficient energy for slow growth only, with a bacterial doubling time of 11 days [20]. It has since emerged that anammox may provide for more rapid growth with a doubling time of 1.8 days, approaching that of ammonium oxidizers [21]:

$$NH_4^+ + NO_2^- \rightarrow N_2 + 2H_2O$$

As biologically available nitrogen is converted back to dinitrogen gas, anammox can be thought of as a form of denitrification. This entire complex pathway is illustrated in Fig. 2.1.

Nitrogen Balance in Mammals

In humans, nitrogen is principally ingested as protein. The average US young adult consumes approximately 90 g of protein daily [22]. This equates to about 14.5 g of nitrogen, nearly all of which is excreted in the urine as urea, creatinine, uric acid, and ammonia. Only a tiny proportion of this nitrogen is converted to nitric oxide via the nitric oxide synthase pathway. In healthy humans, about 1 mmol of nitrate is generated from L-arginine and then nitric oxide oxidation, which represents about one-thousandth of the amount of nitrogen ingested; during illness, such as gastroenteritis, this can increase by as much as eightfold [23].

Ruminants, and other mammals which rely on symbiotic bacteria to metabolize cellulose, can also make use of non-protein nitrogen sources for protein synthesis. Rumen bacteria can convert urea to ammonia which is used to produce amino acids that can be incorporated into mammalian proteins [24].

L-Arginine Nitric Oxide Synthase Pathway

Three distinct NOS isoforms exist in mammals: inducible, neuronal, and endothelial. The loci of each give an indication as to the pluripotent effects of nitric oxide in mammals, with roles in host defense, neuronal and other cellular signaling pathways, and vascular control. L-arginine provides organic nitrogen as a substrate which, with $2O_2$ and NADPH as a cofactor, is converted to nitric oxide and L-citrulline. The nitric oxide produced has a very short half-life, being rapidly oxidized in the presence of superoxide or oxyhemoglobin to nitrate, which enters the mammalian nitrate cycle.

Dietary Sources of Nitrate and Nitrite

Certain foods such as green leafy vegetables and beetroot are particularly rich in nitrate. Consumption of a typical Western diet results in the ingestion of approximately 1–2 mmol nitrate per day. Nitrite, because of its anti-botulism effect, has been used as preservative and colorant for centuries. In addition, humans are also exposed to biologically active nitrogen oxides from the combustion of fossil fuels or inhalation of tobacco smoke.

Enterosalivary Circulation of Nitrate/Nitrite/NO

The remarkable symbiosis between legumes and nitrogen fixing rhizobia finds its counterpoint in the relationship between nitrate-reducing bacteria hidden in crypts in mammalian tongues. Nitrate from the diet is rapidly and completely absorbed from the upper gastro-intestinal tract. This, along with nitrate derived from the oxidation of NO synthesized by the L-arginine NOS pathway, is actively taken up the salivary glands. The resulting salivary nitrate concentration may be ten times greater than the plasma nitrate concentration. In crypts on the dorsum of the tongue, facultative anaerobes e.g., *Vionella* species, utilize nitrate as an alternative electron acceptor [25]. The nitrite released elevates salivary nitrite to levels 1,000 times that of plasma in the resting state. In the presence of acid-generating plaque bacteria, some nitrite is chemically reduced to nitric oxide [26]. The remaining salivary nitrite is then swallowed. In the acidic environment of the stomach, some of this nitrite is further reduced to nitric oxide [27] which has an important role in both protection against enteric pathogens and regulation of gastric blood flow and mucous production [28, 29]. Some of the nitrite is absorbed from the stomach with important consequences for mammalian vascular physiology. This pathway is illustrated in Fig. 2.2.

In 1996, it was discovered that nitric oxide is continually released from the surface of normal human skin [30]. Although it was initially thought that nitric oxide synthase would be responsible, inhibition of this enzyme by infusing monomethyl L-arginine into the brachial artery showed that this was not the case. Further studies showed that nitrate is excreted in human sweat and reduced to nitrite by skin bacteria. As normal skin is slightly acidic (pH around 5.5) this nitrite is reduced to nitric oxide. The function of this NO is thought to be to inhibit skin pathogens – particularly fungi [31] – and, intriguingly, when normal saliva is applied to healthy

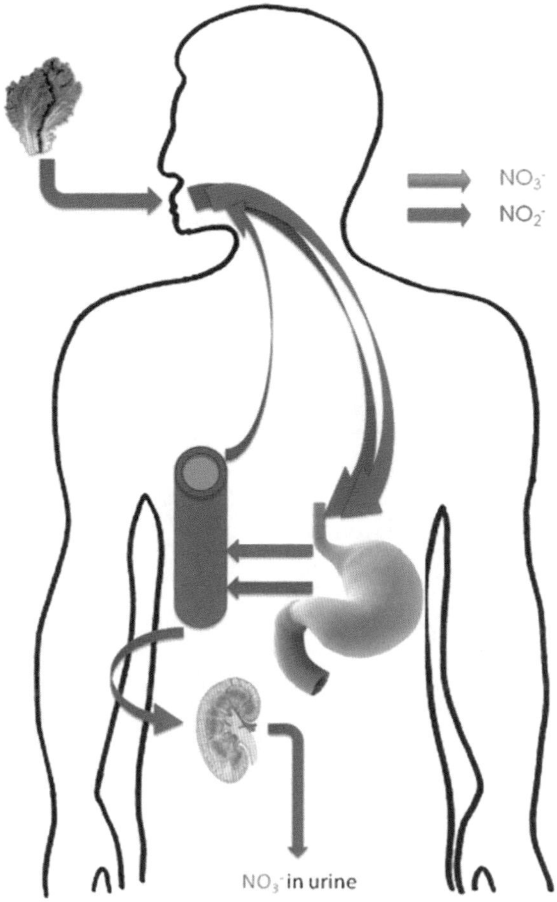

Fig. 2.2 Simplified representation of the entero-salivary circulation of nitrate. Nitrate (represented by the *black arrows*) derived from the diet is swallowed. It is rapidly and completely absorbed in the upper gastro-intestinal tract. Approximately 25% is concentrated in the salivary glands and secreted into the mouth. Here it is reduced to nitrite (represented by the *grey arrows*) by facultative anaerobes on the dorsum of the tongue and swallowed. Some of the nitrite undergoes acidic reduction to nitric oxide in the stomach, with the remainder being absorbed. The fate of nitrite is discussed in depth later. Sixty percent of ingested nitrate is lost in the urine within 48 h

skin, the high concentrations of nitrite considerably increase nitric oxide synthesis, perhaps to protect against infection and encourage wound healing [32].

Breast milk has recently been shown to contain variable amounts of nitrite and nitrate, and it has been suggested that conversion of these anions to more reactive nitrogen oxides may be a factor underlying the protective effect of breastfeeding against infant gastroenteritis [33].

Urinary Nitrate Excretion

Approximately 60% of the nitrate ingested or endogenously synthesized will be lost in the urine within 48 h [34]. The discovery of complete denitrification pathways in human dental plaque flora [14] may offer a clue as to the fate of at least part of the remainder. Nitrate is freely filtered at the glomerulus. Studies in dogs suggest that as much as 90% may be reabsorbed by the renal tubules [35]. Nitrite is not found in human urine under normal physiological conditions. Its presence indicates infection with nitrate-reducing organisms. Detection of urinary nitrite is in widespread use as a simple bedside test for diagnosing urinary tract infection.

Conclusion

When considering the biological nitrogen cycle as part of the global nitrogen cycle, mammalian nitrogen cycling is often considered separately from the processes occurring in plants and micro-organisms. Scientific interest in both these spheres has undergone a resurgence over the last 20–30 years. The complexity of the processes and relationships between animals, plants, and micro-organisms is still being unraveled.

References

1. Feelisch M, Martin JF. The early role of nitric oxide in evolution. Trends Ecol Evol. 1995;10(12):496–9.
2. Addiscott T. Nitrate, agriculture and the environment. Oxford: CABI; 2005.
3. Kiernan VG. Foreign interests in the War of the Pacific. Hisp Am Hist Rev. 1955;35(1):14–36.
4. Harutyunyan EH, et al. The Structure of Deoxy- and Oxy-leghaemoglobin from Lupin. Journal of Molecular Biology. 1995;251(1):104–115.
5. Smil V. Enriching the earth: Fritz Haber, Carl Bosch, and the transformation of world food production. Cambridge: MIT; 2004.
6. Schlesinger WH. On the fate of anthropogenic nitrogen. Proc Natl Acad Sci. 2009;106(1):203–8.
7. Vitousek PM, Aber JD, Howarth RW, Likens GE, Matson PA, Schindler DW, et al. Human alteration of the global nitrogen cycle: sources and consequences. Ecol Appl. 1997;7(3):737–50.
8. Schumann U, Huntrieser H. The global lightning-induced nitrogen oxides source. Atmos Chem Phys. 2007;7(14):3823–907.
9. Barth KR, Isabella VM, Clark VL. Biochemical and genomic analysis of the denitrification pathway within the genus *Neisseria*. Microbiology. 2009;155(12):4093–103.
10. Moir JW, Wood NJ. Nitrate and nitrite transport in bacteria. Cell Mol Life Sci. 2001;58(2):215–24.
11. Cutruzzolà F. Bacterial nitric oxide synthesis. Biochim Biophys Acta. 1999;1411(2-3):231–49.
12. Hendriks J, Oubrie A, Castresana J, Urbani A, Gemeinhardt S, Saraste M. Nitric oxide reductases in bacteria. Biochim Biophys Acta. 2000;1459(23):266–73.

13. Zumft W. Cell biology and molecular basis of denitrification. Microbiol Mol Biol Rev. 1997;61(4):533–616.
14. Schreiber F, Stief P, Gieseke A, Heisterkamp I, Verstraete W, de Beer D, et al. Denitrification in human dental plaque. BMC Biol. 2010;8(1):24.
15. Moreno-Vivián C, Ferguson SJ. Definition and distinction between assimilatory, dissimilatory and respiratory pathways. Mol Microbiol. 1998;29(2):664–6.
16. Berks BC, Ferguson SJ, Moir JWB, Richardson DJ. Enzymes and associated electron transport systems that catalyse the respiratory reduction of nitrogen oxides and oxyanions. Biochim Biophys Acta. 1995;1232(3):97–173.
17. Parham NJ, Gibson GR. Microbes involved in dissimilatory nitrate reduction in the human large intestine. FEMS Microbiol Ecol. 2000;31(1):21–8.
18. Bothe H, Ferguson SJ, Newton WE. Biology of the nitrogen cycle. 1st ed. Amsterdam: Elsevier; 2007.
19. Jetten MSM, Strous M, Pas-Schoonen KT, Schalk J, Dongen UGJM, Graaf AA, et al. The anaerobic oxidation of ammonium. FEMS Microbiol Rev. 1998;22(5):421–37.
20. Strous M, Heijnen JJ, Kuenen JG, Jetten MSM. The sequencing batch reactor as a powerful tool for the study of slowly growing anaerobic ammonium-oxidizing microorganisms. Appl Microbiol Biotechnol. 1998;50(5):589–96.
21. Isaka K, Date Y, Sumino T, Yoshie S, Tsuneda S. Growth characteristic of anaerobic ammonium-oxidizing bacteria in an anaerobic biological filtrated reactor. Appl Microbiol Biotechnol. 2006;70(1):47–52.
22. Fulgoni III VL. Current protein intake in America: analysis of the National Health and Nutrition Examination Survey, 2003–2004. Am J Clin Nutr. 2008;87(5):1554S–1557.
23. Forte P, Dykhuizen RS, Milne E, McKenzie A, Smith CC, Benjamin N. Nitric oxide synthesis in patients with infective gastroenteritis. Gut. 1999;45(3):355–61.
24. Huntington GB, Archibeque SL. Practical aspects of urea and ammonia metabolism in ruminants. J Anim Sci. 2000;77(E-Suppl):1–11.
25. Doel JJ, Benjamin N, Hector MP, Rogers M, Allaker RP. Evaluation of bacterial nitrate reduction in the human oral cavity. Eur J Oral Sci. 2005;113(1):14–9.
26. Duncan C, Dougall H, Johnston P, Green S, Brogan R, Leifert C, et al. Chemical generation of nitric oxide in the mouth from the enterosalivary circulation of dietary nitrate [see comment]. Nat Med. 1995;1(6):546–51.
27. Benjamin N, O'Driscoll F, Dougall H, Duncan C, Smith L, Golden M, et al. Stomach NO synthesis [see comment]. Nature. 1994;368(6471):502.
28. Wallace JL, Miller MJ. Nitric oxide in mucosal defense: a little goes a long way. Gastroenterology. 2000;119(2):512–20.
29. Dykhuizen RS, Frazer R, Duncan C, Smith CC, Golden M, Benjamin N, et al. Antimicrobial effect of acidified nitrite on gut pathogens: importance of dietary nitrate in host defense. Antimicrob Agents Chemother. 1996;40(6):1422–5.
30. Weller R, Pattullo S, Smith L, Golden M, Ormerod A, Benjamin N. Nitric oxide is generated on the skin surface by reduction of sweat nitrate. J Investig Dermatol. 1996;107(3):327–31.
31. Weller R, Price RJ, Ormerod AD, Benjamin N, Leifert C. Antimicrobial effect of acidified nitrite on dermatophyte fungi, *Candida* and bacterial skin pathogens. J Appl Microbiol. 2001;90(4):648–52.
32. Benjamin N, Pattullo S, Weller R, Smith L, Ormerod A. Wound licking and nitric oxide [see comment]. Lancet. 1997;349(9067):1776.
33. Hord NG, Ghannam JS, Garg HK, Berens PD, Bryan NS. Nitrate and nitrite content of human, formula, bovine and soy milks: implications for dietary nitrite and nitrate recommendations. Breastfeed Med 2010; EPub ahead of Print.
34. Wagner DA, Schultz DS, Deen WM, Young VR, Tannenbaum SR. Metabolic fate of an oral dose of 15N-labeled nitrate in humans: effect of diet supplementation with ascorbic acid. Cancer Res. 1983;43(4):1921–5.
35. Godfrey M, Majid DS. Renal handling of circulating nitrates in anesthetized dogs. Am J Physiol. 1998;275(1 Pt 2):F68–73.

Chapter 3
The Nitrate–Nitrite–Nitric Oxide Pathway in Mammals

Jon O. Lundberg, Eddie Weitzberg, Sruti Shiva, and Mark T. Gladwin

Key points

- Accumulating evidence suggests that the molecules nitrite and nitrate can be metabolized in vivo to form NO and other bioactive nitrogen oxides.
- Commensal bacteria play a central role in the bioactivation of nitrate in an entero-salivary bioactivation pathway.
- A number of nitrite reductase enzymes reduce nitrite to bioactive NO along a physiological oxygen and pH gradient.
- Nitrite mediates a number of physiological effects including vasodilation, modulation of mitochondrial function, and protection from ischemia–reperfusion injury.
- Modest dietary intake of nitrate reduces blood pressure, inhibits platelet function, prevents endothelial dysfunction after ischemia–reperfusion, and decreases oxygen cost during exercise.
- Vegetables protect against cardiovascular disease and diabetes type 2; an effect that may be related to the high nitrate content in this food group.

Keywords Vegetables • Endothelial function • Cytoprotection • Saliva • Oral bacteria

Introduction

Inorganic nitrate (NO_3^-) and nitrite (NO_2^-) from dietary sources have been implicated in the development of gastric cancer and other disorders [1, 2]. This has led to very strict regulations of nitrate and nitrite levels in our food and drinking water. In the

J.O. Lundberg (✉)
Institute of Molecular Medicine, University of Texas Health Science Center at Houston, Pressler Street 1825, 77030, Houston, Texas, USA,
e-mail: jon.lundberg@ki.se

N.S. Bryan and J. Loscalzo (eds.), *Nitrite and Nitrate in Human Health and Disease*, Nutrition and Health, DOI 10.1007/978-1-60761-616-0_3,

early 1980s it was shown that, in addition to dietary exposure, nitrate and nitrite are also generated endogenously [3]. Shortly thereafter, the entire L-arginine-nitric oxide synthase (NOS)-system was discovered and was found to be the major endogenous source of nitrate and nitrite, since nitric oxide (NO) is rapidly oxidized to these higher nitrogen oxides [4–6]. Until recently, biologists considered nitrate and nitrite merely as inactive end products of NO metabolism, but this view is rapidly changing. It is now clear that nitrite and nitrate can recycle in vivo and again form bioactive nitrogen oxides, including NO [7–12]. Interestingly, commensal bacteria in the oral cavity play a key role in the bioactivation of nitrate [9]. A picture is emerging suggesting physiological, nutritional, as well as therapeutic roles for the nitrate–nitrite–NO pathway, especially under hypoxic conditions when NO formation from nitrite is greatly enhanced [7]. Thus, instead of simply wasting the products of NO oxidation, mammals store and actively recycle it. Nitrite reduction to NO was first described in the stomach, where salivary nitrite forms NO nonenzymatically via acid-catalyzed reduction [13, 14]. Soon after this observation, Zweier described NOS-independent nitrite reduction in the ischemic and acidic heart [15]. In the last 10 years it has become evident that blood and tissue nitrite is reduced under physiological conditions to form NO and modulate blood flow [16, 17]. Subsequent studies show that a variety of enzymes and proteins can catalyze the one-electron reduction of nitrite to NO in blood and tissues. In this chapter, we discuss the chemical biology of the nitrate–nitrite–NO pathway with special emphasis on mechanisms of bioactivation, biological effects, and therapeutic opportunities. Table 3.1 provides an overview of the many redox reactions that drive the bioactivation and interconversion of nitrate, nitrite, and NO.

NO Synthase and Our Diet are the Two Major Sources of Endogenous Nitrate and Nitrite

The L-arginine–NOS pathway is the predominant source of endogenous nitrate and nitrite. NO, a reactive free radical with a biological half-life in the millisecond range, is rapidly oxidized in vivo to form higher nitrogen oxides including nitrite and nitrate. Nitrate is by far the dominant final NO oxidation product; the levels in blood and tissues exceed those of nitrite by at least two orders of magnitude (μM vs. nM) [6]. In blood, this nitrate forms directly from the NO dioxygenation reaction between NO and oxyhemoglobin: NO reacts with oxyhemoglobin to form nitrate and methemoglobin (Eq. (3.1)) [18–20].

$$NO + Fe^{+2} - O_2 \rightarrow NO_3^- + Fe^{+3} \tag{3.1}$$

Nitrite can also form directly in blood and tissues via NO autooxidation, which involves the reaction of two molecules of NO with oxygen, a reaction that is catalyzed by the plasma protein ceruloplasmin [21].

Table 3.1 Overview of many redox reactions that drive the bioactivation and interconversion of nitrate, nitrite, and NO

Nitrite formation

Auto-oxidation of NO

$$4\,NO^{\bullet} + O_2 + 2H_2O \rightarrow 4NO_2^- + 4H^+$$

Ceruloplasmin

$$NO^{\bullet} + Cu^{2+} \rightarrow NO^+ + Cu^{1+}$$

$$NO^+ + H_2O \rightarrow HNO_2 + H^+$$

$$HNO_2 \leftrightarrow H^+ + NO_2^-$$

Bacterial nitrate reductase

$$NO_3^- + 2e^- + H^+ \rightarrow NO_2^- + H_20$$

Nitrite reduction

Deoxyhemoglobin/myoglobin

$$NO_2^- + Fe^{2+} + H^+ \rightarrow NO^{\bullet} + Fe^{3+} + OH^-$$

Xanthine oxidoreductase

$$NO_2^- + Mo^{2+} + H^+ \rightarrow NO^{\bullet} + Mo^{3+} + OH^-$$

Protons

$$NO_2^- + H+ \rightarrow HNO_2$$

$$2\,HNO_2 \rightarrow 2N_2O_3 + H_2O$$

$$N_20_3 \rightarrow NO^{\bullet} + NO_2^{\bullet}$$

Ascorbate

$$NO_2^- + H^+ \rightarrow HNO_2$$

$$2\,HNO_2 + \text{Asc} \rightarrow 2 + \text{dehydroAsc} + 2H_2O$$

Polyphenols (Ph-OH)

$$NO_2^- + H^+ \rightarrow HNO_2$$

$$Ph - OH + HNO_2 \rightarrow Ph - O^{\bullet} + NO^{\bullet} + H_2O$$

Nitrite oxidation

Hemoglobin

$$4HbO_2 + 4NO_2^- + \rightarrow \text{autcatalytic intermediates } HbFe^{+4} = O \text{ and } NO_2^{\bullet} \rightarrow 4\text{metHb} + 4NO_3^- + O_2 + 2H_2O$$

In addition to the NOS system, our everyday diet represents an equally important source of nitrate and nitrite. Vegetables are the dominant source of nitrate in the diet (>80%) and leafy green vegetables contain particularly high levels [7]. The contribution of nitrate from endogenous versus exogenous sources varies greatly depending on diet, physical activity, disease, and medication. Inflammatory conditions with massive activation of inducible NOS (iNOS) [22] or physical exercise with activation of endothelial NOS (eNOS) [23] will increase endogenous levels of nitrate. Conversely, dietary intake of nitrate can increase systemic nitrate levels dramatically. As an example, one serving of beetroot, lettuce, or spinach contains more nitrate than what is generated by the entire NOS system over a day [24].

Enterosalivary Recirculation of Nitrate and Reduction to Nitrite

Nitrate circulates in plasma, distributes evenly throughout the tissues, and has a half-life of approximately 5 h. For largely unknown reasons, circulating nitrate is actively taken up from blood by the salivary glands and concentrated 10–20-fold in saliva [25]. This concentration of nitrate can result in salivary nitrate levels of several millimolar [26]. Up to 25% of all circulating nitrate enters this peculiar enterosalivary cycle, while a large portion is eventually excreted via the kidney. In the oral cavity, nitrate reductase enzymes in commensal facultative anaerobic bacteria reduce approximately 20% of the nitrate to nitrite [25, 27]. These bacteria use nitrate as an alternative terminal electron acceptor to generate ATP in the absence of oxygen. As a result of bacterial enzymatic activity, the levels of nitrite in saliva also become very high. Salivary-derived nitrite is metabolized to NO and other reactive nitrogen oxides locally in the acidic stomach; a number of physiological roles for gastric NO have been proposed which are discussed below [28–32]. Of particular relevance is the fact that much nitrite also survives gastric passage and enters the systemic circulation [33]. This is of great interest since there are now numerous mechanisms described in blood and tissues for nitrite reduction to NO and other bioactive nitrogen oxides [10, 24, 34, 35]. Thus, a complete reverse nitrate–nitrite–NO pathway exists in mammals. This fact, when revealed, was highly surprising since nitrate had been universally considered to be inert from the time when NO formation in mammals was first discovered in the 1980s (Fig. 3.1).

Stomach NO Generation

The levels of NO generated from acidified nitrite in the stomach are in the range of 10–100 ppm, i.e., several orders of magnitude higher than those required for vasodilation [13, 14, 27]. In such high concentrations, NO and its reaction products are toxic to a variety of microorganisms which suggested a role for gastric NO in host defense. Benjamin and colleagues have tested this hypothesis [13]. They exposed enteropathogens to different combinations of acid and nitrite. *Escherichia coli* and *Candida albicans* species were remarkably resistant when exposed to acid alone but were killed if nitrite was added. Subsequent studies have showed that acidified nitrite inhibits the growth of a variety of enteropathogens,

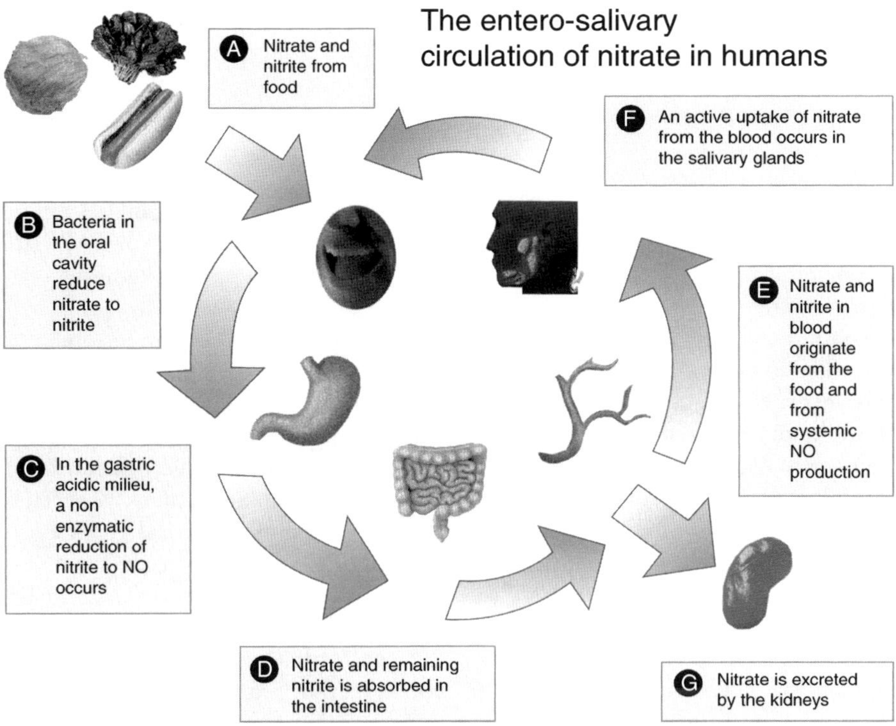

Fig. 3.1 The enterosalivary nitrate–nitrite–NO pathway. Inorganic nitrate from dietary and endogenous sources is partly recycled in our bodies via conversion to nitrite, NO, and other bioactive nitrogen oxides in blood and tissues. This bioactivation of nitrate involves active transportation of nitrate from blood to saliva and reduction to nitrite by commensal bacteria in the oral cavity. Further reduction of nitrite to NO and other nitrogen oxides occurs locally in the acidic gastric lumen as well as systemically after absorption. Modified from ref. [7] and reproduced with permission

including *Salmonella*, *Shigella*, and *Helicobacter pylori* [36, 37]. More recently, Björne et al. tested the antibacterial effects of authentic human gastric juice and saliva, and, again, this combination was effective when salivary nitrite was high [32]. Beside NO, a variety of other reactive nitrogen intermediates (RNIs) are generated from nitrite under acidic conditions, and it is likely that several of these contribute to the antibacterial effects (see below). Altogether, there are good indications for a role of salivary nitrite in primary host defense against swallowed pathogens.

When nitrite is acidified it yields nitrous acid (HNO_2, Eq. 3.2), which spontaneously decomposes to NO and other nitrogen oxides (Eq. 3.3 and 3.4):

$$NO_2^- + H^+ \leftrightarrow HNO_2\,(\text{pKa } 3.2 - 3.4) \tag{3.2}$$

$$2HNO_2 \rightarrow N_2O_3 + H_2O \quad (3.3)$$

$$N_2O_3 \rightarrow NO + NO_2 \quad (3.4)$$

The chemistry of acidified nitrite is very complex and the amount of NO generated from nitrite is dependent not only on pH and nitrite concentrations but also on the presence of other reducing agents (e.g., vitamin C, thiocyanate, polyphenols), proximity to heme groups, proteins, thiols, and the oxygen tension [8].

With very high levels of NO generated luminally, it is not unreasonable to assume that enough NO would reach the gastric mucosa to affect physiological processes. Björne and colleagues used the rat gastric mucosa as a bioassay to test effects of human saliva on gastric mucosal blood flow and mucus generation in vivo [28]. When saliva collected from fasting individuals (low in nitrite) was placed onto the mucosa, no changes in blood flow and mucus generation were noted. In contrast, with nitrite-rich saliva, collected after oral ingestion of nitrate, blood flow and mucus increased greatly. In addition, feeding rats with nitrate in the drinking water for 1 week leads to a sustained increase in gastric mucosal blood flow and a thicker mucus layer [30]. These effects of nitrite are cGMP dependent and associated with NO gas formation. They are also independent of NO synthase and cyclooxygenase activity as evident from experiments using N^{ω}-nitro-L-arginine methyl ester (L-NAME) and indomethacin [28]. Because adequate blood flow and mucus generation are essential for maintaining gastric integrity, it is of interest to study the effects of nitrate and nitrite in models of gastric injury. Indeed, several studies have found a potent protective effect of dietary nitrate in animal models of gastric ulcers [29, 31, 38, 39]. Jansson and colleagues found that a 7-day pretreatment with nitrate in the drinking water protected rats against ulcers induced by a nonsteroidal anti-inflammatory drug (NSAID) [29]. Miyoshi et al. found gastroprotective effects of dietary nitrate in a stress-induced model of gastric ulcers [31]. Interestingly, if they eliminated the oral microflora, this gastroprotection was lost. Sobko and colleagues showed that in contrast to normal conventions rats, gastric NO levels are extremely low in germ-free rats and do not increase after ingestion of nitrate [40]. Together, this demonstrates the central role of bacteria in bioactivation of salivary nitrate to nitrite.

In critically ill patients, endotracheal intubation and sedation interrupt the enterosalivary nitrate cycle, which results in depleted gastric NO, nitrite, and *S*-nitrosothiol levels [41]. It has been hypothesized that the insufficient levels of gastric NO contribute to the gastric lesions and bacterial overgrowth commonly found in these patients [41].

In aggregate, when nitrite-rich saliva meets gastric acid, a variety of biologically active nitrogen oxides are rapidly formed. The enterosalivary circulation of nitrate and its reduction to nitrite in the mouth by commensal bacteria seem to be a beautiful example of symbiosis. The bacteria receive from the host nitrate, a compound necessary for their respiration, and in return they provide us with nitrite, a substrate needed for the generation of gastroprotective NO.

Reactions Between Nitrite and Other Dietary Compounds

The stomach may be regarded as a bioreactor where different dietary compounds react with each other to form novel compounds, some of which are bioactive [42]. As described previously, many vegetables are very rich in inorganic nitrate, which, after ingestion, rapidly accumulates in saliva where it is reduced by commensal bacteria to form nitrite. Nitrous acid (HNO_2) is formed from salivary nitrite in the acidic stomach. As described previously, HNO_2 spontaneously yields NO via disproportionation, and this reduction is greatly enhanced in the presence of dietary polyphenols [42, 43] and ascorbic acid [44]. An example is red wine in which the anthocyanin fraction and wine catechol (caffeic acid) greatly enhance NO generation from nitrite [42]. Interestingly, the ethanol in wine also reacts with nitrite under gastric conditions, thereby forming ethyl nitrite, another bioactive compound [45]. Ethyl nitrite is a potent vasodilator and forms rapidly via O-nitrosation when HNO_2 reacts with ethanol. Ethyl nitrite is also a potent nitrosating compound that can give rise to *S*-nitrosothiols and *N*-nitrosamines.

S-nitrosation reactions give rise to *S*-nitrosothiols [46], and clearly these highly bioactive compounds are formed intragastrically from acidified nitrite [28]. Many of the NO-like effects of nitrite described in the stomach, such as stimulation of mucosal blood flow and mucus generation, might proceed via intermediate formation of *S*-nitrosothiols which can act as stable carriers of NO. In addition, *S*-nitrosation of critical cysteine residues in protein is also important in regulation of protein function [46]. Unsaturated fatty acids (e.g., linoleic and oleic acid) from olive oil or other oils may undergo nitration following reaction with RNIs (in particular the NO_2 radical) giving rise to nitrated lipids [47]. Although the existence of this latter reaction is yet to be proven in vivo, the acidic nitrite-containing stomach is indeed an ideal milieu for nitration. Interestingly, nitrated lipids have anti-inflammatory effects and act as potent agonists of PPAR-gamma, a key regulator of the inflammatory cascade [48, 49].

In aggregate, NO, *S*-nitrosothiols, ethyl nitrite, and possibly nitrated lipids are readily formed from dietary constituents when they react with nitrite under acidic conditions in the stomach. These can all act as transducers of NO-like bioactivity including vasodilation, stimulation of mucus generation, and antimicrobial and anti-inflammatory effects. Many of the compounds formed in the gastric lumen from acidified nitrite are fairly stable and readily absorbed which likely allows also for systemic effects. It is interesting that most nutrients highlighted in the figure may divert the chemistry in the stomach away from nitrosamine formation and instead favor the formation of other chemically related bioactive compounds. Future studies will clarify if the well-known heath promoting effects of a Mediterranean diet is related to this alternative chemical fate of dietary nitrate.

Nitrite in Blood Flow Control and Hypoxic Signaling

Large doses of nitrite given as a treatment for cyanide poisoning have been known for almost 100 years to produce hypotension in humans [50]. Consistent with this observation, high micromolar concentrations of nitrite were shown to vasodilate isolated aortic rings in vitro by Furchgott in 1953 [51], and to activate guanylyl cyclase and vasodilate aortic rings, by Murad and Ignarro [51–54]. However, the very low potency in these systems, requiring high micromolar to millimolar concentrations of nitrite, to mediate these effects contrasted with the low nanomolar concentrations of authentic NO needed to vasodilate aortic rings. The low potency coupled with the very low levels of nitrite present in human blood led to the premature dismissal of nitrite as a regulator of blood flow in vivo. Indeed, studies published by Lauer and colleagues demonstrated that nitrite had no vasodilatory activity when infused into the forearm circulation of three normal volunteers, even at concentrations of 200 μM, a result that was not challenged by the NO field at that time [55]. These studies appeared to preclude a role for nitrite in the regulation of physiological blood flow [56, 57].

Prior to the publication of these negative studies, our group had measured arterial and venous levels of nitrite across the human forearm, and observed that arterial levels were higher than venous levels, suggesting metabolism of nitrite across a vascular bed [16]. In addition, Modin et al. had showed that nitrite dilated aortic ring preparations at physiological concentrations if pH in the medium was reduced to levels seen in metabolically active or ischemic tissues [58]. The concept that nitrite could be utilized biologically rather than serving as an inert end-oxidation product of NO was considered. Normal volunteer studies of NO inhalation showed that the levels of plasma nitrite increased in association with vasodilation at a site distant from the pulmonary circulation [59, 60]. Because the half-life of authentic NO in blood is less than 2 ms due to its rapid reaction with oxyhemoglobin (Eq. 3.1), another species must be forming that accounted for the transport of NO bioactivity in an endocrine fashion from the pulmonary vasculature to the peripheral organs. To test directly whether nitrite could vasodilate in vivo at physiological concentrations, nitrite was infused into the forearm brachial artery of 28 healthy volunteers and resulted in substantial vasodilation, even without exercise stress. Nitrite was potent in humans, increasing blood flow by 170% at 200 μM in the forearm circulation, by 22% at 2.5 μM, and produced vasodilation during exercise stress even at levels of 900 nM [17]. The potent in vivo vasodilating effect of nitrite has now been confirmed by a number of investigators [61–67].

From a physiological standpoint, it is increasingly clear that basal levels of nitrite contribute to blood flow regulation and the cellular resilience to ischemic stress. In studies of nitrite infusion into the human forearm circulation, concentrations of nitrite of only 200 nM significantly increase blood flow [66]. Studies of dietary nitrate manipulation in humans show that increasing nitrate dietary levels by consumption of beet root juice, or by ingestion of authentic nitrate, increases the plasma nitrite levels and lowers blood pressure [68–70]. Indeed, from these studies it appears that even very low levels of nitrite, modulated by the diet, regulate blood pressure.

Mechanisms of Nitrite Bioactivation: Reduction Reactions to NO

Hypoxic vasodilatation is a conserved physiological response to hypoxia that matches blood flow and oxygen delivery to tissue metabolic demand; this process has been characterized for more than 100 years since the initial description by Roy and Brown in 1879 [71, 72]. In mammalian species, the set point for hypoxic vasodilatation occurs as the hemoglobin desaturates from 60 to 40%, around the intrinsic P_{50} of hemoglobin (40–20 mmHg) [73]. During nitrite infusions, iron-nitrosyl-hemoglobin ($HbFe^{+2}$-NO) forms from artery to vein, suggesting that nitrite is being reduced to NO rapidly within one half-circulatory time [17]. The amount of NO formed from artery to vein during infusion studies is inversely correlated with decrease in oxyhemoglobin saturation, so that as hemoglobin deoxygenates more NO is formed. Thus, the mechanism of nitrite-dependent vasodilation appears to be tightly coupled to hemoglobin deoxygenation, indicating a potential role of nitrite as a hypoxic vasodilator, even during physiological deoxygenation from artery to vein [17, 74]. While the suggestion that nitrite is a hypoxic vasodilator has been challenged by other groups [56], this was recently directly tested by Maher and colleagues in normal human volunteers [75]. They reported that under normal physiological conditions, nitrite potently vasodilated the venous circulation. Exposure to 12% oxygen, which reduced arterial hemoglobin oxygen saturation to approximately 85%, the potency of nitrite increased significantly in the arterial circulation but was unchanged in the venous circulation. These experiments provide direct evidence for nitrite-mediated hypoxic vasodilation in humans [75].

The physiological observations that nitrite infusions are tightly coupled to NO-hemoglobin formation are consistent with the classic reaction between nitrite and deoxyhemoglobin to form NO as described in Eq. (3.5) [17, 76–78].

$$NO_2^- + HbFe^{+2} + H^+ \rightarrow NO + HbFe^{+3} + OH^- \quad (3.5)$$

This reaction possesses the sensor and effector properties necessary for hypoxic vasodilatation. The reaction requires deoxyhemoglobin and a proton, providing oxygen and pH sensor chemistry, respectively, and generates NO, a potent vasodilator. Nitrite appears to meet a number of basic criteria for a mediator of hypoxic vasodilatation, as it is naturally occurring, is metabolized or generated in response to tissue hypoxia, and potently vasodilates in response to hypoxia at an oxygen partial pressure of approximately 20–40 mmHg. Nitrite-mediated vasodilatation occurs as hemoglobin unloads oxygen to 50% saturation (P_{50}), and that this vasodilatation is consistent with the biophysical observation of a nitrite reductase activity of hemoglobin allosterically linked to its P_{50} [79–81]. This maximal rate of nitrite reduction at around 50% hemoglobin oxygen saturation occurs because the reactivity of nitrite is highest for the R-state or oxygenated hemoglobin while the reaction will only occur with the deoxygenated hemes, which are more plentiful in the deoxygenated T-state. The balance of more available hemes to react in an oxygenated R-state hemoglobin tetramer occurs around the P_{50}.

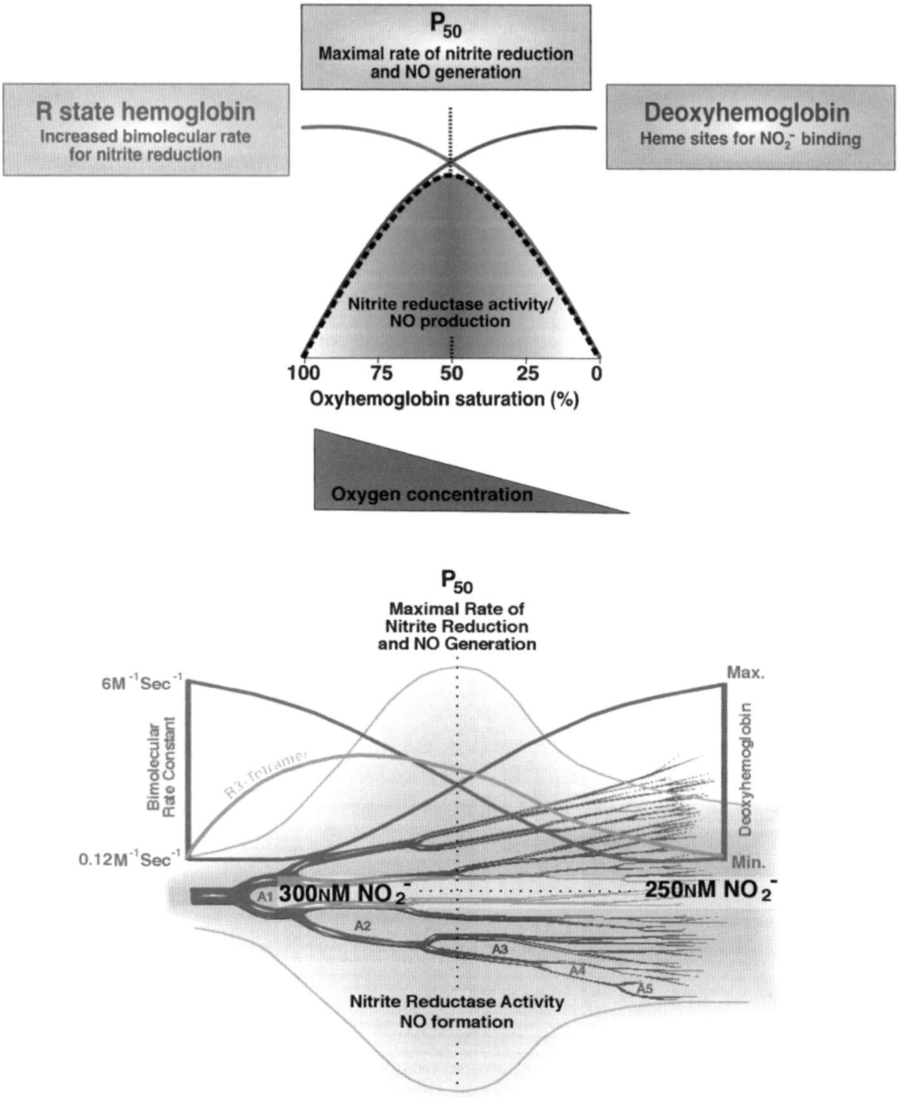

Fig. 3.2 Biochemistry of nitrite-hemoglobin hypoxic vasodilation along the arterioles. There exists a steady state anatomical location within the circulation from artery to vein that has the greatest concentration of R_3 tetramers which possess the maximal nitrite reductase activity. At this location there would always exist an equilibrium rate constant for nitrite reduction and an equilibrium concentration of nitrite and deoxyhemes (maximized in R_3 tetramer). As soon as one red cell moves downstream a new one would replace it, thus preserving the concentration of nitrite and R_3 hemoglobin at that anatomical position. Thus, there will be an increased nitrite reductase rate and increased NO concentration surrounding the blood vessel. The NO concentration should increase in a bell curve distribution from artery to vein according to the predicted rate for

A kinetic model for nitrite regulation of hypoxic vasodilatation and blood flow is illustrated in Fig. 3.2 by the rise in R_3 tetramer during early deoxygenation and the peak rate of nitrite reduction at the P_{50}. This biochemistry allows for rapid nitrite conversion to NO as red blood cells deoxygenate along the arteriolar vascular tree. Indeed, in experimental systems, the faster one deoxygenates the red cell in the presence of nitrite, the faster vasodilatation is observed [79]. Note that in the arterial vasculature, measurements of tissue oxygen tension and hemoglobin oxygen saturation using modern methodologies suggest that much of the oxygen delivery occurs within the resistance arterioles, particularly in the case of skeletal muscle [82]. Thus, in these microvascular beds, the anatomical site of hypoxic sensing may occur in the arterioles and muscularized capillaries. As hemoglobin deoxygenates in these beds, there will be an increased nitrite reductase rate and increased NO concentration surrounding the blood vessel (Fig. 3.2). The anatomical position of this equilibrium NO concentration will be responsive to tissue metabolism and oxygen consumption by moving the R-to-T transition up or downstream.

It is important to note that based on the remarkable potency of NO, very little nitrite must be reduced and escape the red blood cell. Consider that the EC_{50} (the concentration that dilates a vessel 50%) of NO is 5 nM and EC_{20} is less than 0.5 nM, while the concentration of nitrite in a red cell is 250 nM. Therefore, a 20% vasodilation requires only 1/1,000 molecules of NO (or other intermediate such as N_2O_3) to escape. It is also important to recognize in this context that blood flow is proportional to the change in vessel radius to the fourth power. A minimal increase in radius has marked effects on flow so that blood flow regulation occurs below the EC_{20} of NO. The mechanisms surrounding the reactions of nitrite and deoxyhemoglobin and other heme globins discussed previously are more extensively discussed in a recent review by Kim-Shapiro and Gladwin [83].

In addition to hemoglobin, a number of other enzymes can reduce nitrite to NO under hypoxic conditions, particularly in tissues. This includes deoxygenated myoglobin [84, 85], xanthine oxidoreductase (XOR) [65, 86–88], respiratory chain enzymes of mitochondria [89], aldehyde oxidase [90], carbonic anhydrase [91], and even NO synthase [92]. In addition, reducing agents such as vitamin C[44] and polyphenols [42, 45] catalyze nonenzymatic reduction of nitrite. Figure 3.3 shows the bioconversion of nitrate and nitrite to NO.

Fig. 3.2 (continued) nitrite reduction. The anatomical position of this equilibrium NO concentration will be responsive to tissue metabolism and oxygen consumption by moving the R-to-T transition up or downstream. Note that the rate of a second-order reaction is determined by the product of the concentration of two reactants and the bimolecular rate constant. In this case, the nitrite concentration changes only a little as hemoglobin deoxygenates, the deoxyhemoglobin concentration *increases* dramatically, and the bimolecular rate constant *decreases* dramatically as hemoglobin goes from the R-to-T conformation. So the product of bimolecular rate constant and deoxyheme concentration peaks from 60 to 40% hemoglobin oxygen saturation when the most R_3 tetramers are present. Figures modified and reproduced with permision [78, 83]

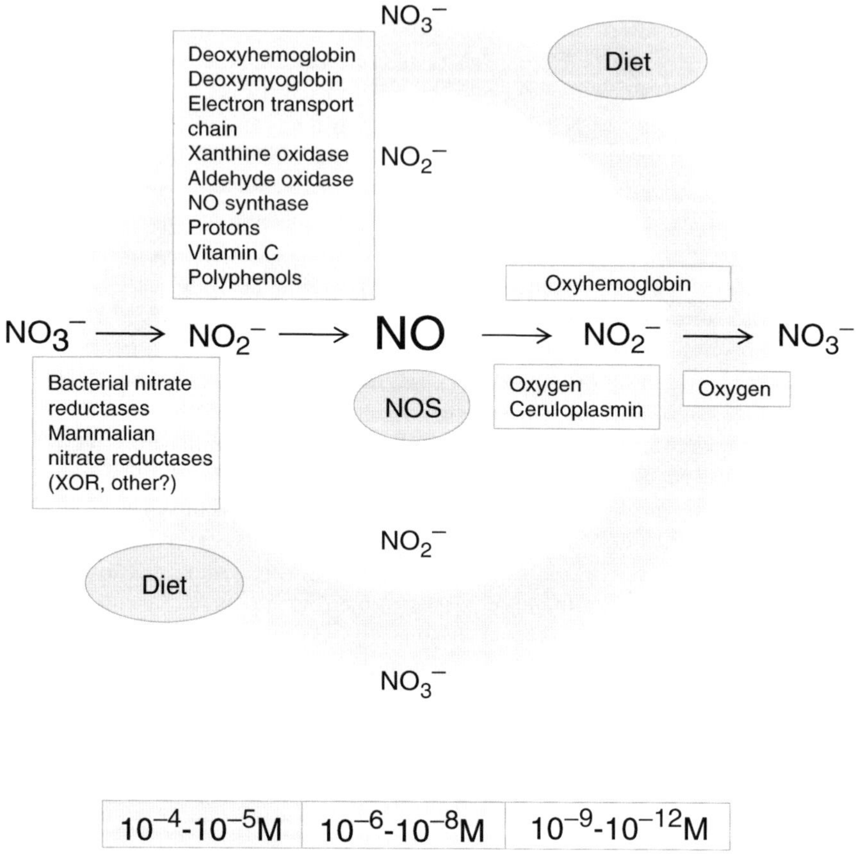

Fig. 3.3 Mechanisms for bioconversion of nitrate and nitrite to NO. Nitric oxide (NO) synthesized by NO synthase (NOS) is a pluripotent biological messenger that controls numerous physiological functions throughout the body. In biological tissues, NO is rapidly inactivated by oxidation to nitrite (NO_2^-) and nitrate (NO_3^-). Nitrate may undergo reduction back to nitrite mainly by commensal bacteria in the oral cavity and to a lesser extent also by mammalian enzymes in tissues. In blood and tissues, numerous pathways exist to further metabolize nitrite to NO and other biologically active nitrogen oxides. NO formation from nitrite is greatly accelerated under hypoxic and acidic conditions. Nitrite reduction represents an alternative to the classical NOS pathway for the generation of NO in blood and tissues. Our diet (mainly green leafy vegetables) is a major contributor to the body pool of nitrate and ingestion of nitrate may fuel the nitrate–nitrite–NO pathway. Dietary nitrate supplementation is associated with robust NO-like effects including a reduction in blood pressure and inhibition of oxygen consumption in humans as well as protection against cardiovascular disease in animal models. Approximate concentration ranges of nitrate, nitrite, and NO in blood and tissues are presented. Figure reproduced with permission [148]

Nitrate–Nitrite and Ischemia–Reperfusion Injury

Treatment of myocardial ischemia due to coronary artery occlusion is aimed at reestablishing perfusion with minimal cardiac injury. However, reperfusion per se may contribute to myocardial damage beyond that induced by ischemia, in a process referred to as ischemia–reperfusion (IR) injury. Several factors are suggested to contribute to the development of myocardial IR injury, including endothelial and microvascular dysfunction, pro-inflammatory activation, and oxidative stress [93]. Decreased NO bioavailability is a central event in IR injury, and importantly contributes to the vascular dysfunction.

In 2004, Webb and coworkers reported protective effects of nitrite in isolated perfused heart preparations subjected to IR injury [65]. Under ischemic conditions both rat and human myocardium generated NO from nitrite. Although NO was produced in a dose-dependent manner (10–100 μM), the degree of protection did not differ, suggesting that the beneficial effect of nitrite treatment was achieved at low concentrations. Furthermore, the authors showed that the conversion to NO was dependent on XOR, since co-administration of XOR inhibitors, allopurinol or BOF-4272, attenuated nitrite dependent NO formation. XOR is generally thought to contribute to IR injuries via production of reactive oxygen species (ROS), including superoxide (O_2^-). However, the findings by Webb et al. suggest that during hypoxic conditions, nitrite supplementation may partly shift the activity of XOR from generation of damaging O_2^- to protective NO.

Duranski and colleagues then demonstrated potent cytoprotective effects of low dose nitrite in vivo in murine models of myocardial infarction and liver ischemia [94]. The effects were independent of NOS and abolished by co-administration of the NO scavenger carboxy-PTIO, suggesting nitrite-derived NO as an active mediator. Furthermore, the efficiency profile of nitrite therapy on liver and heart function was U-shaped, with maximal protection achieved at a dose of 48 nmol nitrite. This supports the idea that small elevations of NO (nano- to micro-molar range) mediate cytoprotection, whereas high, super-physiological levels (high micro- to millimolar range) may promote cellular apoptosis and necrosis [94, 95]. Interestingly, the L-arginine–NOS pathway for NO production is oxygen dependent, whereas the nitrate–nitrite–NO pathway is progressively activated with lower oxygen tensions [7]. Therefore, the nitrate–nitrite–NO pathway can be viewed as a back-up system to ensure sufficient NO levels during ischemic conditions.

A number of subsequent studies in different animal species have confirmed protective effects of low dose nitrite in various settings of IR injury, including models of stroke [96], kidney ischemia [97], lung injury [98, 99], acute myocardial infarction [100], cardiac arrest [101], and chronic limb ischemia [102]. The study by Gonzalez et al. revealed that nitrite dosing during the last 5 min of a 120 min occlusion period reduced myocardial infarction size from 70 to 36% in a canine model [100]. Together, these findings clearly suggest a potential role for nitrite as a useful adjunctive therapy in preventing IR injuries in several organs and tissues.

In 2004, it was demonstrated that ingestion of dietary nitrate resulted in a sustained increase in circulating nitrite levels [33]. This increase could be effectively blocked if the subject avoided swallowing saliva after nitrate ingestion, and rinsing the oral cavity with an antiseptic mouthwash had the same blocking effect [26]. Thus, the acute increase in plasma nitrite after nitrate ingestion is dependent on enterosalivary recirculation of the nitrate and reduction to nitrite by oral commensal bacteria. In addition, Jansson and colleagues more recently reported on the presence of a mammalian nitrate reductase activity in rodent and human tissues [103]. It is intriguing to compare the systemic nitrite load provided by a nitrate-rich meal with the amount of nitrite needed to protect tissues in the setting of IR. In those studies, the maximal protective effects of exogenous nitrite are seen already at a very modest dose [94]. In fact, a similar or even higher systemic load of nitrite is achieved by ingestion of no more than 100 grams of vegetables such as spinach or beetroot [9]. Indeed, several studies have now confirmed that orally administered nitrate and nitrite is protective in IR injury [24, 104, 105].

Although the exact molecular mechanism of nitrite/nitrate-mediated protection is not yet clear, recent studies suggest that modulation of mitochondrial function may play a protective role [101, 105, 106]. To this end, it has been shown that during IR nitrite is able to posttranslationally modify complex I of the mitochondrial respiratory chain resulting in the inhibition of this enzyme's activity and an attenuation of ROS production and apoptotic signaling at reperfusion [105]. Other potential mechanisms include the activation of sGC and the inhibition of deleterious mitochondrial calcium uptake. These ideas are being explored along with the proposal of nitrite as a mediator in the potent ischemic preconditioning cell survival pathway.

Cardioprotective and Blood Pressure Lowering and Effects of Nitrate and Nitrite

Hypertension remains the most common risk factor for cardiovascular morbidity and mortality. NO is a key regulator of cardiovascular function and emerging evidence shows that oxidative stress and subsequent NO deficiency are critically associated with the development of hypertension and other forms of cardiovascular disease [107]. Therefore, treatment modalities that reduce oxidative stress and/or increase NO production may have important implications in preventing and treating cardiovascular disease.

If nitrate can be converted to nitrite and then NO in vivo, it is reasonable that the potent vasodilatory effects of this messenger would lower blood pressure. Larsen and colleagues tested this hypothesis in a double blind, crossover designed study in healthy volunteers [68]. The subjects received sodium nitrate ($NaNO_3$) or placebo sodium chloride (NaCl) for 3 days in a dose of 0.1 mmol kg^{-1} day^{-1}, which corresponds to an intake of 100–300 g of nitrate-rich vegetables. Indeed, diastolic blood pressure (DBP) was reduced by 4 mmHg after nitrate supplementation compared to placebo, which suggests formation of vasodilatory NO. Soon after, the same group repeated a similar study with a greater number of subjects and could show a similar effect also on systolic pressure [108]. Webb and colleagues used beetroot juice as natural source of nitrate to study the same phenomenon in healthy volunteers [69]. Subjects drank 500 mL of either the juice (0.3 mmol nitrate kg^{-1}) or water, and blood pressure was measured repeatedly over a 24-h period. An impressive reduction in both systolic blood pressure (10 mmHg) and DBP (8 mmHg) was noted within 3 h of ingestion, an effect that correlated with peak increases in plasma nitrite concentration. The blood pressure lowering effect was still present after 24 h of a single administration. Interestingly, the blood pressure lowering effects could be completely blocked if the subjects avoided swallowing for a period after drinking the juice, again demonstrating the central role of enterosalivary circulation in bioactivation of nitrate. In the same study, Webb and colleagues also demonstrated inhibitory effects of nitrate on ex vivo platelet aggregation as well as prevention of endothelial dysfunction after a mild ischemic insult in the forearm. In a randomized crossover study, the same group recently demonstrated robust blood pressure lowering effects also with a considerably lower dose of beetroot juice, and effects were similar to those observed with equimolar amounts of potassium nitrate salt [109]. This demonstrates that nitrate is indeed the active component of the juice. In the same study, the authors detected increases in plasma cGMP after nitrate ingestion, which strongly suggests increased NO formation.

In Japan, the occurrence of cardiovascular diseases is low, and Japanese longevity is the highest in the world. In a recent study, Sobko and coworkers examined the blood pressure effects of a 10-day period with traditional Japanese diet rich in vegetables compared with western type diet in 25 healthy volunteers [110]. The traditional Japanese diet is naturally very rich in nitrate and this was reflected in greatly increased levels of nitrate and nitrite in plasma and saliva compared to the control diet. DBP decreased on average 4.5 mmHg with the Japanese diet (~0.3 mmol nitrate kg^{-1} bw^{-1} day^{-1}) compared to the control diet. Again, these findings support the importance of dietary nitrate on blood pressure regulation, and give one possible explanation to the healthy aspects of traditional Japanese food.

Traditional organic nitrates such as nitroglycerine are classically subjected to development of tolerance after repeated administration. To test if dietary nitrate supplementation has sustained blood pressure lowering effects, we measured blood pressure telemetrically in conscious rats during a 5-day period [111]. Nitrate treated (2 mmol nitrate kg^{-1} bw^{-1} day^{-1}) animals displayed a reduction in mean arterial pressure and DBP over the entire observation period. Similar observations have been reported in nonhuman primates with repeated administration of nitrite, indicating no development of tolerance [112].

During the last years, clinical and experimental studies have independently demonstrated that dietary nitrate supplementation reduces blood pressure in healthy normotensive individuals. Considering the important link between NO deficiency and development of hypertension and other forms of cardiovascular disease, it is reasonable to assume that stimulation of the nitrate–nitrite–NO pathway may boost NO production and thus have beneficial effects. In a recent study, Carlström and colleagues tested this hypothesis by studying the effects of dietary nitrate in a rat model of renal and cardiovascular disease, induced by early unilateral nephrectomy in combination with chronic high-salt diet for 10 weeks. Control rats displayed several features of renal and cardiovascular dysfunction, including hypertension, cardiac hypertrophy and fibrosis, proteinuria and histological as well as biochemical signs of renal damage and oxidative stress. In animals receiving nitrate, blood pressure was dose-dependently lowered with no signs of tolerance. Strikingly, proteinuria and histological signs of renal injury were almost completely prevented and the cardiac hypertrophy and fibrosis was attenuated. Mechanistically, dietary nitrate increased or restored tissue levels of bioactive nitrogen oxides and reduced the levels of oxidative stress markers in plasma and urine (M. Carlström, personal communication).

Chronic blockade of NOS with L-NAME results in severe hypertension and progressive kidney damage, and has been used as model for renal and cardiovascular disease. Studies by Tsuchiya and Kanematsu et al. demonstrated that chronic nitrite supplementation (100 mg/L drinking water) attenuated hypertension [113], and that a very low dose of oral nitrite (1 mg/L) protected against L-NAME induced kidney injuries without significant changes in blood pressure [114].

The renal microvasculature plays an important role in blood pressure regulation, and increased preglomerular resistance has been demonstrated in models for hypertension [115, 116]. To further address the role/mechanisms of nitrate–nitrite–NO pathway in the kidney, experiments have been performed with nitrite in isolated renal afferent arterioles. Preliminary results show that nitrite (10 μM) dilates arterioles and attenuates angiotensin II mediated contraction by increasing NO bioavailability. The mechanism for increased NO generation in the renal microvasculature is NOS independent and, rather, involves enzymatic reactions with XOR and activation of cGMP. This finding supports a novel role of nitrite in renal microcirculation and blood pressure control.

In addition to the diet, recent research suggests that the skin can be a considerable source of compounds with NO-like activity [117, 118]. Specifically, it has been proposed that NO-related species from dietary and endogenous sources are stored in the skin and then mobilized by sunlight and delivered to the systemic circulation to exert coronary vasodilator and cardioprotective as well as antihypertensive effects. Opländer and colleagues demonstrated that moderate whole body UVA irradiation (20 J/cm^2) induced formation of S-nitrosothiols in the blood of healthy subjects [117]. Furthermore, UVA irradiation caused a rapid decrease in systolic and DBP which correlated ($R^2=0.74$) with enhanced plasma concentrations of nitrosated species.

Taken together, dietary nitrate may fuel the nitrate–nitrite–NO pathway and partly compensate for disturbances in endogenous NO generation from NOS. The mechanisms for nitrate-mediated antihypertensive effects and renal and cardiovacular protection require further investigations, but modulation of NO bioavailability and reduction in oxidative stress have been suggested. Figure 3.4 highlights the possible role of nitrate and nitrite in signaling and therapeutics.

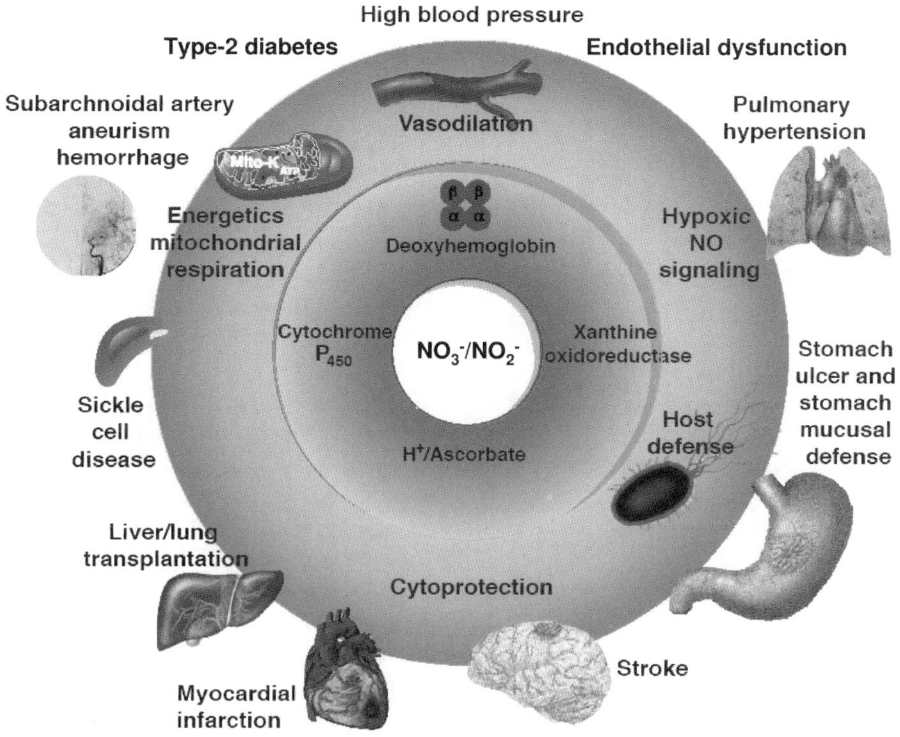

Fig. 3.4 The potential role of the nitrate–nitrite–NO pathway in therapeutics. Modified from ref. [7] and reproduced with permission

Nitrate–Nitrite and Mitochondria and Oxygen Consumption

Recent data suggest that many of the biological effects of nitrite [105, 108] involve interaction with mitochondria. In the last 2 decades, it has been established that the mitochondrion is a physiological target for NO [119]. NO binds reversibly to cytochrome *c* oxidase, the terminal respiratory complex in the mitochondrial electron transport chain, in competition with oxygen. The binding of NO to cytochrome c oxidase, even at concentrations below those that inhibit respiration [120], elicits intracellular signaling events, including elongation of cellular oxygen gradients but also the generation of ROS with potential signaling or damaging effects [121]. These NO-elicited events act as triggers by which mitochondria modulate signal transduction cascades involved in the induction of cellular defense mechanisms and adaptive responses, particularly in response to hypoxia and other environmental stressors [122].

A muscle cell under high metabolic demand is characterized by a very low pO_2 and pH is substantially decreased; conditions which favor nitrite conversion to NO.

With this in mind a study was designed to look at effects of dietary nitrate during physical exercise. Healthy volunteers performed graded exercise on a cycle ergometer while oxygen consumption and circulatory and metabolic parameters were measured. It was found that the metabolic cost of performing standardized constant load exercise was reduced after supplementation with 0.1 mmol kg^{-1} day^{-1} sodium nitrate for three days compared with placebo [108]. This highly surprising effect occurred without any change in lactate concentration, indicating that there was no compensatory increase in glycolytic energy contribution and thus metabolic efficiency seemed to be improved. In subsequent studies, the results have been confirmed and extended using beetroot juice as the nitrate source as well as sodium nitrate salt [123–125]. These studies do not only show that nitrate reduces oxygen cost but intriguingly, the time-to-exhaustion was extended [124, 125], possibly due to a reduced ATP cost of muscle force production [126] or as a direct effect of the improved metabolic efficiency. The molecular mechanisms behind these remarkable effects of nitrate have not been settled in detail but research points towards the mitochondrion as the central target. In a recent study, we examined the effects of dietary nitrate on isolated skeletal muscle mitochondria in healthy volunteers (Larsen et al. in revision). Mitochondria harvested after nitrate supplementation displayed an almost 20% improvement in oxidative phosphorylation efficiency (P/O ratio) and a 45–64% decrease in state 4 respiration and basal oxygen consumption compared to placebo. Moreover, this effect correlated with the improved metabolic efficiency during exercise seen in the same individuals. Mechanistically, expression of the mitochondrial adenine nucleotide translocase (ANT), responsible for a major part of mitochondrial proton leak [127], was reduced by dietary nitrate. In aggregate, nitrate affects fundamental mitochondrial characteristics and the reduction in whole body oxygen consumption seen after ingestion of this anion is likely explained by an increased mitochondrial efficiency involving a decrease in proton leak.

Further evidence that the NO pathway is a fundamental characteristic in cardiovascular health and athletic performance comes from the finding that performance and VO_2 peak correlates robustly with the increase in plasma nitrite after an exercise challenge [128, 129]. Taken together, these findings may have implications, not only for exercise physiology, but also for disease conditions associated with dysfunctional mitochondria and reduced oxygen availability.

Nitrate–Nitrite–Nitric Oxide Pathway and the Metabolic Syndrome

Obesity and hypertension are increasing problems worldwide, resulting in an enormous economic burden to the society [130, 131]. The metabolic syndrome is a combination of medical abnormalities including central obesity, dyslipidemia, hyperglycemia, and hypertension, and it is estimated that around 25% of the world's adult population have the metabolic syndrome [130]. This clustering of metabolic abnormalities that occur in the same individual appears to present a considerably higher cardiovascular risk compared with the sum of the risk associated with each abnormality.

Mice lacking the gene for eNOS do not only develop hypertension, but do also display key features of the metabolic syndrome including dyslipidemia, obesity, and insulin resistance [132–134]. The striking clustering of these particular abnormalities in eNOS-deficient mice has led authors to suggest that a reduced NO bioavailability is a central event in the pathogenesis of metabolic syndrome. In support of this, a polymorphism in the eNOS gene is associated with metabolic syndrome in humans [135]. In a recent study, we investigated the long-term cardiovascular and metabolic effects of nitrate supplementation (0.1 mmol kg^{-1} day^{-1}) in mice lacking eNOS [136]. Previously, Bryan et al. had reported that eNOS-deficient mice have lower levels of nitrite and other nitrogen oxides, which were restored by nitrite supplementation [137]. Remarkably, we could show that inorganic nitrate supplementation for 10 weeks reduced visceral fat accumulation and circulating levels of triglycerides, and reversed the prediabetic phenotype in these animals [136]. These results suggest that stimulation of the nitrate–nitrite–NO pathway can partly compensate for disturbances in endogenous NO generation from eNOS. These findings may have implications for novel nutrition-based preventive and therapeutic strategies against cardiovascular disease and type 2 diabetes.

Interestingly, the dose of dietary nitrate was chosen only to just replace what is being generated by eNOS under normal conditions [138]. The fact that this very modest amount had such profound biological effects supports the intriguing possibility that endogenous nitrate levels are already sufficient to affect cellular processes. Thus, in addition to the second-by-second regulation of vascular tone by eNOS-derived NO, its oxidative end product nitrate may serve as a long-lived reservoir for NO-like bioactivity in tissues. We also noted increased plasma and tissue levels of *S*-nitrosothiols and nitrosylation products with dietary nitrate supplementation [136]. These bioactive compounds may be mediating the observed effects, however, the exact mechanism for this and the signaling pathways involved remains to be elucidated.

Nutritional Implications

It is clear from epidemiological studies that diets rich in fruits and vegetables, such as the Mediterranean or Traditional Japanese diets, protect against development of cardiovascular disease and type 2 diabetes [139–142]. In addition, intervention studies, such as the classical DASH trial, have shown blood pressure lowering effects of such diets [143]; however, the active component(s) responsible for this protection is yet to be pinpointed and trials with single nutrients have generally failed. With the accumulating data on the effects of nitrate in the cardiovascular system, researchers have suggested that nitrate might be one active ingredient in these healthy diets [9, 144–146]. This development is quite astonishing considering that this particular anion is just about the only naturally occurring compound in vegetables that is defined as unwanted and potentially harmful [1]. Although much more research is needed before we can draw any firm conclusions about this, it is striking that the reduction in blood pressure seen by a modest dietary dose of nitrate [68, 69] is similar or even greater than that seen with a vegetable and fruit-rich diet in the DASH study [143]. The possibility of boosting NO production by dietary intervention may have important implications for public health, in particular cardiovascular disease. Future clinical studies will elucidate if nitrate can offer a nutritional approach to prevention and treatment of cardiovascular disease and if such positive effects will outweigh any negative health effects traditionally attributed to this anion. If such investigations point towards an overall protective effect of nitrate we may need to reconsider the current strictly regulated levels of nitrate in food and drinking water.

Summary and Future Perspectives

Accumulating evidence suggests that the supposedly inert inorganic nitrate anion, abundant in vegetables, can be metabolized in vivo to form nitrite, NO, and other bioactive nitrogen oxides. The central role of commensal bacteria in bioactivation of nitrate is intriguing and suggests a symbiotic host–microbial relationship involved in regulation of cardiovascular function [9]. Modest dietary intake of nitrate reduces blood pressure, inhibits platelet function, and prevents endothelial dysfunction after a mild ischemic insult in humans. Moreover, nitrate reduces oxygen cost during exercise and has effects on fundamental mitochondrial functions. In animal studies, nitrate has prolonged blood pressure lowering effects, protects against drug- and salt-induced renal and cardiac injury, enhances postischemic blood flow, and protects against IR injury. While these effects of exogenously delivered nitrate are unequivocal, the physiological relevance of endogenously generated nitrate and nitrite is still to be settled. Nevertheless, experiments using extremely low doses of nitrate or nitrite indicate bioactivity also in the near physiological concentration ranges, and studies with nitrate in amounts titrated to replenish endogenous nitrate generation in eNOS-deficient mice further strengthen this notion.

Although this field of research is still in an early phase it is tempting to speculate that the salutary effects of nitrate and nitrite seen in animal studies and in small trials in healthy volunteers can be utilized also in patients in prevention and treatment of cardiovascular disease. As pharmaceutical agents, inorganic nitrate (NO_3^-) should not be confused with organic nitrates such as nitroglycerine or isosorbide mononitrate which are commonly used in cardiovascular medicine [147]. These latter drugs are lipophilic potent vasodilators that act via rapid generation of NO in smooth muscle. Inorganic nitrate and nitrite also generate NO, but the pharmacokinetic profiles are profoundly different and their potency as vasodilators is orders of magnitude lower. However, a great advantage with nitrate and nitrite, if we view them as potential drug candidates, is that the final step in their bioactivation (formation of NO and other short-lived bioactive nitrogen oxide) is pH and pO_2 dependent and, thus, greatly enhanced where the NO is most needed, i.e., in an ischemic tissue. If nitrate and nitrite were to be used clinically in the future, for example, in treatment and prevention of myocardial infarction or other ischemic events, one can foresee that sufficient formation of active nitrogen oxides locally in the disease area can be achieved without causing generalized NO formation with profound vasodilatation and the risk of hypotension. In addition to these properties, the effects of nitrate and nitrite seem to be devoid of tolerance. Also, one should note that many of the effects of nitrate and nitrite may in fact be unrelated to classical NO-cGMP signaling. As an example, nitrate and nitrite reaction products can directly modulate mitochondrial function via various mechanisms.

Finally, the nutritional implications of nitrate and nitrite biology are among the most intriguing in this area of research. The amounts of these anions needed for the effects on the cardiovascular system, described in this review, are readily achieved via our everyday diet, most easily via a rich intake of vegetables. If the cardiovascular benefits of a diet rich in vegetables turns out to be related to their high nitrate content, we may have to reconsider our current thinking and realize that inorganic nitrate may not necessarily be a threat to human health. Instead, in some years we might even consider it as an essential nutrient.

Acknowledgments Drs. Gladwin and Shiva receive research support from NIH grants RO1HL096973, RC1DK085852 and the Institute for Transfusion Medicine and the Hemophelia Center of Western Pennsylvania. Dr. Shiva also receives research support from the American Heart Association grant 09SDG2150066. Drs. Lundberg and Weitzberg receive research support from Torsten and Ragnar Söderbergs Foundation, The Swedish Reasearch Council, Vinnova (CIDaT), EUs 7th Framework Program (Flaviola), and the Swedish Heart & Lung Foundation.

References

1. Tannenbaum SR, Correa P. Nitrate and gastric cancer risks. Nature. 1985;317:675–6.
2. Mirvish SS. Role of *N*-nitroso compounds (NOC) and *N*-nitrosation in etiology of gastric, esophageal, nasopharyngeal and bladder cancer and contribution to cancer of known exposures to NOC. Cancer Lett. 1995;93:17–48.
3. Green LC, Tannenbaum SR, Goldman P. Nitrate synthesis in the germfree and conventional rat. Science. 1981;212:56–8.
4. Stuehr D, Marletta MA. Mammalian nitrate biosynthesis: mouse macrophages produce nitrite and nitrate in response to *Escherichia coli* lipopolysaccharide. Proc Natl Acad Sci USA. 1985;82:7738–42.
5. Hibbs Jr JB, Taintor RR, Vavrin Z. Macrophage cytotoxicity: role for L-arginine deiminase and imino nitrogen oxidation to nitrite. Science. 1987;235:473–6.
6. Moncada S, Higgs A. The L-arginine-nitric oxide pathway. N Engl J Med. 1993;329:2002–12.
7. Lundberg JO, Weitzberg E, Gladwin MT. The nitrate-nitrite-nitric oxide pathway in physiology and therapeutics. Nat Rev Drug Discov. 2008;7:156–67.
8. Weitzberg E, Lundberg JO. Nonenzymatic nitric oxide production in humans. Nitric Oxide. 1998;2:1–7.

9. Lundberg JO, Weitzberg E, Cole JA, Benjamin N. Nitrate, bacteria and human health. Nat Rev Microbiol. 2004;2:593–602.
10. Lundberg JO, Weitzberg E. NO generation from nitrite and its role in vascular control. Arterioscler Thromb Vasc Biol. 2005;25:915–22.
11. Gladwin MT, Schechter AN, Kim-Shapiro DB, et al. The emerging biology of the nitrite anion. Nat Chem Biol. 2005;1:308–14.
12. Zweier JL, Samouilov A, Kuppusamy P. Non-enzymatic nitric oxide synthesis in biological systems. Biochim Biophys Acta. 1999;1411:250–62.
13. Benjamin N, O'Driscoll F, Dougall H, et al. Stomach NO synthesis. Nature. 1994;368:502.
14. Lundberg JO, Weitzberg E, Lundberg JM, Alving K. Intragastric nitric oxide production in humans: measurements in expelled air. Gut. 1994;35:1543–6.
15. Zweier JL, Wang P, Samouilov A, Kuppusamy P. Enzyme-independent formation of nitric oxide in biological tissues. Nature Med. 1995;1:804–9.
16. Gladwin MT, Shelhamer JH, Schechter AN, et al. Role of circulating nitrite and S-nitrosohemoglobin in the regulation of regional blood flow in humans. Proc Natl Acad Sci USA. 2000;97:11482–7.
17. Cosby K, Partovi KS, Crawford JH, et al. Nitrite reduction to nitric oxide by deoxyhemoglobin vasodilates the human circulation. Nat Med. 2003;9:1498–505.
18. Doherty DH, Doyle MP, Curry SR, et al. Rate of reaction with nitric oxide determines the hypertensive effect of cell-free hemoglobin. Nat Biotechnol. 1998;16:672–6.
19. Dou Y, Maillett DH, Eich RF, Olson JS. Myoglobin as a model system for designing heme protein based blood substitutes. Biophys Chem. 2002;98:127–48.
20. Gladwin MT, Ognibene FP, Pannell LK, et al. Relative role of heme nitrosylation and beta-cysteine 93 nitrosation in the transport and metabolism of nitric oxide by hemoglobin in the human circulation. Proc Natl Acad Sci USA. 2000;97:9943–8.
21. Shiva S, Wang X, Ringwood LA, et al. Ceruloplasmin is a NO oxidase and nitrite synthase that determines endocrine NO homeostasis. Nat Chem Biol. 2006;2(9):486–93.
22. Crawford JH, Chacko BK, Pruitt HM, Piknova B, Hogg N, Patel RP. Transduction of NO-bioactivity by the red blood cell in sepsis: novel mechanisms of vasodilation during acute inflammatory disease. Blood. 2004;104:1375–82.
23. Jungersten L, Ambring A, Wall B, Wennmalm A. Both physical fitness and acute exercise regulate nitric oxide formation in healthy humans. J Appl Physiol. 1997;82:760–4.
24. Lundberg JO, Gladwin MT, Ahluwalia A, et al. Nitrate and nitrite in biology, nutrition and therapeutics. Nat Chem Biol. 2009;5:865–9.
25. Spiegelhalder B, Eisenbrand G, Preussman R. Influence of dietary nitrate on nitrite content of human saliva: possible relevance to *in vivo* formation of N-nitroso compounds. Food Cosmet Toxicol. 1976;14:545–8.
26. Govoni M, Jansson EA, Weitzberg E, Lundberg JO. The increase in plasma nitrite after a dietary nitrate load is markedly attenuated by an antibacterial mouthwash. Nitric Oxide. 2008;19:333–7.
27. Duncan C, Dougall H, Johnston P, et al. Chemical generation of nitric oxide in the mouth from the enterosalivary circulation of dietary nitrate [see comments]. Nat Med. 1995;1:546–51.
28. Bjorne HH, Petersson J, Phillipson M, Weitzberg E, Holm L, Lundberg JO. Nitrite in saliva increases gastric mucosal blood flow and mucus thickness. J Clin Invest. 2004;113:106–14.
29. Jansson EA, Petersson J, Reinders C, et al. Protection from nonsteroidal anti-inflammatory drug (NSAID)-induced gastric ulcers by dietary nitrate. Free Radic Biol Med. 2007;42:510–8.
30. Petersson J, Phillipson M, Jansson EA, Patzak A, Lundberg JO, Holm L. Dietary nitrate increases gastric mucosal blood flow and mucosal defense. Am J Physiol Gastrointest Liver Physiol. 2007;292:G718–24.
31. Miyoshi M, Kasahara E, Park AM, et al. Dietary nitrate inhibits stress-induced gastric mucosal injury in the rat. Free Radic Res. 2003;37:85–90.

32. Bjorne H, Weitzberg E, Lundberg JO. Intragastric generation of antimicrobial nitrogen oxides from saliva–physiological and therapeutic considerations. Free Radic Biol Med. 2006;41:1404–12.
33. Lundberg JO, Govoni M. Inorganic nitrate is a possible source for systemic generation of nitric oxide. Free Radic Biol Med. 2004;37:395–400.
34. Bryan NS. Nitrite in nitric oxide biology: cause or consequence? A systems-based review. Free Radic Biol Med. 2006;41:691–701.
35. van Faassen EE, Bahrami S, Feelisch M, et al. Nitrite as regulator of hypoxic signaling in mammalian physiology. Med Res Rev. 2009;29:683–741.
36. Dykhuizen RS, Frazer R, Duncan C, et al. Antimicrobial effect of acidified nitrite on gut pathogens: importance of dietary nitrate in host defense. Antimicrob Agents Chemother. 1996;40:1422–5.
37. Duncan C, Li H, Dykhuizen R, et al. Protection against oral and gastrointestinal diseases: importance of dietary nitrate intake, oral nitrate reduction and enterosalivary nitrate circulation. Comp Biochem Physiol A Physiol. 1997;118:939–48.
38. Larauche M, Anton PM, Garcia-Villar R, et al. Protective effect of dietary nitrate on experimental gastritis in rats. Br J Nutr. 2003;89:777–86.
39. Larauche M, Bueno L, Fioramonti J. Effect of dietary nitric oxide on gastric mucosal mast cells in absence or presence of an experimental gastritis in rats. Life Sci. 2003;73:1505–16.
40. Sobko T, Reinders C, Norin E, Midtvedt T, Gustafsson LE, Lundberg JO. Gastrointestinal nitric oxide generation in germ-free and conventional rats. Am J Physiol Gastrointest Liver Physiol. 2004;287:G993–7.
41. Bjorne H, Govoni M, Tornberg DC, Lundberg JO, Weitzberg E. Intragastric nitric oxide is abolished in intubated patients and restored by nitrite. Crit Care Med. 2005;33:1722–7.
42. Gago B, Lundberg JO, Barbosa RM, Laranjinha J. Red wine-dependent reduction of nitrite to nitric oxide in the stomach. Free Radic Biol Med. 2007;43:1233–42.
43. Peri L, Pietraforte D, Scorza G, Napolitano A, Fogliano V, Minetti M. Apples increase nitric oxide production by human saliva at the acidic pH of the stomach: a new biological function for polyphenols with a catechol group? Free Radic Biol Med. 2005;39:668–81.
44. Carlsson S, Wiklund NP, Engstrand L, Weitzberg E, Lundberg JO. Effects of pH, nitrite, and ascorbic acid on nonenzymatic nitric oxide generation and bacterial growth in urine. Nitric Oxide. 2001;5:580–6.
45. Gago B, Nystrom T, Cavaleiro C, et al. The potent vasodilator ethyl nitrite is formed upon reaction of nitrite and ethanol under gastric conditions. Free Radic Biol Med. 2008;45:404–12.
46. Stamler JS, Lamas S, Fang FC. Nitrosylation. The prototypic redox-based signaling mechanism. Cell. 2001;106:675–83.
47. O'Donnell VB, Eiserich JP, Chumley PH, et al. Nitration of unsaturated fatty acids by nitric oxide-derived reactive nitrogen species peroxynitrite, nitrous acid, nitrogen dioxide, and nitronium ion. Chem Res Toxicol. 1999;12:83–92.
48. Cui T, Schopfer FJ, Zhang J, et al. Nitrated fatty acids: endogenous anti-inflammatory signaling mediators. J Biol Chem. 2006;281:35686–98.
49. Coles B, Bloodsworth A, Clark SR, et al. Nitrolinoleate inhibits superoxide generation, degranulation, and integrin expression by human neutrophils: novel antiinflammatory properties of nitric oxide-derived reactive species in vascular cells. Circ Res. 2002;91:375–81.
50. Weiss S, Wilkins RW, Haynes FW. The nature of circulatory collapse induced by sodium nitrite. J Clin Investig. 1937;16:73–84.
51. Furchgott RF, Bhadrakom S. Reactions of strips of rabbit aorta to epinephrine, isopropylarterenol, sodium nitrite and other drugs. J Pharmacol Exp Ther. 1953;108:129–43.
52. Mittal CK, Arnold WP, Murad F. Characterization of protein inhibitors of guanylate cyclase activation from rat heart and bovine lung. J Biol Chem. 1978;253:1266–71.
53. Ignarro LJ, Lippton H, Edwards JC, et al. Mechanism of vascular smooth muscle relaxation by organic nitrates, nitrites, nitroprusside and nitric oxide: evidence for the involvement of S-nitrosothiols as active intermediates. J Pharmacol Exp Ther. 1981;218:739–49.

54. Ignarro LJ, Gruetter CA. Requirement of thiols for activation of coronary arterial guanylate cyclase by glyceryl trinitrate and sodium nitrite: possible involvement of S-nitrosothiols. Biochim Biophys Acta. 1980;631:221–31.
55. Lauer T, Preik M, Rassaf T, et al. Plasma nitrite rather than nitrate reflects regional endothelial nitric oxide synthase activity but lacks intrinsic vasodilator action. Proc Natl Acad Sci USA. 2001;98:12814–9.
56. Allen BW, Piantadosi CA. How do red blood cells cause hypoxic vasodilation? The SNO-hemoglobin paradigm. Am J Physiol Heart Circ Physiol. 2006;291:H1507–12.
57. McMahon TJ. Hemoglobin and nitric oxide. N Engl J Med. 2003;349:402–5. author reply -5.
58. Modin A, Bjorne H, Herulf M, Alving K, Weitzberg E, Lundberg JO. Nitrite-derived nitric oxide: a possible mediator of 'acidic-metabolic' vasodilation. Acta Physiol Scand. 2001;171:9–16.
59. Cannon 3rd RO, Schechter AN, Panza JA, et al. Effects of inhaled nitric oxide on regional blood flow are consistent with intravascular nitric oxide delivery. J Clin Invest. 2001;108:279–87.
60. Lang Jr JD, Teng X, Chumley P, et al. Inhaled NO accelerates restoration of liver function in adults following orthotopic liver transplantation. J Clin Invest. 2007;117:2583–91.
61. Kozlov AV, Costantino G, Sobhian B, et al. Mechanisms of vasodilatation induced by nitrite instillation in intestinal lumen: possible role of hemoglobin. Antioxid Redox Signal. 2005;7:515–21.
62. Tsuchiya K, Kanematsu Y, Yoshizumi M, et al. Nitrite is an alternative source of NO in vivo. Am J Physiol Heart Circ Physiol. 2004;288(5):H2163–70.
63. Tsuchiya K, Takiguchi Y, Okamoto M, et al. Malfunction of vascular control in lifestyle-related diseases: formation of systemic hemoglobin-nitric oxide complex (HbNO) from dietary nitrite. J Pharmacol Sci. 2004;96:395–400.
64. Hunter CJ, Dejam A, Blood AB, et al. Inhaled nebulized nitrite is a hypoxia-sensitive NO-dependent selective pulmonary vasodilator. Nat Med. 2004;10:1122–7.
65. Webb A, Bond R, McLean P, Uppal R, Benjamin N, Ahluwalia A. Reduction of nitrite to nitric oxide during ischemia protects against myocardial ischemia-reperfusion damage. Proc Natl Acad Sci USA. 2004;101:13683–8.
66. Dejam A, Hunter CJ, Gladwin MT. Effects of dietary nitrate on blood pressure. N Engl J Med. 2007;356:1590. author reply.
67. Dejam A, Hunter CJ, Tremonti C, et al. Nitrite infusion in humans and nonhuman primates: endocrine effects, pharmacokinetics, and tolerance formation. Circulation. 2007;116:1821–31.
68. Larsen FJ, Ekblom B, Sahlin K, Lundberg JO, Weitzberg E. Effects of dietary nitrate on blood pressure in healthy volunteers. N Engl J Med. 2006;355:2792–3.
69. Webb AJ, Patel N, Loukogeorgakis S, et al. Acute blood pressure lowering, vasoprotective, and antiplatelet properties of dietary nitrate via bioconversion to nitrite. Hypertension. 2008;51:784–90.
70. Kapil V, Milsom AB, Okorie M, et al. Inorganic nitrate supplementation lowers blood pressure in humans: role for nitrite-derived NO. Hypertension. 2010;56:274–81.
71. Roy CS, Brown JG. The blood pressure and its variation in the arterioles, capillaries and smaller veins. J Physiol. 1879;2:323.
72. Tune JD, Gorman MW, Feigl EO. Matching coronary blood flow to myocardial oxygen consumption. J Appl Physiol. 2004;97:404–15.
73. Ross JM, Fairchild HM, Weldy J, Guyton AC. Autoregulation of blood flow by oxygen lack. Am J Physiol. 1962;202:21–4.
74. Nagababu E, Ramasamy S, Abernethy DR, Rifkind JM. Active nitric oxide produced in the red cell under hypoxic conditions by deoxyhemoglobin mediated nitrite reduction. J Biol Chem. 2003;278(47):46349–56.
75. Maher AR, Milsom AB, Gunaruwan P, et al. Hypoxic modulation of exogenous nitrite-induced vasodilation in humans. Circulation. 2008;117(5):670–7.
76. Brooks J. The action of nitrite on haemoglobin in the absence of oxygen. Proc R Soc Med. 1937;123:368–82.

77. Doyle MP, Pickering RA, DeWeert TM, Hoekstra JW, Pater D. Kinetics and mechanism of the oxidation of human deoxyhemoglobin by nitrites. J Biol Chem. 1981;256:12393–8.
78. Huang Z, Shiva S, Kim-Shapiro DB, et al. Enzymatic function of hemoglobin as a nitrite reductase that produces NO under allosteric control. J Clin Invest. 2005;115:2099–107.
79. Crawford JH, Isbell TS, Huang Z, et al. Hypoxia, red blood cells, and nitrite regulate NO-dependent hypoxic vasodilation. Blood. 2006;107:566–74.
80. Huang KT, Keszler A, Patel N, et al. The reaction between nitrite and deoxyhemoglobin. Reassessment of reaction kinetics and stoichiometry. J Biol Chem. 2005;280:31126–31.
81. Huang Z, Shiva S, Kim-Shapiro DB, et al. Enzymatic function of hemoglobin as a nitrite reductase that produces NO under allosteric control. J Clin Invest. 2005;115:2099–107.
82. Tsai AG, Johnson PC, Intaglietta M. Oxygen gradients in the microcirculation. Physiol Rev. 2003;83:933–63.
83. Gladwin MT, Kim-Shapiro DB. The functional nitrite reductase activity of the heme-globins. Blood. 2008;112:2636–47.
84. Shiva S, Huang Z, Grubina R, et al. Deoxymyoglobin is a nitrite reductase that generates nitric oxide and regulates mitochondrial respiration. Circ Res. 2007;100:654–61.
85. Rassaf T, Flogel U, Drexhage C, Hendgen-Cotta U, Kelm M, Schrader J. Nitrite reductase function of deoxymyoglobin: oxygen sensor and regulator of cardiac energetics and function. Circ Res. 2007;100:1749–54.
86. Millar TM, Stevens CR, Benjamin N, Eisenthal R, Harrison R, Blake DR. Xanthine oxidoreductase catalyses the reduction of nitrates and nitrite to nitric oxide under hypoxic conditions. FEBS Lett. 1998;427:225–8.
87. Li H, Cui H, Liu X, Zweier JL. Xanthine oxidase catalyzes anaerobic transformation of organic nitrates to nitric oxide and nitrosothiols: characterization of this mechanism and the link between organic nitrate and guanylyl cyclase activation. J Biol Chem. 2005;280:16594–600.
88. Li H, Samouilov A, Liu X, Zweier JL. Characterization of the magnitude and kinetics of xanthine oxidase-catalyzed nitrite reduction. Evaluation of its role in nitric oxide generation in anoxic tissues. J Biol Chem. 2001;276:24482–9.
89. Kozlov AV, Staniek K, Nohl H. Nitrite reductase activity is a novel function of mammalian mitochondria. FEBS Lett. 1999;454:127–30.
90. Zweier JL, Li H, Samouilov A, Liu X. Mechanisms of nitrite reduction to nitric oxide in the heart and vessel wall. Nitric Oxide. 2010;22:83–90.
91. Aamand R, Dalsgaard T, Jensen FB, Simonsen U, Roepstorff A, Fago A. Generation of nitric oxide from nitrite by carbonic anhydrase: a possible link between metabolic activity and vasodilation. Am J Physiol Heart Circ Physiol. 2009;297:H2068–74.
92. Vanin AF, Bevers LM, Slama-Schwok A, van Faassen EE. Nitric oxide synthase reduces nitrite to NO under anoxia. Cell Mol Life Sci. 2007;64:96–103.
93. Jung C, Gonon AT, Sjoquist PO, Lundberg JO, Pernow J. Arginase inhibition mediates cardioprotection during ischaemia-reperfusion. Cardiovasc Res 2009.
94. Duranski MR, Greer JJ, Dejam A, et al. Cytoprotective effects of nitrite during in vivo ischemia-reperfusion of the heart and liver. J Clin Invest. 2005;115:1232–40.
95. Dimmeler S, Zeiher AM. Nitric oxide and apoptosis: another paradigm for the double-edged role of nitric oxide. Nitric Oxide. 1997;1:275–81.
96. Jung KH, Chu K, Ko SY, et al. Early intravenous infusion of sodium nitrite protects brain against in vivo ischemia-reperfusion injury. Stroke. 2006;37:2744–50.
97. Tripatara P, Patel NS, Webb A, et al. Nitrite-derived nitric oxide protects the rat kidney against ischemia/reperfusion injury in vivo: role for xanthine oxidoreductase. J Am Soc Nephrol. 2007;18:570–80.
98. Zuckerbraun BS, Shiva S, Ifedigbo E, et al. Nitrite potently inhibits hypoxic and inflammatory pulmonary arterial hypertension and smooth muscle proliferation via xanthine oxidoreductase-dependent nitric oxide generation. Circulation. 2010;121:98–109.
99. Dias-Junior CA, Gladwin MT, Tanus-Santos JE. Low-dose intravenous nitrite improves hemodynamics in a canine model of acute pulmonary thromboembolism. Free Radic Biol Med. 2006;41:1764–70.

100. Gonzalez FM, Shiva S, Vincent PS, et al. Nitrite anion provides potent cytoprotective and antiapoptotic effects as adjunctive therapy to reperfusion for acute myocardial infarction. Circulation. 2008;117:2986–94.
101. Dezfulian C, Shiva S, Alekseyenko A, et al. Nitrite therapy after cardiac arrest reduces reactive oxygen species generation, improves cardiac and neurological function, and enhances survival via reversible inhibition of mitochondrial complex I. Circulation. 2009;120:897–905.
102. Kumar D, Branch BG, Pattillo CB, et al. Chronic sodium nitrite therapy augments ischemia-induced angiogenesis and arteriogenesis. Proc Natl Acad Sci USA. 2008;105:7540–5.
103. Jansson EA, Huang L, Malkey R, et al. A mammalian functional nitrate reductase that regulates nitrite and nitric oxide homeostasis. Nat Chem Biol. 2008;4:411–7.
104. Bryan NS, Calvert JW, Elrod JW, Gundewar S, Ji SY, Lefer DJ. Dietary nitrite supplementation protects against myocardial ischemia-reperfusion injury. Proc Natl Acad Sci USA. 2007;104:19144–9.
105. Shiva S, Sack MN, Greer JJ, et al. Nitrite augments tolerance to ischemia/reperfusion injury via the modulation of mitochondrial electron transfer. J Exp Med. 2007;204:2089–102.
106. Raat NJ, Noguchi AC, Liu VB, et al. Dietary nitrate and nitrite modulate blood and organ nitrite and the cellular ischemic stress response. Free Radic Biol Med. 2009;47:510–7.
107. Wilcox CS. Oxidative stress and nitric oxide deficiency in the kidney: a critical link to hypertension? Am J Physiol Regul Integr Comp Physiol. 2005;289:R913–35.
108. Larsen FJ, Weitzberg E, Lundberg JO, Ekblom B. Effects of dietary nitrate on oxygen cost during exercise. Acta Physiol (Oxford). 2007;191:59–66.
109. Kapil V, Milsom AB, Okorie M, et al. Inorganic nitrate supplementation lowers blood pressure in humans: role for nitrite-derived NO. Hypertension. 2010;56:274–81.
110. Sobko T, Marcus C, Govoni M, Kamiya S. Dietary nitrate in Japanese traditional foods lowers diastolic blood pressure in healthy volunteers. Nitric Oxide. 2010;22:136–40.
111. Petersson J, Carlstrom M, Schreiber O, et al. Gastroprotective and blood pressure lowering effects of dietary nitrate are abolished by an antiseptic mouthwash. Free Radic Biol Med. 2009;46:1068–75.
112. Dejam A, Hunter CJ, Tremonti C, et al. Nitrite infusion in humans and nonhuman primates. Endocrine effects, pharmacokinetics, and tolerance formation. Circulation. 2007;116(16): 1821–31.
113. Tsuchiya K, Kanematsu Y, Yoshizumi M, et al. Nitrite is an alternative source of NO in vivo. Am J Physiol Heart Circ Physiol. 2005;288:H2163–70.
114. Kanematsu Y, Yamaguchi K, Ohnishi H, et al. Dietary doses of nitrite restore circulating nitric oxide level and improve renal injury in L-NAME-induced hypertensive rats. Am J Physiol Ren Physiol. 2008;295:F1457–62.
115. Michel FS, Man RY, Vanhoutte PM. Increased spontaneous tone in renal arteries of spontaneously hypertensive rats. Am J Physiol Heart Circ Physiol. 2007;293:H1673–81.
116. Ren Y, D'Ambrosio MA, Liu R, Pagano PJ, Garvin JL, Carretero OA. Enhanced myogenic response in the afferent arteriole of spontaneously hypertensive rats. Am J Physiol Heart Circ Physiol. 2010;298:H1769–75.
117. Oplander C, Volkmar CM, Paunel-Gorgulu A, et al. Whole body UVA irradiation lowers systemic blood pressure by release of nitric oxide from intracutaneous photolabile nitric oxide derivates. Circ Res. 2009;105:1031–40.
118. Feelisch M, Kolb-Bachofen V, Liu D, et al. Is sunlight good for our heart? Eur Heart J. 2010;31:1041–5.
119. Erusalimsky JD, Moncada S. Nitric oxide and mitochondrial signaling: from physiology to pathophysiology. Arterioscler Thromb Vasc Biol. 2007;27:2524–31.
120. Palacios-Callender M, Hollis V, Frakich N, Mateo J, Moncada S. Cytochrome c oxidase maintains mitochondrial respiration during partial inhibition by nitric oxide. J Cell Sci. 2007;120:160–5.
121. Brookes P, Darley-Usmar VM. Hypothesis: the mitochondrial NO(*) signaling pathway, and the transduction of nitrosative to oxidative cell signals: an alternative function for cytochrome C oxidase. Free Radic Biol Med. 2002;32:370–4.

122. Unitt DC, Hollis VS, Palacios-Callender M, Frakich N, Moncada S. Inactivation of nitric oxide by cytochrome c oxidase under steady-state oxygen conditions. Biochim Biophys Acta. 2010;1797:371–7.
123. Larsen FJ, Weitzberg E, Lundberg JO, Ekblom B. Dietary nitrate reduces maximal oxygen consumption while maintaining work performance at maximal exercise. Free Radic Biol Med 2009.
124. Bailey SJ, Winyard P, Vanhatalo A, et al. Dietary nitrate supplementation reduces the O_2 cost of low-intensity exercise and enhances tolerance to high-intensity exercise in humans. J Appl Physiol. 2009;107:1144–55.
125. Vanhatalo A, Bailey SJ, Blackwell JR, et al. Acute and chronic effects of dietary nitrate supplementation on blood pressure and the physiological responses to moderate-intensity and incremental exercise. Am J Physiol Regul Integr Comp Physiol. 2010;299(4):R1121–31.
126. Bailey SJ, Fulford J, Vanhatalo A, et al. Dietary nitrate supplementation enhances muscle contractile efficiency during knee-extensor exercise in humans. J Appl Physiol. 2010;109:135–48.
127. Gouspillou G, Bourdel-Marchasson I, Rouland R, et al. Alteration of mitochondrial oxidative phosphorylation in aged skeletal muscle involves modification of adenine nucleotide translocator. Biochim Biophys Acta. 2010;1797:143–51.
128. Rassaf T, Lauer T, Heiss C, et al. Nitric oxide synthase-derived plasma nitrite predicts exercise capacity. Br J Sports Med. 2007;41:669–73. discussion 73.
129. Allen JD, Stabler T, Kenjale A, et al. Plasma nitrite flux predicts exercise performance in peripheral arterial disease after 3 months of exercise training. Free Radic Biol Med. 2010;49:1138–44.
130. Eckel RH, Grundy SM, Zimmet PZ. The metabolic syndrome. Lancet. 2005;365:1415–28.
131. Despres JP, Lemieux I. Abdominal obesity and metabolic syndrome. Nature. 2006;444:881–7.
132. Huang PL. eNOS, metabolic syndrome and cardiovascular disease. Trends Endocrinol Metab. 2009;20:295–302.
133. Cook S, Hugli O, Egli M, et al. Clustering of cardiovascular risk factors mimicking the human metabolic syndrome X in eNOS null mice. Swiss Med Wkly. 2003;133:360–3.
134. Nisoli E, Clementi E, Paolucci C, et al. Mitochondrial biogenesis in mammals: the role of endogenous nitric oxide. Science. 2003;299:896–9.
135. Monti LD, Barlassina C, Citterio L, et al. Endothelial nitric oxide synthase polymorphisms are associated with type 2 diabetes and the insulin resistance syndrome. Diabetes. 2003;52:1270–5.
136. Carlstrom. Dietary inorganic nitrate reverses features of metabolic syndrome in endothelial nitric oxide synthase deficient mice. PNAS 2010;(in press).
137. Bryan NS, Calvert JW, Gundewar S, Lefer DJ. Dietary nitrite restores NO homeostasis and is cardioprotective in endothelial nitric oxide synthase-deficient mice. Free Radic Biol Med. 2008;45:468–74.
138. Wickman A, Klintland N, Gan LM, et al. A technique to estimate the rate of whole body nitric oxide formation in conscious mice. Nitric Oxide. 2003;9:77–85.
139. Willett WC. Diet and health: what should we eat? Science. 1994;264:532–7.
140. Liese AD, Nichols M, Sun X, D'Agostino Jr RB, Haffner SM. Adherence to the DASH diet is inversely associated with incidence of type 2 diabetes: the insulin resistance atherosclerosis study. Diab Care. 2009;32:1434–6.
141. Joshipura KJ, Ascherio A, Manson JE, et al. Fruit and vegetable intake in relation to risk of ischemic stroke. JAMA. 1999;282:1233–9.
142. Joshipura KJ, Hu FB, Manson JE, et al. The effect of fruit and vegetable intake on risk for coronary heart disease. Ann Intern Med. 2001;134:1106–14.
143. Appel LJ, Moore TJ, Obarzanek E, et al. A clinical trial of the effects of dietary patterns on blood pressure. DASH Collaborative Research Group. N Engl J Med. 1997;336:1117–24.
144. Lundberg JO, Feelisch M, Bjorne H, Jansson EA, Weitzberg E. Cardioprotective effects of vegetables: is nitrate the answer? Nitric Oxide. 2006;15:359–62.

145. Classen HG, Stein-Hammer C, Thoni H. Hypothesis: the effect of oral nitrite on blood pressure in the spontaneously hypertensive rat. Does dietary nitrate mitigate hypertension after conversion to nitrite? J Am Coll Nutr. 1990;9:500–2.
146. Lundberg JO, Weitzberg E. NO generation from inorganic nitrate and nitrite: role in physiology, nutrition and therapeutics. Arch Pharm Res. 2009;32:1119–26.
147. Butler AR, Feelisch M. Therapeutic uses of inorganic nitrite and nitrate: from the past to the future. Circulation. 2008;117:2151–9.
148. Lundberg JO, Weitzberg E. NO-synthase independent NO generation in mammals. Biochem Biophys Res Commun. 2010;396:39–45.

Chapter 4
Sources of Exposure to Nitrogen Oxides

Andrew L. Milkowski

Key points

- There are five primary sources of exposure to nitrite and nitrate:
    1. Endogenous production of nitric oxide
    2. Dietary exposure to nitrite and nitrate in food
    3. Drinking water
    4. Swallowing saliva
    5. Environmental/atmospheric exposure to nitric oxide and nitrogen dioxide
- The majority of exposure to nitrite and nitrate in most people is through dietary intake of vegetables.
- Cured and processed meats contribute very little to the overall burden of exposure.
- Individual levels of exposure are dependent upon where and how one lives and works, what types of food one eats, and lifestyle habits.
- Low to moderate exposure to these sources poses little risk with known benefits while overexposure can lead to health risks.

Keywords Nitrogen dioxide • Nitric oxide • Nitrite • Nitrate • Air pollution • Environmental • Combustion

Introduction

Humans are exposed to nitrogen oxides that include the toxicologically significant gases nitric oxide (NO) and nitrogen dioxide (NO_2), and the non-volatile salts of nitrate and nitrite ions. Nitric oxide is a sharp, sweet-smelling gas at

A.L. Milkowski (✉)
Department of Animal Science, University of Wisconsin, Madison, WI, USA
e-mail: amilkowski@ansci.wisc.edu

N.S. Bryan and J. Loscalzo (eds.), *Nitrite and Nitrate in Human Health and Disease*,
Nutrition and Health, DOI 10.1007/978-1-60761-616-0_4,

room temperature, whereas nitrogen dioxide (NO_2) has a strong, harsh odor and is a liquid at refrigerated and cool room temperatures, becoming a reddish-brown gas above 70°F. Both molecules are free radicals. Before it was realized that NO is produced in our body, they were recognized as toxic air pollutants. The majority of the gaseous nitrogen oxides are metabolized to nitrite and nitrate in the body. Many chapters in this monograph refer to multiple sources of nitrite, nitrate, and nitric oxide in human physiology. In an attempt to summarize the discussion relative to human exposure, they can be grouped into some key areas, and this chapter focuses primarily on exposure of NO, nitrite, and nitrate. In 2006, the International Agency for Research on Cancer (IARC) developed an extensive exposure estimate of nitrite and nitrate during the preparation of their monograph on the carcinogenic potential of nitrate and nitrite under conditions of endogenous nitrosation [1]. This monograph provides an excellent source of exposure information which has been supplemented with additional reports [2, 3]. Because of the chemical reactivity of nitric oxide and nitrogen dioxide, estimates for human exposure have largely defaulted to those of nitrate and nitrite.

Endogenous Production

The demonstration of NO formation by an enzyme in vascular endothelial cells in 1987 has since had profound implications in research and medicine. NO was shown to be a potent vasodilator, inhibitor of platelet aggregation, and active species of nitroglycerin before the discovery of endothelium-derived relaxing factor (EDRF) in 1980 [4]. Subsequent studies revealed that EDRF is NO and is synthesized by mammalian cells from L-arginine through a complex oxidation reaction catalyzed by the flavo-hemoprotein NO synthase (NOS) [5]. NOS catalyzes the NADPH- and oxygen-dependent oxidation of L-arginine to NO plus L-citrulline in a reaction that requires several cofactors in the presence of oxygen, including NADPH, FAD, FMN, tetrahydrobiopterin, heme, reduced glutathione, and calmodulin. Enzyme-bound calmodulin facilitates the transfer of electrons from NADPH to the flavoprotein domain of NOS and also from the flavins to the heme domain of NOS [5, 7]. These electrons are used to reduce the iron to the ferrous state so that it can bind oxygen, which is incorporated into the substrate, arginine, to generate NO plus citrulline. The endogenous production of NO by NOS has been established as playing an important role in vascular homeostasis, neurotransmission, and host defense mechanisms [4]. The major removal pathway for NO is the stepwise oxidation to nitrite and nitrate [6]. Although some NO can be oxidized to nitrogen dioxide, it will quickly nitrosate water to form nitrite. In plasma, NO is oxidized almost completely to nitrite, in which form it will remain stable for several hours [8, 9]. In contrast, NO and nitrite are rapidly oxidized to nitrate in whole blood. NO can also be enzymatically oxidized to nitrite by cerulopasmin or other metal-containing proteins [10]. The half-life of nitrite in human blood is about 110 s [11]. Nitrate, by contrast, has a circulating half life of 5–8 h [12, 13]. Tissue nitrite and nitrate, on the other hand, both have in vivo half-lives of tens of minutes [14]. During fasting conditions with a low previous intake of nitrite/nitrate, enzymatic NO formation accounts for the majority of nitrite [15]. On the basis of these studies, it was believed that NO metabolism acutely terminates by oxidation to nitrite and nitrate. Early studies on nitrogen balance in humans and analyses of fecal and ileostomy samples indicated that nitrite and nitrate are formed de novo in the intestine. It was these findings by Tannenbaum et al. [16] that significantly altered our conceptions of human exposure to exogenous nitrite and nitrates, and represented the original observations that would eventually lead to the discovery of the L-arginine:NO pathway. Prior to these studies, it was thought that steady-state levels of nitrite and nitrate originated solely from the diet and from nitrogen-fixing enteric bacteria. The steady-state concentrations of nitrite in the body are tightly regulated and vary depending on each tissue or compartment and relative NOS activity [17, 18]. However, there is usually more nitrite concentrated in tissues than in the circulatory system [17]. Being limited primarily to blood levels of NO in humans, nitrite and nitrate may be an underestimation of what occurs in specific tissues [19].

The endogenous production of NO plays an important role in vascular homeostasis, neurotransmission, and host defense mechanisms [4]. It has been estimated that our body (70 kg) produces 1.68 mmol NO/day (based on 1 μmol/kg/h NO production) in a healthy individual. The majority of this NO produced will end up as nitrite and nitrate. In plasma, steady-state concentrations of nitrite are conserved across various mammalian species, including humans, in the range of 150–600 nM [20]. Apart from plasma, nitrite can also be transported within red blood cells [21]. The net concentration in plasma is a result of its formation and consumption. It has been shown that up to 70–90% of plasma nitrite is derived from eNOS activity in fasted humans and other mammals [20]. Humans, unlike prokaryotes, are thought to lack the enzymatic machinery to reduce nitrate back to nitrite; however, due to the commensal bacteria that reside within the human body, it has been demonstrated that these bacteria can reduce nitrate, thereby supplying an alternative source of nitrite [22–25]. Plasma nitrite increases after ingestion of large amounts of nitrate. This increase is entirely due to enterosalivary circulation of nitrate (as much as 25% is actively taken up by the salivary glands) and reduction to nitrite by commensal bacteria in the mouth [26]. Therefore, dietary and enzymatic sources of nitrate are actually a potentially large source of nitrite in the human body. Nitrate is rapidly absorbed in the small intestine and readily distributed throughout the body [27]. This nitrate pathway to NO has been shown to help reduce gastrointestinal tract infection, increase mucous barrier thickness and gastric blood flow [27–32]. Nitrite and nitrate are excreted through the kidneys. Nitrate is excreted in the urine as such or after conversion to urea [33]. Clearance of nitrate from blood to urine approximates 40 μmol/h in adults [34], indicating considerable renal tubular reabsorption of this ion. There is little detectable nitrite or nitrate in feces [35]. There is some loss of nitrate and nitrite in sweat, but this is not a major route of excretion [36].

Diet

Diet has been suggested to be the largest source of nitrogen oxides, primarily via nitrite and nitrate intake. Much of the quantified exposure of nitrite and nitrate has been based on the focus of dietary ingestion and endogenous formation of nitrosamines as a cause of human cancers [37–39]. What is often not understood is that nitrogen is essential for growth and reproduction of all plant and animal life. An essential form of that nitrogen is nitrate. Nitrogen is a basic constituent of proteins. The form of nitrogen within plants when consumed by animals has important effects on growth and reproduction. Several different groups of nitrogen-containing compounds may be found in plants. The amount of each form depends on plant species, maturity, and environmental conditions during growth. These nitrogen compounds may be broadly classified as either protein or non-protein compounds. Under normal growing conditions, plants use nitrogen to form plant proteins. When normal growth is altered, protein formation may be slowed and the nitrogen remains in the plant as non-protein nitrogen. Nitrate, nitrite, amides, free amino acids, and small peptides make up most of the non-protein nitrogen fraction. We focus specifically on nitrite and nitrate occurring naturally in plants and that added exogenously in meat processing. First, it is worth a cursory review of how plants accumulate nitrate from the soil in order to understand how they become a significant source of nitrite and nitrate.

Nitrate is a natural material in soils. Adequate supply of nitrate is necessary for good plant growth. Probably more than 90% of the nitrogen absorbed by plants is in the nitrate form. Soils can be supplemented with nitrogen often in the form of ammonium nitrogen (NH_4^+), which is rapidly converted to nitrate (NO_3^-) in the soil. The amount of crop growth is essentially the same whether nitrogen fertilizer is applied as ammonia (NH_3), ammonium, or nitrate (NO_3^-). Chemical fertilizers may be composed of ammonium nitrate, ammonium phosphates, ammonium sulfate, alkaline metal nitrate salts, urea, and other organic forms of nitrogen. Microorganisms must change organic nitrogen to ammonium or nitrate ionic species before assimilation by plants. Usual release of available nitrogen from soil organic matter is 1–4% annually, depending on soil texture and weather conditions. Animal manure is an excellent source of nitrogen and can contribute significantly to soil improvement. Animal manure contains about 10 lb of nitrogen/ton, poultry manure about 20 lb/ton; and legume residues 20–80 lb/ton. About half of this organic nitrogen may be converted to nitrate-nitrogen and become available for plant use the year it is added to the soil; however, it is low in phosphorus content. Excessive manure applications can result in toxic levels of nitrate in forage crops just as excessive use of chemical nitrogen fertilizer. Adding phosphate fertilizer to manure can reduce the nitrate content in the crop produced.

Nitrate content of plants is determined by their inherited metabolic pattern (genetics) and the available nitrate of the soil. Applying fertilizer in amounts beyond the ability of the vegetable crop to use them may result in an accumulation of nitrate, particularly if some other essential nutrient is not adequate. Leafy green vegetables and some root crops naturally contain nitrates and there are wide variations between species. Vegetables produced on high organic soils, and even where no fertilizer nitrogen is applied, frequently have higher nitrate content than the same species grown in fertilized soil. There are also seasonal differences: it is reported that nitrate content in spinach was almost twice as high in autumn–winter harvest than in spring (2,580 mg/kg vs. 1,622 mg/kg) [40]. Disruption of normal plant growth increases the probability for nitrate accumulation in leaves, stems, and stalks. Factors favoring accumulation of nitrate in plants include drought, high temperatures (rapid transpiration of water), shading and cloudiness (reduced protein synthesis), deficiency of other plant growth nutrients (phosphorus, potassium, calcium), excessive soil nitrogen (manures, legume, residues, effluents and solid wastes, and nitrogen fertilizer), damage from insects and certain weed control chemicals, and immaturity at harvest [41, 42]. As a result of all of these factors, nitrate content of vegetables is highly variable but still represents the main dietary source of nitrate.

The National Research Council report *The Health Effects of Nitrate, Nitrite, and N-Nitroso Compounds* [37] reported estimates of nitrite and nitrate intake based on food consumption tables and calculated that the average total nitrite and nitrate intake in the United States was 0.77 and 76 mg, respectively, per day. Meah et al. [43] estimated 1985 dietary intake in the UK at 4.2 mg/day for nitrite and 54 mg/day for nitrate with the largest contribution to both being potatoes. In 1975, White [44] estimated cured meats were a source of 10% of dietary nitrate and 21% of dietary nitrite exposure. The White calculations were based on the relatively higher levels of nitrate (208 ppm residual measured) and nitrite (52.5 ppm residual) that were used prior to concerns about nitrosamine formation. Although the amount of nitrite and nitrate present in cured and processed meats is currently much less than what is found naturally in vegetables, there are a number of epidemiological reports erroneously implicating the nitrite and nitrate in meats with increased risk of some cancers [45, 46]. Past intense investigations into nitrosation chemistry with respect to the use of nitrite and nitrate in meat preservation have resulted in reduction and, in many cases, the elimination of nitrate as a curing salt, reduction in levels of nitrite used for curing, as well as the widespread use of ascorbate or erythorbate salts to better control curing chemistry and greatly reduce the likelihood of nitrosamine formation [47–49]. Thus, changes to cured meat manufacturing procedures over the past 35 years have made earlier estimates of the contribution of cured meats to dietary nitrate and nitrite intake obsolete. A survey of products by Cassens et al. in 1996, representing about one-third of ham, bacon, bologna and frankfurters manufactured in the United States, found residual nitrate at <10 ppm and average nitrite at 10 ppm [50, 51]. This represented more than an 80% reduction in nitrite in cured meats versus earlier reports. Keeton et al. [3] repeated the survey in 2008–2009 and found similar values. In 2009, Hord et al. [2] summarized dietary intake estimates of nitrite and nitrate for the "DASH" (dietary approaches to stop hypertension) diet that included two scenarios for vegetables and fruit consumption. At a recommended level of vegetable and fruit intake that encompassed high nitrate sources, nitrate intake could be over 1,200 mg/day and nitrite would be approximately 0.35 mg/day from food sources exclusive of processed meats. A diet with low nitrate containing vegetable and fruit intake could have as little as 174 mg/day nitrate, but still contain 0.41 mg/day nitrite. This latter estimate was also close to values reported by the IARC committee in their estimates of human diet intake based on data from the UK [1, 52–54]. Adding 75 g/day consumption of cured meats (as done by White in 1975) to the Hord estimates would add 1.5–6 mg/day of nitrate and 0.05–0.54 mg/day of nitrite using the mean residual levels of nitrate and nitrite reported in a wide range of cured meats collected at American retail outlets [3]. Thus, cured meats contribute negligible nitrate compared to other dietary sources and amounts of nitrite equivalent to other dietary sources such as vegetables and fruits. Other foods such as dairy items, eggs, fish and grains have been reported to supply relatively minor fractions of total intake of nitrite and nitrate [1, 2, 43, 55–57].

The indirect exposure to nitrite from dietary nitrate, however, must also be considered because of the enterosalivary recirculation of nitrate and biochemical reduction of salivary nitrate to nitrite by commensal bacteria in the oral cavity. Hord et al. [2] estimated the nitrite exposure from this source to be approximately 5.2 mg/day, about half of which was a result of recycling of ingested nitrate, and the other half from oxidation and recycling of endogenously produced nitric oxide. Thus, oral nitrite ingestion can be summarized as 6–7% directly from fruits and vegetables, 0.5–6.5% directly from cured meats, approximately 50% indirectly from vegetables and fruits and from 40 to 50% from endogenous production.

Water

Even though drinking water usually contains very low levels of nitrate and nitrite, when large volumes are consumed, this can be a significant source for ingestion. The U.S. Environmental Protection Agency (EPA) regulates nitrate and nitrite contamination levels in approximately 160,000 different water supplies, and does periodic monitoring [58–60]. The regulatory maximum contaminant level (MCL) for nitrate is 10 mg/L N from nitrate which equates to 44 mg nitrate ion/L. For nitrite, the MCL is 1 mg/L as N or 3.3 mg/L nitrite ion. Using these maximum values as an estimate for intake may provide a crude estimate of an upper limit for the drinking water contribution to ingestion of nitrite and nitrate in the United States. Recommended dietary consumption of water and beverages is listed at 2.7 L/day for adult males and 2.2 L/day for females [61]. However, IARC has used an average water intake of 1.4 L/day on which to base their estimates [1]. Thus, as much as 60–119 mg/day nitrate ingestion from drinking water may occur in drinking water meeting legal standards. Likewise as much as 4.6–8.9 mg/day of nitrite may also be ingested from drinking water.

The natural background level of nitrate and nitrite are very low in water, generally <10 mg/L for the former and rarely over 3 mg/L for the latter. The increased use of nitrogen fertilizers in agriculture during the second half of the twentieth century contributed to potential contamination of water supplies by nitrate containing runoff of farms into watersheds that ultimately became sources of drinking water. Due to their very high solubility, nitrates can enter groundwater. Levels of nitrate can be significantly higher in shallow wells that are connected to agricultural runoff. EPA monitoring in 2009 revealed violations of federal standards affecting 4.5 million people with over 70% of these violations for population centers of less than 100,000 [62]. Much less is known about private wells which in the United States are usually required to be tested only when the well is constructed or when the property is sold. Tentative estimates for Western Europe and the United States are that 2–3% of the population is potentially exposed to drinking water nitrate exceeding 50 mg/L [63]. Thus, higher exposures from drinking water are not an uncommon event. Excessive levels of nitrate and nitrate in drinking water have been associated with illness and death in newborns and young infants. Thus today, the regulations exist for nitrate and nitrite in drinking water in many parts of the world based on this related toxic effect. Newborn infants are particularly sensitive to oxidation of hemoglobin in erythrocytes and, if the level approaches 20% of total hemoglobin, methemoglobinemia or cyanosis can develop which interferes with blood transport of oxygen. Methemoglobinemia became better understood in the mid twentieth century through medical and epidemiological investigations which resulted in special cautions being applied to diet and water sources for this segment of the human population. It is noteworthy that the few reported human nitrate and nitrite exposure studies, including those in both children and adults, have not produced methemoglobinemia. Infants exposed to 175–700 mg nitrate/day did not experience methemoglobin levels above 7.5%, suggesting that nitrate alone is not causative for methemoglobinemia [64]. A more recent randomized three-way crossover study exposed healthy volunteer adults to single doses of sodium nitrite that ranged from 150–190 to 290–380 mg/volunteer [65]. Observed methemoglobin concentrations were 12.2% for volunteers receiving the higher dose of nitrite ion and 4.5% for those receiving the lower dose. Recent nitrite infusion studies of up to 110 μg/kg/min for 5 min induced methemoglobin concentrations of only 3.2% in adults [66]. These data have led to alternative explanations for the observed methemoglobinemia in infants, including gastroenteritis and associated iNOS-mediated production of NO induced by bacteria-contaminated water [67, 68]. Experts have questioned the veracity of the evidence supporting the hypothesis that nitrates and nitrites are toxic for healthy post-infant populations [69–71]. It appears that the earlier biologically plausible hypothesis of nitrite toxicity (e.g., methemoglobinemia) has essentially been transformed into sacrosanct dogma [69] despite a lack of proof [67, 68]. These studies call into question the mechanistic basis for exposure regulations for nitrate and nitrite. At best, these findings highlight a serious, but context-specific, risk associated with nitrite overexposure in infants.

Saliva

Similar to the environmental nitrogen cycle, there is now a recognized human nitrogen cycle whereby nitrate is reduced to nitrite and to NO [72]. Nitrate, after its absorption in the upper gastrointestinal tract, reaches the salivary glands via the blood circulation where it is secreted into the oral cavity and partially reduced to nitrite by the oral microflora. There is a linear relationship between the amount of nitrate ingested and nitrate and nitrite found in saliva. The ability of the oral microflora to reduce nitrate to nitrite depends on the individual's age. Mean salivary nitrite concentration was found to increase from well below 1 ppm in infants of up to 6 months to about 7 ppm or higher in adults [73], but levels over 500 mg/L have been reported to occur subsequent to consumption of nitrate rich foods [26].

Early work by Tannenbaum et al. [24, 74] reported that nitrite in saliva of healthy individuals is 6–10 mg/L and higher levels in the hundreds of milligrams/liter could form under conditions of very high nitrate intake and active bacterial reduction in saliva. Values of 97 mg/L nitrate and 10 mg/L nitrite were reported in a Chinese study after an overnight fasting period [75]. More recently, Bjorne et al. [76] reported 1.9 mg/L fasting levels of salivary nitrite which were increased to near 100 mg/L after a standardized nitrate load of 4.6 mg/kg body weight was administered. Certainly, the high variation in many reports reflects biological variability and dietary interactions. Thus, it is extremely difficult to accurately quantify the contribution of saliva to nitrate and nitrite ingestion but it must be considered to be a major source. The current consensus is that as much as 25% of dietary, water-solubilized, and endogenously generated nitrate is recycled in the body via the enterosalivary route and that up to 25% of that nitrate pool is reduced in the saliva by commensal bacteria [77]. The earlier discussed Hord values for salivary nitrite contribution to ingestion were based on a 2.0 mg/L concentration of nitrite in saliva and secretion of 1.5 L/day. Thus, saliva can commonly be a source of as much 15 mg ingested nitrite per day and under conditions of sustained nitrate intake from vegetables, this could even be higher.

Non-Dietary Sources of Nitrogen Oxide Exposure: Occupational, Smoking, Toothpaste

On a global scale, quantities of nitric oxide and nitrogen dioxide produced naturally by bacterial growth, volcanic action and by lightning by far outweigh those generated by man's activities; however, as they are distributed over the entire earth's surface, the resulting background atmospheric concentrations are very small. The major source of man-made emissions of nitrogen oxides into the atmosphere is the combustion of fossil fuels in stationary sources (heating, power generation) and in motor vehicles (internal combustion engines). Other contributions to the atmosphere may come from non-combustion industrial processes, such as the manufacture of nitric acid and explosives. Indoor sources include smoking, gas-fired appliances, and oil stoves. Differences in the nitrogen dioxide emission of various countries are mainly due to differences in fossil fuel combustion. Worldwide emissions of oxides of nitrogen in 1970 were estimated at approximately 53 million tons. Nitrogen dioxide is a respiratory pollutant and the US EPA first set standards for NO_2 in 1971 at 0.053 ppm (53 ppb) averaged annually. While reviewed periodically, they have not been changed since then and all areas of the United States currently meet these standards [78]. Nitrogen oxides are released to the air from the exhaust of motor vehicles, the burning of coal, oil, or natural gas, and during processes such as arc welding, electroplating, engraving, and dynamite blasting. When it is NO that is released, it quickly reacts with oxygen to form NO_2, which is the brown gas formed over cities with high pollution [77]. Photo catalysis in the troposphere results in reformation of nitric oxide and because of this cycle, they are both typically referred to as NO_x in air pollution literature (not to be confused with nitrite+nitrate in other literature).

Nitrogen oxides are used in the production of nitric acid, lacquers, dyes, and other chemicals and also used in rocket fuels, nitration of organic chemicals, and the manufacture of explosives. As a result, humans are chronically exposed to these nitrogen oxides, including nitric oxide. Exposure is primarily dependent on where people live and the industry in which they work. Everybody is exposed to small amounts of nitrogen oxides in ambient air. Exposure to high levels of nitrogen oxides, particularly nitrogen dioxide, can damage the respiratory airways. Contact with the skin or eyes can cause burns. Nitrogen dioxide and nitric oxide have been found in at least 9 and 6 of the 1,585 National Priorities List sites identified by the EPA, respectively. People who live near combustion sources such as coal burning power plants or areas with heavy motor vehicle use may be exposed to higher levels of nitrogen oxides. Households that burn wood or use kerosene heaters and gas stoves tend to have higher levels of nitrogen oxides in air when compared to houses without these appliances. Exposure from indoor sources such as home appliances and smoking should not be underestimated. In the immediate proximity of domestic gas-fired appliances, nitrogen dioxide concentrations of up to 2,000 $\mu g/m^3$ (1.1 ppm) have been measured. Workers employed in facilities that produce nitric acid or certain explosives, like dynamite and trinitrotoluene (TNT), as well as workers involved in the welding of metals may breathe in nitrogen oxides during their work. Some of the human toxicological studies of nitrate have relied on workers in explosives and fertilizer manufacturing and usage occupations. While they are not representative of the general human population they can be a significant source for some people. Ammonium and potassium nitrate are important fertilizers. There have been estimates of nitrate exposure as high as >10 mg/m^3 from dust [1, 79]. Much literature in this area examines nitrate containing dust without actual measurements of the dust composition, hence precise data is lacking [80, 81]. This is understandable given the complexity of crystalline forms of nitrate salts in fertilizers, changes with storage and mixtures with other fertilizer components [82, 83]. Farmers who were exposed to silo gases from the fermentation of harvested crops were acutely affected by nitrogen oxides, some of them fatally. It has been estimated that exposure to nitrogen dioxide levels of 560–940 mg/m^3 (300–500 ppm) may result in fatal pulmonary edema or asphyxia and that levels of 47–140 mg/m^3 (25–75 ppm) can cause bronchitis or pneumonia. Miners who used explosives repeatedly in their work were reported to develop chronic respiratory diseases. Analysis of the products of explosion showed the presence of oxides of nitrogen at concentrations of 88–167 ppm [84].

Nitric oxide and nitrogen dioxide are found in tobacco smoke, exposing people who smoke or breathe in second-hand smoke to nitrogen oxides. Nitrogen oxides in cigarette smoke have been implicated in the genesis of chronic obstructive pulmonary disease (COPD) [85] and as precursors to carcinogenic *N*-nitrosamines [86]. Jenkins and Gill [87] report a range of 3–47 μmol of NO_x which is highly variable depending on type and brand of cigarette. Total NO_x produced from smoking one cigarette is on the order of 0.5–2 mg [88] which can be a significant exposure for heavy smokers, being potentially equal to that from endogenous production. Tobacco smoke has been reported to contain nitric oxide levels of about 98–135 mg/m^3 (80–110 ppm) with up to 1,000 ppm reported [89, 90], and nitrogen dioxide levels of about 150–226 mg/m^3 (80–120 ppm), but these levels may fluctuate considerably with the conditions of combustion. Normal, metabolic production of small amounts of nitric oxide causes airways to expand. The large amount of nitric oxide in tobacco smoke acutely impacts individuals in two ways: (1) when smokers are smoking, it expands their airways even further, making it easier for their lungs to absorb nicotine and other smoke constituents; (2) when they are not smoking, it shuts off endogenous nitric oxide production, causing their airways to constrict. This may be one reason why regular smokers often have difficulty with normal breathing.

Although it is not widely realized, potassium nitrate is used in many toothpaste formulations to reduce pain sensitivity. Potassium nitrate is the active anti-sensitivity ingredient that penetrates the exposed tubules (tiny holes) of the teeth to relieve the pain from sensitive nerves inside teeth. Additionally, it is one of the ADA-approved ingredients to treat tooth hypersensitivity. Toothpaste designed for sensitive teeth is formulated with up to 5% (5,000 ppm) potassium nitrate, which is higher than the concentration found in vegetables [88, 91–94]. Since toothpaste is not commonly swallowed, the net contribution to nitrate and nitrite exposure cannot be accurately estimated. But, nitrate and nitrite as water soluble ions are likely to be extracted from the toothpaste and mixed with saliva. The total exposure could therefore be equivalent to or greater than that contributed by water or cured processed meats.

Nitrosative Chemistry of Nitrite and Nitrate

Regardless of the source of exposure the primary cause for concern of exposure to nitrite, nitrate or other nitrogen oxides is the nitrosative chemistry that can occur. The risk–benefit spectrum from nitrite and nitrate may very well depend upon the specific metabolism and the presence of other components that may be concomitantly ingested. The stepwise reduction to nitrite and NO may account for the benefits while pathways leading to nitrosation of low molecular weight amines or amides may account for the health risks of nitrate exposure. Understanding and affecting those pathways will certainly help in mitigating the risks. The discussion below will describe these two pathways.

Nitrate itself is generally considered to be harmless at low concentrations. Nitrite, on the other hand, is reactive especially in the acid environment of the stomach where it can nitrosate other molecules including proteins, amines, and amides. Nitrite is occasionally found in the environment but most human exposure occurs through ingested nitrate that can be chemically reduced to nitrite by commensal bacteria often found in saliva [95]. The specter of a cancer risk posed by nitrite and nitrate is invariably accompanied by concern about exposure to preformed *N*-nitrosamines or *N*-nitrosamines formed in the stomach from ingesting foods enriched in nitrite and nitrate. This is because some low molecular weight amines can be converted (nitrosated) to their carcinogenic *N*-nitroso derivatives by reaction with nitrite (see reaction below) [96]. Nitrosamines are a class of chemical compounds that were first described in the chemical literature over 100 years ago, but not until 1956 did they receive much attention. In that year two British scientists, John Barnes and Peter Magee, reported that dimethylnitrosamine produced liver tumors in rats [97]. This discovery was made during a routine screening of chemicals that were being proposed for use as solvents in the dry cleaning industry. Magee and Barnes' landmark discovery caused scientists around the world to investigate the carcinogenic properties of other nitrosamines. Approximately 300 of these compounds have been tested, and 90% of them have been found to be carcinogenic in a wide variety of experimental animals. Most nitrosamines are mutagens and a number are transplacental carcinogens. Most are organ specific. For instance, dimethylnitrosamine causes liver cancer in experimental animals, whereas some of the tobacco specific nitrosamines cause lung cancer. Nitrosamines can occur because their chemical precursors – amines and nitrosating agents – occur commonly, and the chemical reaction for nitrosamine formation is quite facile. The reactions below illustrate the nitrosation reactions:

$$R_2NH(\text{amines}) + NaNO_2(\text{sodium nitrite}) \rightarrow R_2N - N = O(\text{nitrosamine}).$$

In the presence of acid (such as in the stomach) or high temperature (such as via frying), nitrosamines are converted to diazonium ions.

$$R_2N - N = O(\text{nitrosamine}) + (\text{acid or heat}) \rightarrow R - N^+ - N = O(\text{diazonium ion}).$$

Certain nitrosamines such as dimethylnitrosamine and *N*-nitrosopyrrolidine form carbocations that react with biological nucleophiles (such as DNA or an enzyme) in the cell.

$$R - N^{+} - N = O(\text{diazonium ion}) \rightarrow R^{+}(\text{carbocation}) + N_2(\text{leaving group})$$
$$+ : Nu(\text{biological nucleophiles}) \rightarrow R - Nu.$$

If this reaction occurs at a crucial site in a biomolecule, it can disrupt normal cell function leading to cancer or cell death.

About 1970 it was discovered that ascorbic acid inhibits nitrosamine formation [98]. Another antioxidant, α-tocopherol (vitamin E), has also been shown to inhibit nitrosamine formation [99]. Ascorbic acid, erythorbic acid, and α-tocopherol inhibit nitrosamine formation due to their oxidation–reduction properties. For example, when ascorbic acid is oxidized to dehydroascorbic acid, nitrous anhydride, a potent nitrosating agent formed from sodium nitrite, is reduced to nitric oxide, which is not a nitrosating agent. Most vegetables which are enriched in nitrate are also rich in antioxidants such as vitamins C and E that can act to prevent the unwanted nitrosation chemistry. These compounds are also now almost universally added to cured processed meats. Nitrate in drinking water on the other hand has no such protective nitrosation inhibitor present and may be cause for concern. Controlling the metabolic fate of nitrate and nitrite away from nitrosation and toward reduction to NO may provide a strategy to promote health benefits while mitigating the health risks. Adverse health effects may be the result of a complex interaction of the amount of nitrite and nitrate ingested, the concomitant ingestion of nitrosation cofactors and precursors, and specific medical conditions such as chronic inflammation that increase nitrosation. Controlling for such factors is essential for defining the safety of nitrite and nitrate.

Summary

Human exposure to nitrate and nitrite can vary widely based on both dietary habits and geographical location of their water supply as well as individual exposure to atmospheric nitrogen oxides. A summary of estimates is shown in Table 4.1. The largest source of nitrogen oxides in the body are derived from plant sources in the diet. While drinking water can be the next highest source, endogenous production is likely higher. Salivary nitrate and nitrite reflect a recycling of the large dietary nitrate consumption and endogenous nitric oxide synthesis and in some cases may be the largest source of nitrate and nitrite that is swallowed. Processed meats, dairy products, and poultry products provide only minor contributions to the human exposure. Overall human exposure to nitrate is primarily from exogenous sources and nitrite is primarily from endogenous sources. Nitric oxide exposure is exclusively from endogenous synthesis or smoking. Nitrogen oxide exposure may also occur from administered pharmaceuticals which have not been considered in this analysis. Safety and health concerns can be addressed by inhibiting unwanted nitrosation reactions.

Table 4.1 Ranges of nitrate, nitrite, and nitric oxide exposure form diet, endogenous synthesis, and recycling for adult humans expressed as mg/day

Source	Nitrate (NO_3^-)	Nitrite (NO_2^-)	Nitric oxide (NO)
Diet (excluding cured processed meat)[a]	50–220	0–0.7	–
From 75 g/day cured processed meat intake[b]	1.5–6	0.05–0.6	–
Water[c]	0–132	0–10	–
Saliva[d]	>30–1,000	5.2–8.6	–
Endogenous synthesis[e]	–	–	70

[a]Based on International Agency for Research on Cancer (IARC) Table1.8 [1]
[b]Based on Keeton et al. [3] average values and intake described in White [43]
[c]Based on none present to US Environmental Protection Agency (EPA) maximum contaminant level (MCL) for water and 2.7 L consumption/day
[d]Based on White [43] and Hord et al. [2] and includes both recycling of diet derived nitrate via the enterosalivary route and that from endogenous NO
[e]Based on 1 mg/kg/day endogenous synthesis for 70 kg adults

References

1. International Agency for Research on Cancer. Ingested nitrate and nitrite and cyanobacterial peptide toxins, Monograph. 94. Lyon: IARC Press; 2010.
2. Hord NG, Tang Y, Bryan NS. Food sources of nitrates and nitrites: the physiologic context for potential health benefits. Am J Clin Nutr. 2009;90(1):1–10.
3. Keeton JT, Osburn WN, Hardin MD, Bryan NS, Longnecker MT. A national survey of the nitrate/nitrite concentrations in cured meat products and non-meat foods available at retail. 2009, National Pork Broad Project #08–124. Report accessed at http://www.amif.org/ht/d/sp/i/26883/pid/26883#sodium note: 2 manuscripts sumbitted and currently under peer review.
4. Moncada S, Palmer RMJ, Higgs A. Nitric oxide: physiology, pathophysiology and pharmacology. Pharmacol Rev. 1991;43(2):109–42.
5. Abu-Soud HM, Yoho LL, Stuehr DJ. Calmodulin controls neuronal nitric-oxide synthase by a dual mechanism. J Biol Chem. 1994;269(51):32047–50.
6. Palmer, RMJ, Ashton DS, Moncada S. Vascular endothelial cells synthesize nitric oxide from L-arginine. Nature. 1988;333(6174):664–666.
7. Yoshida K, Kasama K, Kitabatake M, Imai M. Biotransformation of nitric oxide, nitrite and nitrate. Int Arch Occup Environ Health. 1983;52:103–15.
8. Grube R, Kelm M, Motz W, Strauer BE. The biology of nitric oxide. In: Moncada S, Feelisch M, Busse R, Higgs EA, editors. Enzymology, biochemistry, and immunology, vol. 4. London, UK: Portland Press; 1994. p. 201–4.
9. Kelm M, Feelisch M, Grube R, Motz W, Strauer BE. The biology of nitric oxide. In: Moncada S, Marletta MA, Hibbs JB, Higgs EA, editors. Physiological and clinical aspects, vol. 1. London, UK: Portland Press; 1992. p. 319–22.
10. Shiva S, Wang X, Ringwood LA, Xu X, Yuditskaya S, Annavajjhala V, et al. Ceruloplasmin is a NO oxidase and nitrite synthase that determines endocrine NO homeostasis. Nat Chem Biol. 2006;2(9):486–93.
11. Kelm M. Nitric oxide metabolism and breakdown. Biochim Biophys Acta. 1999;1411:273–89.
12. Tannenbaum SR. Nitrate and nitrite: origin in humans. Science. 1994;205:1333–5.
13. Kelm M, Yoshida K. Metabolic fate of nitric oxide and related N-oxides. In: Feelisch M, Stamler JS, editors. Methods in nitric oxide research. Chichester, UK: Wiley; 1996.p. 47–58.

14. Bryan NS, Fernandez BO, Bauer SM, Garcia-Saura MF, Milsom AB, Rassaf T, et al. Nitrite is a signaling molecule and regulator of gene expression in mammalian tissues. Nat Chem Biol. 2005;1(5):290–7.
15. Rhodes P, Leone AM, Francis PL, Struthers AD, Moncada S. The L-arginine:nitric oxide pathway is the major source of plasma nitrite in fasted humans. Biochem Biophys Res Commun. 1995;209:590–6.
16. Tannenbaum SR, Fett D, Young VR, Land PD, Bruce WR. Nitrite and nitrate are formed by endogenous synthesis in the human intestine. Science. 1978;200:1487–8.
17. Bryan NS, Rassaf T, Maloney RE, Rodriguez CM, Saijo F, Rodriguez JR, et al. Cellular targets and mechanisms of nitros(yl)ation: an insight into their nature and kinetics in vivo. Proc Natl Acad Sci U S A. 2004;101(12):4308–13.
18. Rodriguez J, Maloney RE, Rassaf T, Bryan NS, Feelisch M. Chemical nature of nitric oxide storage forms in rat vascular tissue. Proc Natl Acad Sci U S A. 2003;100:336–41.
19. Bryan NS. Nitrite in nitric oxide biology: cause or consequence? A systems-based review. Free Radic Biol Med. 2006;41(5):691–701.
20. Kleinbongard P, Dejam A, Lauer T, Rassaf T, Schindler A, Picker O, et al. Plasma nitrite reflects constitutive nitric oxide synthase activity in mammals. Free Radic Biol Med. 2003;35(7):790–6.
21. Dejam A, Hunter CJ, Pelletier MM, Hsu LL, Machado RF, Shiva S, et al. Erythrocytes are the major intravascular storage sites of nitrite in human blood. Blood. 2005;106(2):734–9.
22. Goaz PW, Biswell HA. Nitrite reduction in whole saliva. J Dent Res. 1961;40:355–65.
23. Ishiwata H, Tanimura A, Ishidate M. Nitrite and nitrate concentrations in human saliva collected from salivary ducts. J Food Hyg Soc Jpn. 1975;16:89–92.
24. Tannenbaum SR, Sinskey AJ, Weisman M, Bishop W. Nitrite in human saliva. Its possible relationship to nitrosamine formation. J Natl Cancer Inst. 1974;53(1):79–84.
25. van Maanen JM, van Geel AA, Kleinjans JC. Modulation of nitrate-nitrite conversion in the oral cavity. Cancer Detect Prev. 1996;20(6):590–6.
26. Lundberg JO, Govoni M. Inorganic nitrate is a possible source for systemic generation of nitric oxide. Free Radic Biol Med. 2004;37(3):395–400.
27. Walker R. The metabolism of dietary nitrites and nitrates. Biochem Soc Trans. 1996;24(3):780–5.
28. McKnight GM, Smith LM, Drummond RS, Duncan CW, Golden M, Benjamin N. Chemical synthesis of nitric oxide in the stomach from dietary nitrate in humans. Gut. 1994;40(2):211–4.
29. Pique JM, Whittle BJ, Esplugues JV. The vasodilator role of endogenous nitric oxide in the rat gastric microcirculation. Eur J Pharmacol. 1989;174(2–3):293–6.
30. Brown JF, Hanson PJ, Whittle BJ. Nitric oxide donors increase mucus gel thickness in rat stomach. Eur J Pharmacol. 1992;223(1):103–4.
31. Bjorne HH, Petersson J, Phillipson M, Weitzberg E, Holm L, Lundberg JO. Nitrite in saliva increases gastric mucosal blood flow and mucus thickness. J Clin Invest. 2004;113(1):106–14.
32. Lundberg JO, Weitzberg E, Lundberg JM, Alving K. Intragastric nitric oxide production in humans: measurements in expelled air. Gut. 1994;35(11):1543–6.
33. Green LC, Ruiz de Luzuriaga K, Wagner DA, Rand W, Istfan N, Young VR, et al. Nitrate biosynthesis in man. Proc Natl Acad Sci U S A. 1981;78(12):7764–8.
34. Wennmalm A, Benthin G, Edlund A, Jungersten L, Kieler-Jensen N, Lundin S, et al. Metabolism and excretion of nitric oxide in humans. An experimental clinical study. Circ Res. 1993;73(6):1121–7.
35. Bednar C, Kies C. Nitrate and vitamin C from fruits and vegetables: impact of intake variations on nitrate and nitrite excretions in humans. Plant Foods Hum Nutr. 1994;45(1):71–80.
36. Weller R, Pattullo S, Smith L, Golden M, Ormerod A, Benjamin N. Nitric oxide is generated on the skin surface by reduction of sweat nitrate. J Invest Dermatol. 1996;107(3):327–31.
37. National Research Council. The health effects of nitrate, nitrite and N-nitroso compounds. Washington, DC: National Academy Press; 1981. ISBN 0-309-03230-X.
38. National Research Council. Alternatives to the current use of nitrite in foods. Washington, DC: National Academy Press; 1982. ISBN 0-309-03277-6.

39. National Toxicology Program. Toxicology and carcinogenesis studies of sodium nitrite (CAS NO. 7632-00-0) in F344/N rats and B6C3F1 mice (drinking water studies). NTP Report #495 NIH Publication No. 01-3954. National Institutes of Health; 2001. p. 7–273.
40. Santamaria P, Elia A, Serio F, Todaro E. A survey of nitrate and oxalate content in fresh vegetables. J Sci Food Agric. 1999;79:1882–8.
41. Salomez J, Hofman G. Nitrogen nutrition effects on nitrate accumulation of soil-grown greenhouse butterhead lettuce. Soil Sci Plant Anal. 2009;40(1):620–32.
42. Anjana SU, Iqbal M. Nitrate accumulation in plants, factors affecting the process, and human health implications. A review. Agronomy. 2007;27(1):45–57.
43. Meah MN, Harrison N, Davies A. Nitrate and nitrite in foods and the diet. Food Addit Contam. 1994;11(4):519–32.
44. White Jr JW. Relative significance of dietary sources of nitrate and nitrite. J Agric Food Chem. 1975;23(5):886–91.
45. Cross AJ, Ferrucci LM, Risch A, Graubard BI, Ward MH, Park Y, et al. A large prospective study of meat consumption and colorectal cancer risk: an investigation of potential mechanisms underlying this association. Cancer Res. 2010;70(6):2406–14.
46. Sinha R, Cross AJ, Graubard BI, Leitzmann MF, Schatzkin A. Meat intake and mortality: a prospective study of over half a million people. Arch Intern Med. 2009;169(6):562–71.
47. Honikel KO. The use and control of nitrate and nitrite for the processing of meat products. Meat Sci. 2008;78:68–76.
48. Sebranek JG, Fox JB. A review of nitrite and chloride chemistry: interactions and implications for cured meats. J Sci Food Agric. 1985;36:1169–82.
49. Cassens RG. Use of sodium nitrite in cured meats today. Food Technol. 1995;49:72–81.
50. Cassens RG. Residual nitrite in cured meat. Food Technol. 1997;51:53–5.
51. Cassens RG. Composition and safety of cured meats in the USA. Food Chem. 1997;59:561–6.
52. Sebranek JG, Bacus JN. Cured meat products without direct addition of nitrate or nitrite: what are the issues? Meat Sci. 2007;77(1):136–47.
53. Food Standards Agency. Survey of nitrite and nitrate in bacon and cured meat products. Food Surveillance Information Sheet No. 142. London; 1998.
54. Food Standard Agency. UK monitoring programme for nitrate in lettuce and spinach. Food Survey Information Sheet 74/05; 2004.
55. Food Standards Agency. Total diet study, nitrate and nitrite. Food Surveillance Information Sheet No. 163. London; 1998.
56. World Health Organization. WHO Food additives series: 50 Nitrate and Nitrite 2007 http://www.inchem.org/documents/jecfa/jecmono/v50je07.htm (accessed 08-02-10).
57. Jakszyn PG, Ibáñez R, Pera G, Agudo A, García-Closas R, Amiano P, González CA. Food content of potential carcinogens. Catalan Institute of Oncology. Barcelona. 2004 European Prospective Investigation on Cancer, http://epic-spain.com/libro.html (accessed 08-02-10).
58. US Environmental Protection Agency, Technical fact sheet on nitrate/nitrite, http://www.epa.gov/safewater/pdfs/factsheets/ioc/tech/nitrates.pdf. (accessed 08-02-10).
59. US Environmental Protection Agency, EPA community water regulations handbook March 2010 edition, http://www.epa.gov/safewater/publicnotification/pdfs/Revised-Public-Notification-Handbook-CWS.pdf. (accessed 08-02-10).
60. US Environmental Protection Agency, Listing of local drinking water contaminant results, http://www.epa.gov/safewater/dwinfo/index.html. (accessed 08-02-10).
61. Food Nutrition Board National Academies of Sciences. Institute of Medicine dietary reference intakes: water, potassium, sodium, chloride, and sulfate. Washington, DC: National Academies Press; 2004.
62. US Environmental Protection Agency, Water Quality monitoring report 2009, http://www.epa.gov/safewater/databases/pdfs/data_factoids_2009.pdf. (accessed 08-02-10).
63. van Grinsven HJ, Ward MH, Benjamin N, de Kok TM. Does the evidence about health risks associated with nitrate ingestion warrant an increase of the nitrate standard for drinking water? Environ Health. 2006;5:26.

64. Cornblath M, Hartmann AF. Methhemoglobinaemia in young infants. J Pediatr. 1948; 33:421–5.
65. Kortboyer J, Olling M, Zeilmaker MJ. The oral bioavailability of sodium nitrite investigated in healthy adult volunteers. Bilthoven, The Netherlands: National Institute of Public Health and the Environment; 1997.
66. Dejam A, Hunter CJ, Tremonti C, et al. Nitrite infusion in humans and nonhuman primates: endocrine effects, pharmacokinetics, and tolerance formation. Circulation. 2007;116(16):1821–31.
67. L'Hirondel JL, Avery AA, Addiscott T. Dietary nitrate: where is the risk? Environ Health Perspect. 2006;114(8):A458–9; author reply A459–61.
68. Powlson DS, Addiscott TM, Benjamin N, Cassman KG, de Kok TM, van Grinsven H, et al. When does nitrate become a risk for humans? J Environ Qual. 2008;37(2):291–5.
69. McKnight GM, Duncan CW, Leifert C, Golden MH. Dietary nitrate in man: friend or foe? Br J Nutr. 1999;81(5):349–58.
70. L'Hirondel JL. Nitrate and man: toxic, harmless or beneficial? Wallingford, UK: CABI Publishing; 2001.
71. European Food Safety Authority. Nitrate in vegetables: scientific opinion of the panel on contaminants in the food chain. EFSA J. 2008;289:1–79.
72. Lundberg JO, Weitzberg E, Gladwin MT. The nitrate-nitrite-nitric oxide pathway in physiology and therapeutics. Nat Rev Drug Discov. 2008;7(8):156–67.
73. Eisenbrand G, Spiegelhalder B, Preussmann R. Nitrate and nitrite in saliva. Oncology. 1980;37(4):227–31.
74. Tannenbaum SR, Weisman M, Fett D. The effect of nitrate intake on nitrite formation in human saliva. Food Cosmet Toxicol. 1976;14(6):549–52.
75. Xia D, Deng D, Wang S. Alterations of nitrate and nitrite content in saliva, serum, and urine in patients with salivary dysfunction. J Oral Pathol Med. 2003;32(2):95–9.
76. Bjorne HH, Petersson J, Phillipson M, Weitzberg E, Holm L, Lundberg JO. Nitrite in saliva increases gastric mucosal blood flow and mucus thickness. J Clin Invest. 2004;113(1):106–14.
77. Lawrence MG, Chameides WL, Kasibhatla PS, Levy H, Moxim W. Lightning and atmospheric chemistry: the rate of atmospheric NO production, in Handbook of Atmospheric Dynamics, ed Volland H, CRC Press, Boca Raton 1995;1:189–203.
78. US Environmental Protection Energy, Nitrogen Dioxide Primary and Secondary Standards, 2010, http://www.epa.gov/ttn/naaqs/standards/nox/s_nox_index.html (accessed 08-05-10).
79. Al-Dabbagh S, Forman D, Bryson D, Stratton I, Doll R. Mortality of nitrate fertiliser workers. Br J Ind Med. 1986;43(8):507–15.
80. Fandrem SI, Kjuus H, Andersen A, Amlie E. Incidence of cancer among workers in a Norwegian nitrate fertiliser plant. Br J Ind Med. 1993;50(7):647–52.
81. Addiscott TM. Fertilizers and nitrate leaching. Issues Environ Sci Technol. 1996;5:1–26.
82. Miller P, Saeman WC. Properties of monocrystalline ammonium nitrate. Fertilizer Indus Eng Chem. 1948;40:154–9.
83. European Fertilizer Manufacturers' Association, Production of Ammonium Nitrate and Calcium Ammonium Nitrate, Brussels, Belgium, 2000, Booklet No. 6 of 8.
84. World Health Organization. Environmental health criteria 4: oxides of nitrogen. Geneva, Switzerland: World Health Organization; 1977.
85. Anderson HR, Spix C, Medina S, et al. Air pollution and daily admissions for chronic obstructive pulmonary disease in 6 European cities: results from the APHEA project. Eur Respir J. 1997;10(5):1064–71.
86. Hoffmann D, Rivenson A, Hecht SS. The biological significance of tobacco-specific N-nitrosamines: smoking and adenocarcinoma of the lung. Crit Rev Toxicol. 1996; 26(2):199–211.
87. Jenkins RA, Gill BE. Determination of oxides of nitrogen (NOx) in cigarette smoke by chemiluminescent analysis. Anal Chem. 1980;52:925–8.
88. Lofroth G, Burton RM, Forehand L, Hammond SK, Seila RL, Zweidinger RB, et al. Characterization of environmental tobacco smoke. Environ Sci Technol. 1989;23:610–4.

89. Norman V, Keith CG. Nitrogen oxides in tobacco smoke. Nature. 1965;205:915–6.
90. Cueto R, Pryor WA. Cigarette smoke chemistry: conversion of nitric oxide to nitrogen dioxide and reactions of nitrogen oxides with other smoke components as studied by Fourier transform infrared spectroscopy. Vib Spectrosc. 1994;7:97–111.
91. Manochehr-Pour M, Bhat M, Bissada N. Clinical evaluation of two potassium nitrate toothpastes for the treatment of dental hypersensitivity. Periodontal Case Rep. 1984;6(1):25–30.
92. Orchardson R, Gillam DG. The efficacy of potassium salts as agents for treating dentin hypersensitivity. J Orofac Pain. Winter 2000;14(1):9–19.
93. Poulsen S, Errboe M, Hovgaard O, Worthington HW. Potassium nitrate toothpaste for dentine hypersensitivity. Cochrane Database Syst Rev. 2001;(2):CD001476.
94. Wara-aswapati N, Krongnawakul D, Jiraviboon D, Adulyanon S, Karimbux N, Pitiphat W. The effect of a new toothpaste containing potassium nitrate and triclosan on gingival health, plaque formation and dentine hypersensitivity. J Clin Periodontol. 2005;32(1):53–8.
95. Lundberg JO, Weitzberg E, Cole JA, Benjamin N. Nitrate, bacteria and human health. Nat Rev Microbiol. 2004;2(7):593–602.
96. Spiegelhalder B, Eisenbrand G, Preussmann R. Influence of dietary nitrate on nitrite content of human saliva: possible relevance to in vivo formation of N-nitroso compounds. Food Cosmet Toxicol. 1976;14(6):545–8.
97. Magee PN, Barnes JM. The production of malignant primary hepatic tumours in the rat by feeding dimethylnitrosamine. Br J Cancer. 1956;10(1):114–22.
98. Mirvish SS. Blocking the formation of N-nitroso compounds with ascorbic acid in vitro and in vivo. Ann N Y Acad Sci. 1975;258:175–80.
99. Mirvish SS. Inhibition by vitamins C and E of in vivo nitrosation and vitamin C occurrence in the stomach. Eur J Cancer Prev. 1996;5 Suppl 1:131–6.

Part II
Food and Environmental Exposures to Nitrite and Nitrate

Chapter 5
History of Nitrite and Nitrate in Food

Jimmy T. Keeton

Key points

- Saltpeter (potassium nitrate), a natural contaminant of salt, contributed historically to the pinkish-red color in salted meats
- Nitric oxide, derived from nitrate/nitrite reduction, when combined with myoglobin produces as the color pigment in cured meat products
- Nitrite prevents sporulation of *Clostridium boutlinum* in cured meats
- In 1925, the United States Department of Agriculture (USDA) approved nitrite's use in curing brines and formulas at a maximum ingoing level of 200 ppm
- Residual nitrite declined in cured meats from 1930s to 1970s and has remained low since the 1980s mitigating the formation of *N*-nitrosamines
- USDA regulations restrict ingoing levels of nitrite and nitrate (if allowed) to specific levels in meat product categories
- Residual nitrite levels in conventional and "organic" cured meats is <10 ppm in the United States
- Naturally occurring nitrate in raw vegetables at retail in the United States ranges from 200 to 3,000 ppm
- Nitrates in the municipal water supply of 25 metropolitan U.S. cities is <5 mg/L
- Nitrite contributes to the safety of cured meats and currently no suitable alternative is available

Keywords Nitrate • Nitrite • Food born pathogens • Cured meats • Food safety • Diet • Exposure

J.T. Keeton (✉), Ph.D
Head, Professor, AgriLife Faculty Fellow
Department of Nutrition and Food Science, Texas A&M University,
College Station, Texas 77843-2253 USA
Office: Kleberg Center, Room 122
e-mail: jkeeton@tamu.edu

N.S. Bryan and J. Loscalzo (eds.), *Nitrite and Nitrate in Human Health and Disease*,
Nutrition and Health, DOI 10.1007/978-1-60761-616-0_5,

Introduction

Salting as means of preserving meat, poultry, fish, seafood, and vegetables predates written history and was essential in ancient times for providing nutrient-dense foods during scarcity or population migration. The application of rock salt to animal tissues led to the curing practices of today that include brine injection, marination, dry rubs, or a combination of each that preserve, protect, and favor present-day cured meat products. Meat curing is historically defined as the addition of salt (sodium chloride) to fresh meat cuts to remove moisture and reduce the water activity of the tissues to prevent spoilage. In ancient times, salt was obtained from crystalline deposits by mining directly from the earth, or evaporating water from brine pools or seawater. As a consequence, it often contained natural contaminants such as sodium or potassium nitrate that contributed directly to the curing reaction and the preservation process. These contaminants, it was later learned, were the primary components in curing reactions that allowed reduction of nitrate salts to gaseous nitric oxide and its subsequent reaction with myoglobin, a cellular transport protein.

The fundamental utility of nitrite has been its ability to inhibit the growth of a number of aerobic and anaerobic microorganisms and especially suppress the outgrowth of spores from *Clostridium botulinum.* Nitrite in combination with salt and other curing factors may also control the growth of other pathogens such as *Bacillus cereus*, *Staphylococcus aureus* and *Clostridium perfringens.* However, under conditions of prolonged temperature fluctuations in a range that promote bacterial growth, it does not prevent pathogen outgrowth and may allow toxin production or spoilage. Other properties of nitrite include its reduction and reaction with myoglobin to produce the characteristic reddish-pink cured meat color, its ability to inhibit lipid oxidation (and reduce oxidative rancidity or warmed over flavor), and its contribution to the cured flavor of meat products.

Meat curing today is understood to mean the incorporation of salt, nitrite, and occasionally nitrate into meat products. Additional ingredients in a curing mixture may include water, potassium chloride, sodium ascorbate (or its isomer erythorbate), sweeteners (sucrose, glucose, honey), alkaline phosphates, lactic acid, citric acid, acetic acid, sodium or potassium lactate, sodium citrate, glucono delta lactone, lactic acid starter cultures, spices and seasonings, antioxidants, nonmeat binders and extenders, hydrolyzed proteins, gelatin, modified starch, hydrocolloids, and liquid or natural smoke. Some of these ingredients (saltpeter, vinegar, sugar, spices, natural smoke, lactic acid bacteria) were used historically to impart flavor, functional properties, and textural characteristics to cured meats, but were self-limiting in most cases and unregulated in the United States prior to the turn of the twentieth century. These ingredients, and especially nitrates and nitrites, have been studied to identify their specific functions in the curing process and their role in the potential formation of *N*-nitrosamines, some of which are known carcinogens in several animal species. The purpose of this chapter is to provide a brief historical perspective on the use of nitrates and nitrites in preserving meat products, the regulatory limits of cure ingredients as established by the United States Department of Agriculture – Food Safety

Inspection Service (USDA-FSIS) or Food and Drug Administration (FDA), and a synopsis of the residual levels of nitrates and nitrites in present-day cured meat products. A brief overview of the concentrations of naturally occurring nitrates in selected vegetables and municipal water supplies in the United States are also presented since these contribute to dietary nitrate and nitrite load.

History of Nitrate and Nitrite in Meat Curing

In a review of the history of nitrate and nitrite in meat curing, Binkerd and Kolari [1] concluded that salting as a means of meat preservation was first practiced in the deserts of Asia. Saline salts from this area contained impurities such as nitrates that contributed to the characteristic red color of cured meats. As early as 3,000 BC in Mesopotamia, cooked meats and fish were preserved in sesame oil and dried, salted meat and fish were part of the Summerian diet. Salt from the Dead Sea was in common use by Jewish inhabitants around 1,600 BC, and by 1,200 BC, the Phoenicians were trading salted fish in the Eastern Mediterranean region. By 900 BC, salt was being produced in "salt gardens" in Greece and dry salt curing and smoking of meat were well established. The Romans (200 BC) acquired curing procedures from the Greeks and further developed techniques to "pickle" various kinds of meats in a brine marinade. It was during this time that the reddening effect of salting was noted. Saltpeter (potassium nitrate) is mentioned as being gathered in China and India prior to the Christian era for use in meat curing along with "wall" saltpeter (calcium nitrate), which is formed by nitrifying bacteria and deposited on the walls of caves and stables. In Medieval times, the application of salt and saltpeter as curing ingredients was commonplace and the reddening effect on meat was attributed to saltpeter.

As early as 1835, saltpeter was cited as imparting juiciness and flavor to bacon when applied at 0.5 lb per 100 lb of meat (5,000 ppm or mg/kg). In 1873, Edward Smith described salt as "the oldest and best known of preserving agents…its chief action appears to be due to its power of attracting moisture, and thus extracting fluid to harden the tissues." He further described the development of a "reddish color throughout" in meat preserved with saltpeter as compared to preservation by salt alone, which allowed the meat color to fade. Meat curing was more of an art than a science in the early nineteenth century, but as a greater understanding of the curing process evolved in the late 1800s and early 1900s, the role of nitrate and nitrite in the formation of cured meat color and flavor became apparent. In 1891, Polenske first reported finding nitrite in cured meat and reused curing pickle [1]. He concluded that nitrite was the result of the bacterial reduction of nitrate added to the pickle [2]. In 1899, Lehman and Kisskalt (cited in [1]) demonstrated that the color development in cured meats was actually due the presence of nitrite and not nitrate. Later, Haldane [3] and Hoagland [4] explained the chemical reactions involved in the development of red color in cooked, cured meats in a series of studies with hemoglobin. Haldane demonstrated the formation of nitrosylhemoglobin when nitrite was combined with hemoglobin, and the subsequent conversion of nitrosylhemoglobin to its red pigment form, nitrosylhemochromogen, with heating. Hoagland, also using "hemoglobin", later described the reduction of nitrate to nitrite and nitrite to nitrous acid with subsequent formation of nitric oxide by bacterial and/or enzymatic reactions [5]. Conversion of nitrous acid to nitric oxide allowed reaction with "hemoglobin" resulting in the formation of nitrosylhemoglobin and nitrosohemochromogen. Additional studies by Hoagland in 1914 [6] with cooked salted meats showed that nitrous acid, or its metabolite nitric oxide, reacted with myoglobin to form nitrosylmyoglobin and nitrosylprotohaem, the characteristic color compounds observed in uncooked and cooked cured meat products, respectively [7, 8].

Regulatory Restrictions of Nitrate and Nitrite

Following the work of Haldane and Hoagland, the German government in 1909 recommended the use of partially reduced nitrate in curing mixes and marketed these across Europe. Up until this time, U.S. processors had limited control, if any, over the production of nitrite in a nitrate brine or pickle. Curing establishments, prior to the early 1900s, frequently used the same lot of pickle more than once, adding more salt and nitrate to maintain the original concentration in the pickle, failing to realize that the pickle now contained more nitrite derived from the reduction of the original nitrate in the brine. The effect was a pickle relatively rich in nitrite, as well as highly contaminated with bacteria. Products produced using these brines were susceptible to spoilage, quite variable in nitrite concentration (often ranging from 2 to 960 ppm), and in some cases had exceptionally high levels of residual nitrite [2]. To resolve this problem, the Bureau of Animal Industry of the USDA authorized a series of experiments in 1923 to study the direct use of nitrite as a curing agent rather than nitrate. Kerr et al. [9], under the supervision of Bureau of Animal Industry inspectors, cured hams with a closely controlled amount of nitrite (~2,000 ppm) in a pickle, analyzed the hams for residual nitrite and compared these results to the residual nitrite in nitrate-only cured products. They established that the addition of approximately 2,000 ppm of sodium nitrite in the curing pickle resulted in an average nitrite content of 42–150 ppm in hams with no more than 200 ppm in any part of the product. Nitrate-cured hams, on the other hand had a maximum residual nitrite content of 45 ppm. Subsequent evaluation of the flavor and quality of the nitrite cured hams indicated that they were equivalent to traditionally cured (nitrate) hams. Based on these results, a tentative limit of 200 ppm residual nitrite was established as the maximum nitrite level allowed in finished cured meat products. Additional commercial scale experiments in 17 meat curing plants with shoulders, loins, tongues, hams, bacon, corned beef, dried beef, and sausages enabled Kerr et al. to conclude that "from 1/4th to 1 oz. of sodium nitrite is sufficient to fix the color in 100 lb (of meat), the exact quantity depending on the meat to be cured and the process employed." Based on these experiments, sodium nitrite was allowed as a curing ingredient in federally inspected establishments in 1925 [10, 11]. The Bureau authorization later stated that "Extended experiments have demonstrated that successful curing may be accomplished by the addition of as small a quantity as one-fourth of an ounce of sodium nitrite to each 100 pounds of meat; therefore, pending further ruling by the Bureau, the finished product shall not contain sodium nitrite in excess of 200 parts per million." U.S. meat packers found that by using nitrite in place of nitrate in curing pickles, they gained more control over the process with more uniform results. The direct addition of nitrite also shortened pickling times and resulted in a curing process that did not require the presence of reducing microorganisms for nitrate, thereby allowing for a process that limited the presence of spoilage bacteria.

In 1931, the USDA issued a rule-limiting curing solutions, containing both nitrite and nitrate, to deliver no more than 156 ppm nitrite and 1,716 ppm nitrate per 100 lbs of pumped, brine cured meats. Additional regulations for cured meat were issued in 1970 that allowed sodium or potassium nitrate at 7 lb per 100 gal pickle, 3.5 oz per 100 lb meat in dry cure, or 2.75 oz per 100 lb of chopped meat. Sodium nitrite was permitted at 2 lb per 100 gal pickle assuming a 10% pump level, while 1 oz could be applied to 100 lb meat as a dry cure or 0.25 oz to 100 lb of chopped meat or meat product that would result in not more than 200 ppm nitrite in the finished product.

Current U.S. regulations allow the use of nitrite and selective use of nitrate in meat products based on product category and method of curing. Immersion cured, massaged, or pumped products, such as hams or pastrami, are limited to a maximum ingoing level of 200 ppm of sodium or potassium nitrite and/or 700 ppm of nitrate, respectively, based on the raw product weight [12]. Dry-cured products, however, are allowed a maximum ingoing level of 625 ppm nitrite and/or 2,187 ppm nitrate since these products have longer curing times that allow for immediate nitrite reaction with myoglobin and longer term conversion of nitrate to nitrite. If a combination of nitrite and nitrate are used, the combination must not result in more than 200 ppm sodium nitrite in the finished product. Comminuted products, such as frankfurters, bologna, and other cured sausages, are limited to a maximum input level of 156 ppm of sodium or potassium nitrite based on the raw weight of the meat block. Nitrate may be added to these products at 1,718 ppm regardless of the type of salt used. In 1978, USDA regulations lowered the input sodium nitrite in bacon to 120 ppm (148 ppm potassium nitrite), required addition 547 ppm sodium ascorbate or its isomer erythorbate, and eliminated the use of nitrates in bacon [13]. Bacon regulations were again changed in 1986 to give processors three alternatives that would allow lower levels of nitrite in combination with other processing procedures. Skinless bacon was required to have 120 ppm of sodium nitrite (148 ppm potassium nitrite) in combination with 547 ppm of sodium ascorbate or erythorbate to reduce the ingoing nitrite level and the potential for *N*-nitrosamine formation. These regulations also allowed for a ± 20% (96–144 ppm) variance from the specified concentration of nitrite at the time of injection or massaging. Other exceptions to these regulations include reducing sodium nitrite to 100 ppm (123 ppm potassium nitrite) with an "appropriate partial quality control program" [14], or 40–80 ppm of sodium nitrite (49–99 ppm potassium nitrite) if sugar and lactic acid starter culture are included in the curing brine. Dry-cured bacon was limited to 200 ppm sodium nitrite or 246 ppm potassium nitrite.

In other countries, the levels of nitrite and nitrate in cured meats are regulated, but vary depending upon the maximum input levels allowed by each regulatory authority and the specific processing procedures followed by manufacturers. The European Union rules, as specified by Directive No. 95/2/EC [15], fix the input nitrite level in bacon at 150 ppm and residual amounts between 50 and 175 ppm. In Denmark, a lesser amount is allowed (60–150 ppm) in semi-preserved products and special cured hams. Directive No. 95/2/EC allows only 250 ppm residual nitrate (sodium) in salted meats, but for unheated products 150 ppm nitrite+150 ppm nitrate are allowed. Wiltshire or dry cured bacon may have 175 ppm residual nitrite+250 ppm residual nitrite [7]. In Canada, the maximum allowable limit of nitrite in meat products, such as hams, loins, shoulders, cooked sausages, and corned beef, is 200 ppm, but further reductions to 120–180 ppm have been taken by the Canadian meat industry [8]. Bacon may not contain more than 120 ppm nitrite (input) to reduce the risk of *N*-nitrosamine formation.

Residual Nitrate and Nitrite Levels in Meat Products

From 1925 until 1970, many meat processors continued to use nitrate in their curing formulations. In 1930, Mohler (cited in [1]) reported that 54% of meat processors were still using sodium nitrate, 17% used sodium nitrite, and 30% a combination of nitrate and nitrite. By 1970, approximately 50% of meat processors were reported using nitrate in canned, shelf-stable products, but by 1974, all manufactures surveyed had discontinued the use of nitrate in these product categories as well as cured bacon, hams, canned sterile meats, and frankfurters. Reasons for this change have been attributed to a technological shift to more rapid curing and processing procedures, careful control of nitrite quantities that in turn eliminated the need for nitrate, and concern over nitrate being a precursor to the formation of volatile *N*-nitrosamines during processing or after consumption.

Early surveys of nitrate and nitrite levels in U.S. cured meats by Mighton [16] and Lewis [17] showed lower levels of nitrite and nitrate than were previously observed by Kerr et al. [9] just 10 years earlier. By 1972, no nitrates were found in the cured products such as hams, bacon, corned beef, bologna, or frankfurters (Table 5.1) [18]. As noted in Table 5.1, residual nitrite levels were lower in bologna (36 ppm) and frankfurtrers (38 ppm) in the 1972 survey compared to mean nitrite levels of 61–72 ppm and 84–102 ppm observed in two studies in the mid-1930s. In the 1970s, the use of nitrite to cure meat was questioned due to the potential formation of *N*-nitrosamines and the level of residual nitrite that was being contributed to the dietary load [19]. Recommendations derived from this concern and the subsequent publication of additional studies on the use of nitrite in curing led to a change in USDA regulations in 1978 and 1986 that reduced the allowable levels of nitrite and nitrate in meat products and made provision for the use of reductants such as sodium ascorbate to decrease the level of residual nitrite and reduce the potential for *N*-nitroso compound formation [20]. The use of nitrate was restricted to only certain

Table 5.1 Residual nitrite and nitrate levels (mg/kg) in cured meat products surveyed at retail

	Nitrite (ppm)		Nitrate (ppm)	
Product	Average	Range	Average	Range
Hams	52 M	7–145	360	120–610
Smoked	80 L	34–184	600	100–1,200
Boiled	48 K	16–100	–	–
	59 M	11–87	370	30–910
	49 L	31–63	700	300–1,100
Bacon (raw)	13 M	4–22	1,400	700–2,100
	16 L	11–29	1,200	300–1,900
	96 K	24–170	–	–
Corned beef	3 M	3–5	200	169–270
	75 L	1–216	1,700	500–3,000
Bologna	61 M	44–86	380	30–790
	72 L	60–114	650	100–1,020
	36 K	0–76	–	–
Frankfurters	84 M	55–146	710	570–880
	102 L	13–195	513	100–1,090
	38 K	15–80	–	–

M Mighton [16]; *L* Lewis [17]; *K* Kolari and Aunan (1972) [18]

long-term cured products (dry-cured hams, prosciutto) and stricter handling controls of nitrites were instituted in the manufacturing process. In 1980, The National Academy of Sciences (NAS) entered into a contract with USDA and the FDA and established the *Committee on Nitrate and Alternative Curing Agents in Food*. The committee was charged with the task of assessing the health risks associated with overall exposure to nitrate, nitrate, and *N*-nitroso compounds derived from both natural sources and the nitrate and nitrite added to foods. They also were to review the status of research and future prospects for developing feasible alternatives to the use of nitrite as a preservative. In 1981, a comprehensive report entitled, "The Health Effects of Nitrate, Nitrite and *N*-Nitroso Compounds," was published by the NAS to assess the human health risk of these compounds. Eleven recommendations, which are shortened below [21, 22], were made to reduce the risk associated with consumption of nitrites, nitrates and *N*-nitroso compounds in cured meats.

1. Nitrate is neither carcinogenic nor mutagenic. Some human population studies have indicated an association of exposure to high nitrate levels with certain cancers. Future studies were therefore recommended.
2. Nitrite does not act directly as a carcinogen in animal studies. Further testing may be warranted.
3. Most *N*-nitroso compounds are carcinogenic in laboratory animals, mutagenic in microbial and mammalian test systems, and some are teratogenic in laboratory animals. Future work should emphasize quantitative assessment of potency and outcome.
4. Because nitrate and nitrite can exert acute toxic effects and contribute to the total body burden of *N*-nitroso compounds, it was recommended that exposure

to these agents be reduced. Reduction in nitrate use should not compromise protection against botulism. The use of nitrate salts in curing should be eliminated, with the exception of a few special products. Attention should be given to reducing nitrate content of vegetables and drinking water.

5. Sources of exposure to *N*-nitroso compounds in various environmental media should be determined so that it can be reduced. Analytical procedures should be improved, especially for nonvolatile compounds and for both free and bound nitrite.
6. The exposure of humans to amines and nitrosamines can be reduced in some instances by modifying manufacturing practices, such as with certain pesticides and drugs.
7. Additional studies are needed to increase the understanding of the metabolism and pharmacokinetics of nitrate in humans.
8. Further study of inhibitors of nitrosation is needed.
9. Further studies should be made to determine the mechanism(s) whereby nitrite controls the outgrowth of *C. botulinum* spores, and also its effect against other spoilage and pathogenic microorganisms.
10. Although it is not possible to estimate the potential morbidity nor mortality from *C. botulinum* in the absence of nitrite as a curing agent in certain products, the prudent approach to protecting public health requires consideration of the possibility that certain preserved food items may be contaminated and may be abused.

A subsequent report entitled, *"Alternatives to the Current Use of Nitrite in Foods"* [23], made recommendations for reducing nitrite and nitrate levels in cured meats and recommended alternatives that could serve as partial or complete replacements for nitrate, and agents that block the formation of nitrosamines in products containing conventional concentrations of nitrite. The following alternatives were found by the committee to be the most promising: a combination of ascorbate, α-tocopherol, and nitrite; irradiation (with or without nitrate); lactic acid-producing organisms (with or without nitrite); potassium sorbates with low concentrations of nitrite; sodium hypophosphite (with or without nitrite); and several fumarate esters. Combined, these reports alleviated public concern about cured meat as a human health risk. However, an epidemiological report that processed meats (specifically hot dogs) caused cancer in children [24] led to additional concerns about the potential risk of *N*-nitroso compounds and the levels of nitrite in these products. Following the publication of the NAS reports and other studies linking nitrite as a curing ingredient to cancer, a significant amount of research was initiated. However, a suitable alternative to nitrite with the same protective effect against *Clostridium botulinum* has not been identified.

To substantiate that levels of nitrite were lower in cured meat products since USDA regulatory changes were instituted in 1978 and 1986, a multi-city survey of U.S. cured meat products (bacon, ham bologna, and wieners from major manufacturers) was conducted by Cassens [19] to assess the levels of residual nitrite and ascorbate in these products. The overall mean residual nitrite and ascorbate levels in 164 samples were 10 and 209 ppm, respectively. In comparison, the average nitrite value was lower than those reported by White [25] and those given in the NAS [21] report. White's mean values for nitrite and nitrate were 52.5 and 235 ppm (361 ppm for canned items), respectively, for a broad range of U.S. cured meat products taken from different databases in the early 1970s. Overall mean residual nitrite levels in Cassens' survey indicated that the residual nitrite in cured meats had been reduced significantly from the levels reported in the 1970s.

A national U.S. retail survey of the residual nitrate and nitrite concentrations in cured meat products was conducted by Keeton in 2009 to verify that levels contributed by cured meats have remained low, and are minor contributors to the total human dietary intake of nitrate and nitrite. A comparison of the mean nitrite and nitrate concentrations in the major cured meat categories are shown in Table 5.2. Nitrite means from the NAS [21] study derived from databases amassed in the 1970s ranged from 6 to 42 ppm except for dry-cure products, which had nitrite levels of 280 ppm. Compared to mean nitrite values (3–15 ppm) in three product categories reported by Cassens [19], residual nitrite levels have decreased significantly since the NAS study was conducted. Keeton [26] likewise evaluated 467 cured meat products and found comparable mean nitrite values (0.6–7 ppm) to those of Cassens [19]. Keeton found much lower nitrate levels (15–79 ppm) within each product category compared to those in the NAS survey (24–640 ppm). The nitrite and nitrate concentrations of cured meat products classified as conventional or organic are given in Table 5.3. In general, all residual nitrite levels were not different between conventional and organic categories. However, the nitrate concentrations in organic labeled cured meat products tended to be lower in dried sausage, cooked sausage, fermented sausage, and dry-cured products when compared to conventional products. Thus, it appears that implementation of regulatory restrictions by the USDA-FSIS in 1978 and 1986, the use of reductants such as ascorbate, and controlled manufacturing procedures have dramatically reduced the levels of residual nitrite, and nitrate, in all meat product categories.

Nitrates and Nitrites Occurring Naturally in Food

Vegetables

Nitrates are intrinsic, naturally occurring constituents of plants and as a consequence are consumed with vegetables and fruits. Nitrates are produced in the soil by microorganisms acting on manure, urine and vegetable waste, thus making them available to growing plants [27]. Several factors can affect the nitrate content of vegetables [21, 28–30]. These include:

- Related plant strains (cultivars) that systematically differ in nitrate content.
- Different levels and timing of nitrogen fertilizer application. Nitrate accumulation increases as the amount of nitrogen fertilizer used increases and if the fertilizer is applied shortly before harvest.

Table 5.2 Mean nitrite and nitrate concentrations (mg/kg) of cured meat products in the United States from the 1970s to present

	Keeton et al. [26][a]	Keeton et al. [26][a]	Cassens [19]	NAS [21] Table 5.1	NAS [21][b] Table 5.1
Product category	Nitrite ppm	Nitrate ppm	Nitrate ppm	Nitrite ppm	Nitrate ppm
Cured, dried sausages (uncooked)	0.7	79	–	13–17	78–89
German air-dry sausage, chorizo, Italian dry sausage					
Cured sausages (cooked)	7	28	8–15	10–31	32–110
Bologna, frankfurters, polish sausage					
Fermented/acidified sausages (cooked)	0.6	36	–	6–17	78–89
Pepperoni, summer sausage, snack sticks					
Whole-muscle brine cured (uncooked)	7	26	3–5	12–42	33–96
Bacon					
Whole-muscle brine cured (cooked)	7	15	4–7	16–37	140–150
Hams, bacon (precooked), cured poultry, pastrami					
Corned beef					
Whole muscle dry-cured (uncooked)	2	67	–	280	24–640
Dry-cure country style hams, dry-cure bacon, proscuitto					

[a]Combined means of the nitrite and nitrate concentrations in conventional and organic cured meat products at retail

[b]National Academy of Sciences (1981)

Table 5.3 Mean nitrite and nitrate concentrations (mg/kg) of retail cured meat categories classified as conventional or organic

		Conventional				Organic			
Product category		*N*	Mean	Min	Max	*N*	Mean	Min	Max
Cured dried uncooked sausage	Nitrite	40	0.8	0.03	10	24	0.6	0.01	8
	Nitrate	40	113	0.1	2,289	23	19	0.1	137
Cured cooked sausage	Nitrite	59	8	0.1	29	30	5	0.05	35
	Nitrate	59	32	0.8	541	30	18	0.4	74
Fermented cooked sausage	Nitrite	52	0.8	0	27	19	0.1	0	0.5
	Nitrate	52	46	2	320	19	7	0	57
Whole-muscle brine cured uncooked	Nitrite	20	7	1	36	19	8	0.1	36
	Nitrate	20	14	3	32	18	38	2	462
Whole-muscle brine cured cooked	Nitrite	97	7	0.03	28	44	7	0.1	22
	Nitrate	97	16	0.2	108	44	13	0	143
Whole-muscle dry-cured uncooked	Nitrite	39	1	0.02	16	24	3	0.02	29
	Nitrate	38	106	0.4	1,367	24	7	0.2	50

Keeton et al. [26]. See Table 5.2 for a more detailed explanation and examples of actual products in each category

- Nitrate levels tend to increase as daytime temperatures drop below an optimal temperature; thus, geographical region and season of harvest affect nitrate content.
- Greenhouse plants tend to accumulate higher levels of nitrate than do plants grown outdoors perhaps because nitrogen fertilizers are used more heavily indoors.
- Plants grown in shade, at high latitudes with limited sunlight, and during drought accumulate higher levels of nitrate than do plants grown under optimal conditions of water and light supply.
- Leafy plants harvested on a sunny afternoon often contain less nitrate than those harvested in the morning or during inclement weather.
- Some plant diseases, insect damage, or exposure to herbicides, such as those used in weed control, often increase nitrate accumulation.
- Soil deficiencies, such as insufficient molybdenum or potassium, or acidic, organically rich (peat) soils lead to elevated nitrate content.

These factors may alter nitrate levels in vegetables by affecting one or more plant processes such as nitrogen uptake, nitrogen transport, or nitrate reduction and assimilation. Processes such as canning and blanching can reduce the nitrate levels by 20–50% in some vegetables, while storage, particularly at higher temperatures, can result in an increase of nitrite due to the reduction of nitrate to nitrite by reductase enzymes present in plant tissues and by contaminating bacteria. While the nitrate and nitrite content of vegetables varies greatly, the initial nitrate content can be modified by certain modifications in growing conditions including water source, soil conditions, time of harvest, plant-specific factors, and by the amount, kind and timing of nitrogen fertilization [21].

In general, the nitrate concentration in raw vegetables is extremely variable and also varies from country to country and region to region due to the factors previously mentioned. Results from a recent national U.S. survey of nitrite and nitrate

Table 5.4 Mean nitrite and nitrate concentrations (mg/kg) of raw vegetables at retail classified as conventional or organic

Product category		Conventional				Organic			
		N	Mean	Min	Max	*N*	Mean	Min	Max
Broccoli	Nitrite	20	0.6	0.01	9	20	0.1	0.01	0.7
	Nitrate	20	394	29	1,140	19	204	3	683
Cabbage	Nitrite	20	0.1	0.01	0.4	20	1	0.01	11
	Nitrate	20	418	37	1,831	20	552	2	2,114
Celery	Nitrite	18	0.1	0.02	0.5	20	1	0.02	9
	Nitrate	19	1,495	20	4,269	20	912	0.7	3,589
Lettuce	Nitrite	19	0.6	0.01	10	20	0.1	0	0.3
	Nitrate	19	850	79	2,171	20	844	58	2,013
Spinach	Nitrite	19	8	0	137	20	1	0	17
	Nitrate	19	2,797	65	8,000	20	1,318	16	4,089

Keeton et al. [26]

concentrations of raw vegetables at retail classified as conventional or organic are given in Table 5.4. In general, the concentrations of nitrate, and nitrite, were not different between conventional or organic labeled broccoli, cabbage, celery, and lettuce. One exception was the nitrate level in organic spinach (1,318 ppm), which was lower than its conventional counterpart (2,797 ppm). The concentrations (ppm) of nitrate from highest to lowest in raw vegetables shown in Table 5.4 were: spinach (1,318–2,797), celery (912–1,495), lettuce (844–850), and broccoli (204–394). Other vegetables such as beets, radishes, eggplant, celery, lettuce, collards, and turnip greens also contain similar high concentrations of nitrates.

Drinking Water

The nitrate ion (NO_3^-), the oxidized form of dissolved nitrogen, is the most common groundwater contaminant worldwide and is highly soluble making it easily leached from soils into the available groundwater. It is derived from both natural and human sources and is a relatively stable form of nitrogen in oxygen-rich soils and aquifers. Nitrogen is an inert gas that makes up 78% of the atmosphere and through natural processes is converted into a variety of common bioavailable compounds, the most common being ammonia (NH_3), ammonium (NH_4^+), nitrate nitrogen (NO_3–N), nitrite (NO_2^-), *N*-nitrosamines, or organic nitrogen (R-N_2H) [31]. Natural sources of nitrate from nitrogen include fixation by lightning, bacterial conversion in plants and to a lesser extent igneous rocks, deep geothermal fluids and dissolution of some evaporite minerals. During the decomposition of plants, stored nitrogen can be released to the soil where it is then converted to nitrate and incorporated into aquifers by precipitation.

The primary sources of nitrates that contaminate groundwater are derived from human activity and also include waste from farm animals, fertilizers, manure applied to soils, human waste from septic tanks, and waste water treatment systems (discharge) [32]. In the Southwestern United States and other agricultural areas, inorganic fertilizer and animal manure are the most common nitrate sources while urban areas without proper sewer containment contribute to the nitrate levels in groundwater. Ammonia used in fertilizers may volatilize, be used by plants, or may be denitrified by microbial action, thus releasing gaseous nitrogen.

The U.S. Environmental Protection Agency (EPA) established the maximum contaminant level (MCL) for nitrate-nitrogen (NO_3^-N) at 10 mg/L in 1975 to regulate nitrate levels in drinking water and protect human health [31]. It has been known for more than 50 years that high concentrations of nitrate (>20 mg/L as nitrogen) could cause methemoglobinemia ("blue baby disease") in infants less than 6 months of age and was part of the impetus for the EPA regulation. Other conditions such as hypertension, central nervous system birth defects, certain cancers, non-Hodgkins lymphoma, and diabetes mellitus have been linked to nitrate in drinking water. Having similar concerns, the World Health Organization in 1993 established the MCL for nitrate at 11.3 mg/L nitrate-nitrogen which has been the limit adopted by many other countries throughout the world. In the United States, a survey conducted by Nolan and Stoner [33] of 33 regional aquifers found that more than 15% of the wells drawing from the aquifers had nitrate concentrations above the EPA maximum limit of 10 mg/L nitrate-N. In other studies, nitrate was the most frequently reported groundwater contaminant in over 40 states [34, 35].

The permissible concentration of nitrate in drinking water is 50 mg nitrate/L in the European Union and 44 mg/L in the United States [36]. The U.S. EPA MCL or exposure level for drinking water is 10 mg/L nitrate and 1 mg/L nitrite. The Joint Food and Agricultural Organization/World Health Organization set the Acceptable Daily Intake (ADI) for nitrate ion at 3.7 mg/kg body weight and for the nitrite ion at 0.6 mg/kg body weight. The EPA has set a Reference Dose for nitrate at 1.6 mg nitrate ion/kg-day body weight and for the nitrite ion a level of 0.1 mg/kg-day body weight. Lower doses are set for infants and children due to the potential for methemoglobinemia.

A survey of EPA reports giving the nitrate and nitrite concentrations in the municipal water supply of 25 major cities across the United States indicated that the water from each city was compliant for nitrate and nitrite content [26]. Philadelphia, PA, Atlantic City, NJ, and Los Angeles, CA, reported the highest levels of nitrate of the 25 cities at 4.9, 4.6 and 2.2–9.2 ppm (mg/L), respectively. It is interesting to note that the highest nitrate concentrations in Los Angeles came from groundwater that was taken from wells. All drinking water sources were below the allowable limits for nitrate (10 mg/L) and nitrite (1 mg/L) established by the EPA.

Current and future treatment options to lower nitrate levels in drinking water include: (1) blending high-nitrate water with low-nitrate water; (2) ion exchange of potable water (most widely use nitrate removal method, but may contribute very low levels of nitrosamines or their precursors from the membrane resins); (3) membrane separation (reverse osmosis and electrodialysis are used in small communities, but require high energy inputs); (4) biological denitrification with selective microorganisms that convert NO_3^- to N_2; and (5) chemical denitrification (under development) that reduces nitrate. Both denitrification systems require low-dissolved oxygen levels.

Conclusion

Nitrite, and in some cases nitrate, are functional food ingredients that serve as effective antimicrobials to inhibit pathogens such as *Clostridium botulinum*, impart a distinctive reddish-pink cured color to meat products, provide antioxidant properties to retard lipid oxidation, and extend the shelf-life of cured meat products. Some concern has been raised about the use of these ingredients and whether their level of use truly poses a sufficient health risk to warrant their restriction or removal from cured meats.

Historically, meat curing consisted of the addition of salt to fresh meat cuts to remove moisture and reduce the water activity of the tissues to prevent spoilage. In ancient times, salt was obtained from crystalline deposits left by evaporating water from brine pools, seawater or mining directly from the earth. As a consequence, it often contained natural contaminants such as sodium or potassium nitrate (saltpeter) that contributed directly to the curing reaction and the preservation process. Written records of meat curing are available as early as 3,000 BC and dry salt curing and smoking of meat were well established by 900 BC in Greece. The Romans (200 BC) acquired curing procedures from the Greeks, developed "pickling" techniques of meats with a brine marinade, and noted a reddening effect when salt extracted from natural sources was used. Meat curing was more of an art than a science in the early nineteenth century, but as a better understanding of the curing process evolved in the late 1800s and early 1900s, the role of nitrate, and specifically nitrite, in the formation of a distinct cured meat color and its suppression of the outgrowth of *Clostritidum botulinum* spores became apparent.

In 1925, for the first time the U.S. Department of Agriculture, Bureau of Animal Industry allowed the use of nitrite salts in meat curing formulas to reduce product spoilage and high levels of residual nitrite in cured products. They also established a maximum limit of 200 ppm nitrite in finished cured meat products. From the 1930s to the mid-1970s, the use of nitrate declined in cured meat products, and today, nitrate use is limited to only dry-cured and speciality meat products. With the implementation of additional regulatory restrictions in 1978 and 1986, the use of nitrite in cured meat products is now allowed at no more than 120–200 ppm depending upon the category. These restrictions have dramatically reduced the levels of residual nitrite, and nitrate, in all meat product categories and their levels have remained low since the implementation regulatory restrictions. The most recent survey indicates an average of 0.6–7 ppm residual nitrite and 15–67 ppm nitrate across the major categories of cured meat products consumed in the United States. This could be significant when evaluating cured meat's contribution to the total dietary load of nitrite and nitrate and their contribution to increased risk of gastrointestinal cancer, or their potential benefits to cardiovascular health and gastrointestinal immune function.

The nitrate content of vegetables varies greatly due to cultivar, variety, growing conditions, water source, soil conditions, time of harvest, plant-specific factors, the amount, kind and timing of nitrogen fertilization, storage period and conditions, and processing procedures. Although vegetables are considered a major source of nitrate in the diet, the fact that the nitrate content of vegetables are variable poses a potential dilemma in determining nitrate's actual contribution to the nitrate/nitrite load in the diet. This variation might be of sufficient magnitude to alter epidemiological predictions if not considered appropriately by region and vegetable category.

EPA reports of the nitrate and nitrite levels in 25 of the largest municipal water supplies across the United States indicates that water from these sources meets acceptable nitrate and nitrite levels based on EPA limits. However, some municipalities were close to the EPA maximum indicating that water in certain regions of the United States should be given consideration when evaluating dietary nitrate and nitrite intake.

Although nitrite and nitrate have only recently received public scrutiny since the 1960s, these two molecules have a rich history dating back thousands of years. They are a natural part of our diet through the consumption of vegetables and are essential for food safety when used as an additive. Understanding the rich history of the use of nitrite and nitrate will hopefully help in educating consumers and scientists alike on the safety and potential risks associated with their ingestion.

References

1. Binkerd EF, Kolari OE. The history and use of nitrate and nitrite in the curing of meat. Food Cosmet Toxicol. 1975;13(6):655–61.
2. Jones O. Nitrite in cured meats. Analyst. 1933;58(684):140.
3. Haldane J. The red colour of salted meat. J Hyg (Lond). 1901;1(1):115–22.
4. Hoagland R. The action of saltpeter upon the color of meat. In the 25th Annual Report of the Bureau of Animal Industry, U.S. Department of Agriculture. 1908:301–314.
5. Pegg RB, Shahidi F. History of the curing process. Nitrite curing of meat, the N-nitrosamine problem and nitrite alternatives. Trumbull: Food & Nutrition Press, Inc.; 2000. p. 7–21.
6. Hoagland R. Coloring matter of raw and cooked salted meats. J Agric Res. 1914;3:211–25.
7. Honikel K-O. The use and control of nitrate and nitrite for the processing of meat products. Meat Sci. 2008;78:68–76.
8. Shahidi F, Samaranayaka AGP. Brine. In: Jensen WK, Devine C, Dikeman M, Jensen WK, Devine C, Dikeman M, editors. Encyclopedia of meat sciences. Oxford: Elsevier Ltd; 2004. p. 366–74.
9. Kerr RH, Marsh CTN, Schroeder WF, Boyer EA. The use of sodium nitrite in curing meat. J Agric Res. 1926;33:541–51.
10. (USDA) UDoA. Service and regulatory announcements. In: Industries BoA, ed. Washington, DC; 1925:102.
11. (USDA) UDoA. Service and regulatory announcements. In: Industries BoA, ed. Washington, DC; 1926:2.
12. USDA USDoA. Processing inspectors' calculations handbook (FSIS Directive 7620.3); 1995.
13. (IFT) IoFT. Nitrate, nitrite and nitroso compounds in foods. Food Technol. 1987;41(4):127–36.
14. Sebranek JG, Bacus JN. Cured meat products without direct addition of nitrate or nitrite: what are the issues? Meat Sci. 2007;77(1):136–47.
15. Leth T, Fagt S, Nielsen S, Andersen R. Nitrite and nitrate content in meat products and estimated intake in Denmark from 1998 to 2006. Food Addit Contam Part A Chem Anal Control Expo Risk Assess. 2008;25(10):1237–45.
16. Mighton CJ. An analytical survey of cured meats. Chicago: Institute of American Meat Packers; 1936.

17. Lewis WL. The use of nitrite of soda in curing meat. Chicago: Institute of American Meat Packers; 1937.
18. Kolari OE, Aunan WJ. The residual levels of nitrite in cured meat products. Meat Sci Rev. 1972;6:27.
19. Cassens RG. Residual nitrite in cured meat. Food Technol. 1997;51:53–5.
20. Milkowski AL. The controversy about sodium nitrite. Savoy: Meat Industry Research Conference: American Meat Science Association; 2006.
21. (NAS) NAoS. The health effects of nitrate, nitrite, and N-nitroso compounds. Washington, DC; 1981.
22. Cassens RG. Use of Sodium Nitrite in cured meats today. Food Technol. 1995;49:72–81.
23. (NAS) NAoS. Alternatives to the current use of nitrite in foods. Washington, DC; 1982.
24. Peters JM, Preston-Martin S, London SJ, Bowman JD, Buckley JD, Thomas DC. Processed meats and risk of childhood leukemia (California, USA). Cancer Causes Control. 1994;5(2):195–202.
25. White Jr JW. Relative significance of dietary sources of nitrate and nitrite. J Agric Food Chem. 1975;23(5):886–91.
26. Keeton JT, Osburn W, Hardin MD, Longnecker MT, Bryan NS. A national survey of the nitrite/nitrate concentrations in cured meat products and non-meat foods available at retail. Des Moines: American Meat Institute Foundations; 2009.
27. Barnum DW. Some history of nitrates. J Chem Ed. 2003;80(12):1393–6.
28. Corre WJ, Breimer, T. Nitrate and nitrite in vegetables. In: Documentation TNCfAPa, ed. Wageningen Pudoc; 1979.
29. Maynard DN, Barker AV. Regulation of nitrate accumulation in vegetables. Acta Hortic. 1979;93:153–62.
30. Maynard DN. Nitrate accumulation in vegetables. Adv Agron. 1976;28:71–118.
31. Rosen MR, Kropf C. Nitrates in Southwest groundwater. Southwest Hydrol. 2009;8(4):20–1; 34.
32. Harder T. Agricultural impacts on groundwater nitrate. Hydrology. 2009;8(4):22–3; 35.
33. Nolan BT, Stoner JD. Nutrients in ground waters of the conterminous United States, 1992–1995. Environ Sci Technol. 2000;34(7):1156–65.
34. Fetter CS. Contaminant hydrology. New York: MacMillan Publishing Company; 1993.
35. (EPA) UEPA. Nitrates and nitrites. TEACH chemical summary. In: EPA U, ed; 2007.
36. Hord NG, Tang Y, Bryan NS. Food sources of nitrates and nitrites: the physiologic context for potential health benefits. Am J Clin Nutr. 2009;90(1):1–10.

Chapter 6
Nutritional Epidemiology of Nitrogen Oxides: What do the Numbers Mean?

Martin Lajous and Walter Willett

Key points

- When evaluating epidemiologic data on nutrition, careful consideration of the study design, implementation and accuracy of dietary assessment is required.
- Prospective cohort studies on diet may avoid biases commonly observed in case-control studies and may be less problematic than large randomized experiments.
- Food-frequency questionnaires provide useful information about intake of major nutrients and are a practical option for dietary assessment in large studies.
- Biochemical measurements of nutrients are particularly important when a specific nutrient is poorly measured by other methods but it is essential to evaluate its sensitivity and the validity of this measure as an indicator of long-term intake.
- The great variability in nitrate and nitrite contents of foods and the short half-lives of their biochemical indicators complicate exposure measurement; however, some studies suggest that nitrate and nitrite intake in epidemiologic studies is feasible.
- The epidemiologic evidence for the relation between dietary nitrate and nitrite and cancer is weak.
- There is no direct epidemiologic evidence of the relation between nitrite and nitrate intake and cardiovascular disease.

Keywords Nutrition • Diet • Epidemiology • Food frequency questionnaire • Cancer • Cardiovascular disease

M. Lajous (✉)
Department of Epidemiology, Harvard School of Public Health, 665 Huntington Avenue, Boston, MA 02115, USA
and
Centro de Investigación en Salud Poblacional, Instituto Nacional de Salud Pública, Av. Universidad # 655, Col. Santa Maria Ahuacatitlán, Cuernavaca Morelos 62508, México
e-mail: mlajous@hsph.harvard.edu; mlajous@insp.mx

N.S. Bryan and J. Loscalzo (eds.), *Nitrite and Nitrate in Human Health and Disease*, Nutrition and Health, DOI 10.1007/978-1-60761-616-0_6,

Introduction

Understanding the role of diet in health is important because dietary variables are potentially modifiable risk factors on which preventive efforts may focus. New insights into nitrogen oxide metabolism and beneficial effects of nitrite intake on cardiovascular health in experimental models have opened new areas of investigation for prevention of cardiovascular disease [1]. When evaluating epidemiologic data on nutrition, careful consideration of the study design and the implementation and accuracy of dietary assessment is required. Case-control studies of nutrition may afford important insights, but as compared to prospective cohort studies these are more often susceptible to bias. Affected individuals may associate their disease with foods perceived to be poor in nutritional value and over-report them relative to unaffected controls. Thus, prospective studies are strongly preferred.

Dietary exposures can be assessed using questionnaires or biochemical indicators. These methods have different strengths and limitations which need to be considered for a particular application. The current review is a summary of the available epidemiologic evidence relating dietary nitrates and nitrites and cancer and cardiovascular disease that will focus on methods and study designs commonly used in nutritional epidemiology.

Nutritional Epidemiology and Dietary Assessment

Study Designs

The complex nature of human diets has posed difficult challenges for the evaluation of diet–disease relations [2]. For other exposures, such as cigarette smoking or aspirin use, individuals can easily recall the presence or absence, level and pattern of life-time exposure. In contrast, diet comprises a set of strongly correlated variables where all individuals usually have some level of intake; everyone eats fat, carbohydrates, and vitamin C. In addition, individuals rarely make abrupt changes in their diet and are unaware of the nutrient content of the foods they eat; thus nutrient intake is determined indirectly based on reports of foods. Hence, the major limitation in research on nutrition and health has been the accurate measurement of dietary factors.

Initially, nutritional epidemiology relied on ecologic or correlation studies to investigate dietary factors and health. This strategy relies on comparisons of disease rates and per capita consumption of specific dietary factors across populations. Ecologic studies in nutritional epidemiology have been profoundly important in developing hypotheses on diet and health, but they are considered to provide insufficient evidence for decision making by individuals or for public health policy. The primary reason is the potential for confounding by factors that are difficult to measure. There may be determinants of the disease other than the dietary factor under study that vary between areas with high and low disease incidence and could explain the observed relation. In addition to this important limitation, ecologic studies rely on national dietary data that may vary in quality.

Two study designs commonly employed in epidemiology have been used to avoid the limitations of ecologic studies. In case-control studies, information about previous diet is obtained from diseased individuals and compared with that from individuals free of disease. In contrast, prospective cohort studies obtain data on diet from healthy individuals who are then followed to determine disease rates according to their level of intake of the dietary factors under study. Both study types allow for control of confounding effects of other factors when these have been measured. Investigators may limit the potential effect of these factors by matching individuals to be compared on the basis of known risk factors or by restriction to a particular level of that factor. Alternatively, the effect of these factors may be taken into account in the analysis stage of the study through the use of multivariate modeling. However, case-control and cohort studies have limitations that also need to be considered when evaluating the evidence for diet–disease relations. Case-control studies in nutritional epidemiology may be at a particular risk for recall bias. Diseased individuals associate their disease to an unhealthy diet and over-report consumption of foods considered unhealthy, whereas disease-free individuals may not have this selective recall of certain foods. Thus, recall bias may result in illusory associations between disease and dietary factors considered to be unhealthy [3]. In addition, like in other nondiet applications, case-control studies are subject to selection bias which results from inappropriate enrollment of control individuals. Typically control individuals should represent the population from which the diseased individuals arose and their participation in the study should be unrelated to their level of exposure to the dietary factor considered. However, in practice investigators often include as controls individuals who have a different disease and make strong assumption that this disease is unrelated to the dietary factor of interest. Even when individuals free of disease from the general population are invited to participate, problems may arise. Participation rates for epidemiologic studies are often low, and because diet is particularly associated to health consciousness, the diet of those who decide to participate may be very different from those who do not. There is mounting evidence that case-control studies, even when consistent throughout different populations, may be biased.

Prospective cohort studies may avoid both recall and selection bias because diet is assessed before the diagnosis of disease and even if few individuals decide to participate from the eligibility pool, this low participation rate will not distort observed associations as long as the enrolled persons are not lost to follow-up. For this reason, evidence from prospective cohort studies is considered to be stronger than that from case-control studies. However, prospective cohorts are limited for one strong logistical reason: a large number of people need to be enrolled over a relatively long period of follow-up in order to accrue sufficient cases. Thus, in instances where a rare disease is under study there will never be sufficient cases to conduct analyses of diet when the true effect is relatively small.

The strongest evidence for establishing causal relations in epidemiology comes from double-blind randomized trials; the primary advantage being that randomization and assignment to an exposure by the investigator results in equal distribution of factors that may distort the observed relation. With the exception of studies on dietary supplements, randomized experiments of diet are problematic. Individuals who enroll must change their diet substantially and follow a dietary regime for several years. Only recently a large randomized experiment in women aimed at reducing fat consumption [4] showed the difficulty in maintaining the intervention in women who had consumed a high fat diet most of their life, making results difficult to interpret [5]. Specifically, in this large trial, there was no difference in plasma HDL cholesterol levels between the groups assigned to low fat or usual diets, despite abundant information from controlled feeding trials and other studies showing that reducing the percentage of calories from fat decreases HDL levels.

Diet Assessment Methods

In addition to the methodological concerns related to study design, the main hurdle in nutritional epidemiology is the accurate measurement of diet. In epidemiologic settings, measurement of diet has taken two general paths: assessment by structured questionnaire or interview or measurement by biochemical methods. Today, the food-frequency questionnaire (FFQ) is the primary method for measuring diet in nutritional epidemiology research. As compared to other methods like short-term recall (i.e., 24-h recalls) and diet records, the FFQ focuses on intake over a long period of time, which is the relevant exposure period for chronic disease etiology, rather than in a few specific days. The FFQ has two components: a food list, typically comprising 50–200 food items, and a frequency response section that requests information on the average frequency of intake over the last year (typically in nine or ten categories, ranging from never to 6 or more times a day). Food intake is then transformed to nutrients using a food composition database. The FFQ captures food preferences and average frequencies of consumption reasonably well; however, because it has a restricted list of food items it may miss important foods that were consumed, resulting in error in nutrient calculations. Further error in classification of nutrient intakes may occur if the nutrient content of a particular nutrient in food varies substantially. Thus, it is essential to evaluate the performance of a questionnaire in terms of reproducibility and validity to determine the impact of these sources of error.

The reproducibility of FFQs is often assessed by repeating the same questionnaire at 2 time points within a realistic period of time, typically 1–4 years. To assess the validity of FFQs, most studies compare the FFQ with a series of 24-h recalls or dietary records, or with biomarkers for specific nutrients. One early study evaluating the reproducibility and validity of a FFQ administered the questionnaire twice to 173 women 1 year apart [6]. Within that year, participants also completed four 1-week dietary records. When evaluating the reproducibility, the correlation coefficients for nutrients assessed 1 year apart using the FFQs ranged between 0.49 and 0.71. After adjustment for energy, a comparison between the second FFQ and the average of the dietary records showed correlation coefficients ranging from 0.36 for vitamin A without supplements to 0.75 for vitamin C. Correlation coefficients for carbohydrate, protein, and fat intake were within this range. Plasma levels of folate and an N-3 fatty acid in subcutaneous adipose tissue have been used to validate dietary folate and N-3 fatty acid intake from the FFQ. Adjusted correlation coefficients were 0.63 for folate and 0.47 for the N-3 fatty acid [7, 8]. An extensive literature has developed on the validation of FFQs, which have become more detailed and extensive [9]. Using the average of three such questionnaires over a period years, the correlations with dietary intakes can be as high as 0.8 or 0.9, as found for total and saturated fat. Taken together, these results show that the questionnaire provides useful information about intake of major nutrients and is a practical option for dietary assessment that is easy for individuals to complete, as well as being simple and inexpensive to process.

Biochemical indicators of dietary intake in blood or urine seem at first glance particularly attractive for their objectiveness as compared to dietary questionnaires. However, these measures are subject to the same problems of measurement error bias due to confounding factors as questionnaires, and careful consideration needs to be given regarding their use for particular applications. First, it is important to know the degree to which the marker is sensitive to dietary intake. Certain measurable markers in blood or urine may be strongly affected by factors other than diet like genetics or homeostatic mechanisms, so that large changes in diet are necessary to produce any effect. Serum sodium provides an extreme example; the levels can be measured precisely, but are not informative regarding dietary intake. Serum cholesterol levels are also largely determined by factors other than dietary intake. The second consideration is whether the measure provides an indication of time-integrated intake. Similar to the FFQ, the conceptually important exposure is long term rather than a few hours, the previous meal, or a couple of days. The third consideration to be made is that blood levels of a biochemical indicator may be affected by other factors and that adjustment for these may improve the estimation of dietary intake. The last, but important consideration is the laboratory error in measuring the biochemical parameter. However, as compared to other applications, in epidemiological research consistency in the laboratory is more important than accuracy on an absolute scale.

Biochemical measurements of nutrients are particularly important when a specific nutrient is poorly measured by other methods due to large within-food variation in nutrient content. However, similar to FFQs, it is essential to evaluate the sensitivity of a particular nutrient biomarker to intake of that nutrient and the validity of this measure as an indicator of long-term intake. One important strategy to evaluate biomarkers is repeating measurements of the biomarker among the same group of people on more than one occasion to assess the indicator's stability over time, usually over an interval of several months to several years. If the correlation is high, and the biomarker has been shown to be responsive to intake, the biomarker may be a well-integrated measure of long-term intake.

Assessment of Nitrate and Nitrite Intake

Nitrite and nitrite oxide are biologically active compounds involved in vasodilatation, inhibition of endothelial inflammation, and platelet aggregation. Nitrates and nitrites in diet can serve as substrate for a stepwise reduction of nitrate to nitrite to nitric oxide and are the relevant compounds for nutritional epidemiology. However, as compared to other nutrients, quantifying nitrate and nitrite intake is challenging. The major sources of nitrate exposure are foods and drinking water. Large within-food variation of nitrates and nitrites and large geographical and seasonal differences in nitrate content in drinking water complicate accurate estimation of exposure [10]. Assessment methods for the two major sources of intake are different. Assessment of exposure to nitrates and nitrites from food relies for the most part on 24-h recalls and FFQs; while drinking water nitrate exposure assessment is based on environmental epidemiologic methods that often rely on drinking water quality registries.

The validity of nutrient intake assessment using 24-h recalls, dietary records, or FFQs relies on the availability of an accurate nutrient content database. Most epidemiologic studies rely on food composition values from the U.S. Department of Agriculture's National Nutrient Database. Currently, there is no standard database for the nitrate and nitrite content of foods. There have been some efforts to develop food databases [11, 12]. However, most investigators use ad hoc databases based on publications judged to be most relevant for a particular population and on drinking water quality registries [13–16]. In general, the main sources of nitrate intake are green leafy vegetables, while for nitrite intake the major sources are processed meats to which nitrite is added as a preservative [10]. Relying on literature-based nutrient content database, in the Nurses' Health Study (NHS) a prospective cohort of 78,191 women and in similar study, the Health Professional's Health Study, among 31,537 men the main contributors of nitrate intake were lettuce, spinach, celery, and broccoli (Table 6.1). Nitrate and nitrite values can vary substantially within foods.

Table 6.1 Percent contribution of different foods to nitrate intake from a food-frequency questionnaire in two large epidemiologic studies

Nurses' Health Study		Health Professionals' Health Study	
Food	% contribution	Food	% contribution
Iceberg lettuce	28.2	Iceberg lettuce	27.8
Romaine lettuce	26.1	Romaine lettuce	22.4
Cooked spinach	6.1	Cooked spinach	6.5
Celery	5.5	Broccoli	5.2
Broccoli	5.1	Celery	4.6
Potatoes	3.1	Kale	3.2
Raw spinach	2.1	Potatoes	3.0
Tomato sauce	2.0	Raw spinach	2.5
Kale	1.9	Tomato sauce	2.3
String beans	1.7	Cabbage	1.6

Based on 2002 nitrate estimates for both cohorts

The level of nitrate intake from vegetables depends importantly on soil quality, nitrate content of the fertilizer used, nitrate level on the water supply, and storage and transport practices [10]. Variation in nitrite depends on meat processing practices. Nitrate content of spinach collected from three different markets varied from 71 to 387 mg/100 g [17]. Comparing estimated nitrate and nitrite content in foods from a convenience sample in the United States [10] to a recent literature-based nutrient content database showed dramatic differences [11]. Measured nitrate in spinach was 741 mg/100 g while the nutrient database estimated 210 mg/100 g. For broccoli, the convenience sample estimated 39.5 mg/100 g of nitrate and the database reported 34.2 mg/100 g. Laboratory analysis of nitrites in bacon showed 0.38 mg/100 g, while the database reported 2.91 mg/100 g. Similarly, for hot dog the nitrite estimates varied substantially: measured nitrite was 0.05 mg/100 g and literature estimated nitrite 2.7 mg/100 g.

The great variability in nitrate and nitrite contents of foods highlights the value of identifying valid and practical blood or urine biomarkers and of validating questionnaire-based assessments of nitrate and nitrite intake. There are several factors that make the estimation of nitrate and nitrite exposure in vivo challenging. Serum and urinary nitrate levels are affected by dietary nitrate intake and from endogenous nitric oxide production [18]; however, nitrate and nitrite have short half-lives: 5–8 h for nitrate and 1–5 min for nitrite [19]. In addition, other factors like exercise have been shown to affect circulating nitrite and nitric oxide levels [20]. Thus, plasma and urinary nitrate and nitrite may not accurately reflect long-term exposure, and measured levels may be affected by other factors that need to be taken into account in any analysis. One strategy is assessing overnight nitrate and nitrite metabolite excretion in urine. However, this measurement may still not be reflective of long-term exposure and its logistical difficulty limits its applicability in large-scale population studies. A recent pilot study that evaluated the feasibility of using plasma nitrate and nitrite in epidemiologic studies found this composite biomarker to be stable and the within-person variability was modest but comparable to commonly used biomarkers (personal communication Tianying Wu).

Repeated measurements of urine nitrate may be useful to account for within-person variation and have been used to evaluate the validity of nitrate intake assessed by FFQ [21]. This type of validation study was conducted among 59 individuals who responded to a FFQ; the correlation between estimated intake and nitrate excretion was estimated. The crude correlation was 0.20, but after correction for within-person variation and adjustment for sex, gender, and body mass intake the correlation coefficient was 0.59. Thus, FFQ assessment of nitrate intake may provide useful information on usual nitrate intake [21]. This study also suggests that the within-food variation in nitrate content is not so large as to preclude useful assessment of intake by dietary questionnaires. This is possible if the average nitrate content of a food does not vary substantially among individuals. For example, even though the nitrate content of spinach may vary from sample to sample, over a year the average nitrate content of spinach consumed by one person may not be substantially different from the average nitrate content of spinach consumed by other persons. Because spinach consumption does vary greatly among individuals, it can be possible to distinguish among persons according to their nitrate intake.

Nitrate and nitrite are widely distributed in fruits, vegetables, and processed meats and drinking water, however, evaluating exposure to these substances is challenging. Their wide distribution in the food and water supply makes clinical studies of endogenous nitrogen oxides difficult as designing nitrate-free diets is complex. Also, estimates of dietary and drinking water nitrate and nitrites are subject to important measurement error that may complicate the interpretation of epidemiological studies. However, the report by van den Brandt et al. does suggest that assessment of nitrate intake in epidemiologic studies is feasible. In the context of extensive measurement error that is independent of the outcome, the absence of an association should be interpreted with caution.

Nitrate and Nitrite Intake and Cancer

Nitrate and nitrite intake and their relation to cancer, and more specifically gastric and other gastrointestinal cancers, have been extensively studied [22]. Interest in nitrates and nitrites as risk factors for cancer stems from the generation of *N*-nitroso compounds, potential carcinogens, from nitrites (inorganic nitrates are not subject to nitrosation rxns) in the stomach [23]. For gastric cancer, data are still inconclusive as very few prospective cohorts have evaluated this relation and results from these studies show no association [24]. More recently, prospective studies have also evaluated relations with glioma, bladder cancer, and thyroid cancer, but results are still preliminary and inconclusive [25–27]. The World Cancer Research Fund's (WCFR) report on Nutrition, Physical Activity and Prevention of Cancer did evaluate nitrate and nitrite intake; however, data were too limited to allow conclusions. Nevertheless, the report suggests that they may be potential carcinogens "under conditions that promote nitrosation" [28]. A more recent International Agency for Research in Cancer monograph concluded that, based on the association with gastric cancer, there was limited evidence of carcinogenicity of nitrites in food and inadequate evidence for dietary nitrates [29]. Nevertheless, as mentioned earlier, the epidemiologic evidence supporting these reports is based on nitrate and nitrite assessments with potential important measurement error.

An additional approach to evaluate these relations is to determine associations between the main food sources of nitrates and nitrites. The WCFR report concluded that the evidence from epidemiologic studies that vegetables or foods protect against cancer was not as impressive as originally thought [28]; the initial suggestions of benefit came mainly from case-control studies, and these had generally not been supported by prospective cohort studies. Green leafy vegetables, the major source of dietary nitrates, were independently evaluated for several cancers. Evidence from cohort studies showed a slightly reduced risk of lung cancer when comparing high to low intake groups (meta-analysis RR = 0.91 (95% CI 0.89–0.93) per serving/day). One important consideration when analyzing dietary data is that, as compared to other exposures, nutrients are not consumed independently because many different nutrients are found in the same food. Thus, it may be difficult to

Table 6.2 Correlation coefficient between dietary nitrate and other nutrients from two large epidemiologic studies

Nutrient	Nurses' Health Study	Health Professional's Health Study
Carbohydrates	-0.01	0.04
Saturated fat	-0.20	-0.20
Protein	0.18	0.20
Animal protein	0.12	0.08
Fiber	0.45	0.43
Vitamin C	0.20	0.19
Beta-carotene	0.66	0.60
Folate	0.64	0.63

Based on the 2002 diet from both cohorts. Nutrients were log transformed and energy adjusted

separate the effects of nutrients that are highly correlated. Table 6.2 shows the correlation coefficients of nitrates with other nutrients in two large prospective studies of women and men. Nitrate intake is highly correlated with intake of folate, which is considered to have anticarcinogenic effects [30], so it is possible that this and other nutrients may explain the observed inverse relation with lung cancer. The WCRF report concluded there is convincing evidence that processed meat, to which nitrite is added, increases the risk of colorectal cancer, as well as red meat where no nitrite is added (red meat meta-analysis RR = 1.29 (95% CI 1.04–1.60) per 100 g/day; processed meat meta-analysis RR = 1.21 (95% CI 1.04–1.42) per 50 g/day) [28]. There are several proposed mechanisms for these robust associations: formation of *N*-nitroso compounds through exposure to nitrites, which are added as preservatives; accumulation of heme iron free iron that can lead to production of free radicals; and exposure to heterocyclic amines and polycyclic aromatic hydrocarbons produced during cooking.

Nitrate and Nitrite Intake and Cardiovascular Disease

Organic nitrates at pharmacologic doses have been used as vasodilators for a long time. Inorganic salts of nitrate have been used for centuries to preserve food. More recently, data supporting the vasoprotective, blood-pressure lowering and anti-platelet aggregation effects of low-dose dietary nitrate intake have emerged [31–33]. To date there is no epidemiologic evidence relating nitrate and nitrite intake and cardiovascular outcomes such as coronary heart disease (CHD) and stroke. Nevertheless, intake of fruits and vegetables, green leafy vegetables, red and processed meats, and cardiovascular disease have been extensively studied.

A meta-analysis of nine cohort studies found a significant inverse relation between fruits and vegetables intake and CHD: the risk decreased by 4% (RR 0.96 (95% CI 0.93–0.99)) for every additional serving [34]. One of the studies included in the meta-analysis, the NHS with 14 years of follow-up and repeated dietary measures reported the same lower 4% risk of CHD for one daily serving increase in fruit and vegetable intake and observed a significant 22% lower risk of CHD for an increase of one serving a day of green leafy vegetables [35]. For stroke, a meta-analysis of seven prospective cohorts observed an inverse relation that was similar in magnitude to CHD (RR = 0.95 (95% CI 0.92–0.97) for an increase in one serving per day) [36] and the NHS reported a significant 21% lower risk of stroke for an increase in one daily serving of green leafy vegetables [37].

Processed meats are the main contributors of nitrite intake (fresh red meat has little or no residual nitrite). However, a recent meta-analysis found a significant direct association with CHD for processed meat (RR = 1.42 (95% CI 1.07–1.89)) but not for red (unprocessed) meat (RR = 1.00 (95% CI 0.81–1.23)), although the data for unprocessed meats were extremely limited [38]. No significant association for processed and unprocessed meats and stroke was observed. In the largest prospective study included in the meta-analysis, with 322,263 men and 223 390 women, a direct association for processed meats was observed [39]. There was a 9% increase in CHD risk comparing extreme quintiles in men (RR = 1.09 (95% CI, 1.03–1.15)), and a 38% increase in women (RR = 1.38 (95% CI, 1.26–1.51)). The increased risks with processed meats may be due in part to the content of preservatives in these foods, however, a significant association was also observed for red meat to which no preservatives are added (RR = 1.27 (95% CI, 1.20–1.35) in men and RR = 1.50 (95% CI, 1.37–1.65) in women).

Conclusion

Elucidating the relationship between nitrate and nitrite intake and different health outcomes is of great importance because intakes of these molecules may be amenable to intervention. When evaluating the evidence to establish these relationships it is important to take into account both the strengths and limitations of individual studies with respect to the design, validity of the dietary assessment of nitrates and nitrites, adjustment for other dietary and nondietary factors that may account for the observed relation, and the distinction of different sources of nitrates and nitrites. In one validation study using urinary nitrate excretion as the gold standard, there did appear to be sufficient validity of nitrate intake assessed by FFQ to be useful for evaluating hypotheses in epidemiologic studies. Evaluating foods and food groups that are main contributors to intakes of nitrate and nitrite in relation to health outcomes may be helpful in the absence of a standard nitrate and nitrite content database and wide within-food variability. However, the intake of a particular food of food group may be closely correlated to intake of other foods or health-related behaviors. Therefore, care should be taken in the interpretation of results.

To date, the epidemiologic evidence for the relation between dietary nitrate and nitrite and cancer is weak. However, the evidence for the relation between red and processed meat and colorectal cancer is robust. There is no direct epidemiologic evidence of the relation between nitrite and nitrate and cardiovascular disease. The inverse association with green leafy vegetables, the main source of nitrates, and the observed direct relation with processed meats, the main contributors on nitrites, is perplexing as experimental evidence on nitrate and nitrite intake would predict otherwise. It is possible that the risks associated with meat intake may be related to sodium and other compounds that may be altered or added during processing or cooking. Future studies, should more thoroughly investigate nitrate and nitrite intake from both vegetable and meat sources.

References

1. Lundberg JO, Gladwin MT, Ahluwalia A, Benjamin N, Bryan NS, Butler A, et al. Nitrate and nitrite in biology, nutrition and therapeutics. Nat Chem Biol. 2009;5(12):865–9.
2. Willett W. Nutritional epidemiology: issues and challenges. Int J Epidemiol. 1987;16(2):312–7.
3. Giovannucci E, Stampfer MJ, Colditz GA, Manson JE, Rosner BA, Longnecker M, et al. A comparison of prospective and retrospective assessments of diet in the study of breast cancer. Am J Epidemiol. 1993;137(5):502–11.
4. Prentice RL, Caan B, Chlebowski RT, Patterson R, Kuller LH, Ockene JK, et al. Low-fat dietary pattern and risk of invasive breast cancer: the Women's Health Initiative Randomized Controlled Dietary Modification Trial. JAMA. 2006;295(6):629–42.
5. Willett WC. The WHI joins MRFIT: a revealing look beneath the covers. Am J Clin Nutr. 2010;91(4):829–30.
6. Willett WC, Sampson L, Stampfer MJ, Rosner B, Bain C, Witschi J, et al. Reproducibility and validity of a semiquantitative food frequency questionnaire. Am J Epidemiol. 1985;122(1):51–65.
7. Hunter DJ, Rimm EB, Sacks FM, Stampfer MJ, Colditz GA, Litin LB, et al. Comparison of measures of fatty acid intake by subcutaneous fat aspirate, food frequency questionnaire, and diet records in a free-living population of US men. Am J Epidemiol. 1992;135(4):418–27.
8. Jacques PF, Sulsky SI, Sadowski JA, Phillips JC, Rush D, Willett WC. Comparison of micronutrient intake measured by a dietary questionnaire and biochemical indicators of micronutrient status. Am J Clin Nutr. 1993;57(2):182–9.
9. Willett WC. Nutritional epidemiology. 2nd ed. New York: Oxford University Press; 1998.
10. Hord NG, Tang Y, Bryan NS. Food sources of nitrates and nitrites: the physiologic context for potential health benefits. Am J Clin Nutr. 2009;90(1):1–10.
11. Griesenbeck JS, Steck MD, Huber JC, Jr., Sharkey JR, Rene AA, Brender JD. Development of estimates of dietary nitrates, nitrites, and nitrosamines for use with the Short Willet Food Frequency Questionnaire. Nutr J. 2009;8:16.
12. Jakszyn P, Agudo A, Ibanez R, Garcia-Closas R, Pera G, Amiano P, et al. Development of a food database of nitrosamines, heterocyclic amines, and polycyclic aromatic hydrocarbons. J Nutr. 2004;134(8):2011–4.
13. Griesenbeck JS, Brender JD, Sharkey JR, Steck MD, Huber JC, Jr., Rene AA, et al. Maternal characteristics associated with the dietary intake of nitrates, nitrites, and nitrosamines in women of child-bearing age: a cross-sectional study. Environ Health. 2010;9:10.
14. Ward MH, Heineman EF, McComb RD, Weisenburger DD. Drinking water and dietary sources of nitrate and nitrite and risk of glioma. J Occup Environ Med. 2005;47(12):1260–7.
15. Dich J, Jarvinen R, Knekt P, Penttila PL. Dietary intakes of nitrate, nitrite and NDMA in the Finnish Mobile Clinic Health Examination Survey. Food Addit Contam. 1996;13(5):541–52.
16. Michaud DS, Holick CN, Batchelor TT, Giovannucci E, Hunter DJ. Prospective study of meat intake and dietary nitrates, nitrites, and nitrosamines and risk of adult glioma. Am J Clin Nutr. 2009;90(3):570–7.
17. Anjana S. Are nitrate concentrations in leafy vegetables within safe limits? Curr Sci. 2007;92(3):355–60.
18. Pannala AS, Mani AR, Spencer JP, Skinner V, Bruckdorfer KR, Moore KP, et al. The effect of dietary nitrate on salivary, plasma, and urinary nitrate metabolism in humans. Free Radic Biol Med. 2003;34(5):576–84.

19. Lundberg JO, Weitzberg E. NO generation from nitrite and its role in vascular control. Arterioscler Thromb Vasc Biol. 2005;25(5):915–22.
20. Rassaf T, Lauer T, Heiss C, Balzer J, Mangold S, Leyendecker T, et al. Nitric oxide synthase-derived plasma nitrite predicts exercise capacity. Br J Sports Med. 2007;41(10):669–73; discussion 73.
21. Van den Brandt PA, Willett WC, Tannenbaum SR. Assessment of dietary nitrate intake by a self-administered questionnaire and by overnight urinary measurement. Int J Epidemiol. 1989;18(4):852–7.
22. Mirvish SS. Role of N-nitroso compounds (NOC) and N-nitrosation in etiology of gastric, esophageal, nasopharyngeal and bladder cancer and contribution to cancer of known exposures to NOC. Cancer Lett. 1995;93(1):17–48.
23. IARC. Supplement 7. Overall evaluations of carcinogenicity: an updating of IARC monographs volumes 1 to 42. Lyon: International Agency for Research on Cancer; 1987.
24. Jakszyn P, Gonzalez CA. Nitrosamine and related food intake and gastric and oesophageal cancer risk: a systematic review of the epidemiological evidence. World J Gastroenterol. 2006;12(27):4296–303.
25. Dubrow R, Darefsky AS, Park Y, Mayne ST, Moore SC, Kilfoy B, et al. Dietary components related to *N*-nitroso compound formation: a prospective study of adult glioma. Cancer Epidemiol Biomarkers Prev. 2010 Jun 22.
26. Ward MH, Kilfoy BA, Weyer PJ, Anderson KE, Folsom AR, Cerhan JR. Nitrate intake and the risk of thyroid cancer and thyroid disease. Epidemiology. 2010;21(3):389–95.
27. Zeegers MP, Selen RF, Kleinjans JC, Goldbohm RA, van den Brandt PA. Nitrate intake does not influence bladder cancer risk: the Netherlands cohort study. Environ Health Perspect. 2006;114(10):1527–31.
28. WCRF/AICR. Food, nutrition, physical activity, and the prevention of cancer: a global. Washington: AICR; 2007.
29. AICR. Volume 94. Ingested nitrate and nitrite, and cyanobacterial peptide toxins. Lyon: IARC; 2010.
30. Mason JB. Folate, cancer risk, and the Greek god, Proteus: a tale of two chameleons. Nutr Rev. 2009;67(4):206–12.
31. Larsen FJ, Ekblom B, Sahlin K, Lundberg JO, Weitzberg E. Effects of dietary nitrate on blood pressure in healthy volunteers. N Engl J Med. 2006;355(26):2792–3.
32. Webb AJ, Patel N, Loukogeorgakis S, Okorie M, Aboud Z, Misra S, et al. Acute blood pressure lowering, vasoprotective, and antiplatelet properties of dietary nitrate via bioconversion to nitrite. Hypertension. 2008;51(3):784–90.
33. Bryan NS, Calvert JW, Elrod JW, Gundewar S, Ji SY, Lefer DJ. Dietary nitrite supplementation protects against myocardial ischemia-reperfusion injury. Proc Natl Acad Sci USA. 2007;104(48):19144–9.
34. Dauchet L, Amouyel P, Hercberg S, Dallongeville J. Fruit and vegetable consumption and risk of coronary heart disease: a meta-analysis of cohort studies. J Nutr. 2006;136(10):2588–93.
35. Joshipura KJ, Hu FB, Manson JE, Stampfer MJ, Rimm EB, Speizer FE, et al. The effect of fruit and vegetable intake on risk for coronary heart disease. Ann Intern Med. 2001;134(12):1106–14.
36. Dauchet L, Amouyel P, Dallongeville J. Fruit and vegetable consumption and risk of stroke: a meta-analysis of cohort studies. Neurology. 2005;65(8):1193–7.
37. Joshipura KJ, Ascherio A, Manson JE, Stampfer MJ, Rimm EB, Speizer FE, et al. Fruit and vegetable intake in relation to risk of ischemic stroke. JAMA. 1999;282(13):1233–9.
38. Micha R, Wallace SK, Mozaffarian D. Red and processed meat consumption and risk of incident coronary heart disease, stroke, and diabetes mellitus: a systematic review and meta-analysis. Circulation. 2010;121(21):2271–83.
39. Sinha R, Cross AJ, Graubard BI, Leitzmann MF, Schatzkin A. Meat intake and mortality: a prospective study of over half a million people. Arch Intern Med. 2009;169(6):562–71.

Chapter 7
Nutritional Impact on the Nitric Oxide Pathway

Wing Tak Wong and John P. Cooke

Key points

- The endothelial nitric oxide synthase (eNOS) pathway is highly modulated by nutrition.
- Dietary choices and interventions may modulate endothelial function.
- The Mediterranean diet enhances endothelial vasodilator function and reduces major adverse cardiovascular events (MACE).
- Increased dietary consumption of fish and other sources of omega-3 fatty acids improve NO activity, and also reduce MACE in those with cardiovascular disease.
- A number of nutritional supplements may improve endothelial function by reducing oxidative stress, by restoring elements of the NOS pathway, or by ameliorating insulin resistance, hypertension, or hyperlipidemia; however, their long-term effects on MACE are unknown.
- Some dietary supplements known to enhance the NOS pathway and improve endothelial vasodilator function in humans (Vitamin E, the B vitamins, and L-arginine) were ineffective at reducing MACE in large randomized clinical trials.
- Thus, dietary interventions that improve endothelial vasodilator function in the short term might not necessarily have long-term benefit.

Keywords Diet • Nutrition • Mediterranean diet • Vitamins • Cardiovascular disease • Functional foods

J.P. Cooke (✉)
Department of Medicine, Stanford School of Medicine, Cardiovascular Medicine Clinic, 300 Pasteur Drive A260 MC 5319, Stanford, CA 94305-2200, USA
e-mail: John.Cooke@stanford.edu

N.S. Bryan and J. Loscalzo (eds.), *Nitrite and Nitrate in Human Health and Disease*, Nutrition and Health, DOI 10.1007/978-1-60761-616-0_7,

Vascular Homeostasis and Its Modulation by the Diet

In this chapter, we review the effect of nutritional interventions on vascular homeostasis. We focus on the effect of diet and dietary supplements on endothelial vasodilator function, and in particular, the activity and expression of endothelial nitric oxide synthase (eNOS). Endothelium-derived nitric oxide (NO) is a powerful regulator of vascular homeostasis. By virtue of its ability to activate soluble guanylyl cyclase and increase intracellular cyclic GMP, NO relaxes the underlying vascular smooth muscle to improve vascular compliance and to reduce vascular resistance. In addition, endothelium-derived NO inhibits platelet adhesion and aggregation, suppresses leukocyte adhesion and vascular inflammation, and limits the proliferation of the underlying vascular smooth muscle cells. Furthermore, NO is mitogenic for endothelial cells and increases the regeneration of the endothelial monolayer. In large conduit vessels such as the coronary artery, NO plays a critical role in defending against vascular inflammation and lesion formation.

Diet has a profound and immediate impact upon the bioactivity of NO generated by the endothelium. Chronic nutritional interventions, by modulating the expression and activity of NOS, can dramatically affect vascular homeostasis. This knowledge may be useful to the clinician in management of patients with or at risk of cardiovascular disease. We begin by considering the points along the NOS pathway that are most susceptible to modification by diet.

Regulation of the NOS Pathway

eNOS metabolizes L-arginine to NO and L-citrulline (Fig. 7.1). Endothelial shear stress, as well as a variety of humoral or paracrine factors such as acetylcholine, adenosine diphosphate, thrombin, and vasopressin, is known to induce vasodilation, secondary to phosphorylation and activation of eNOS [1, 2]. A major cause of endothelial vasodilator dysfunction is impairment of this pathway. The ability of the endothelium to respond to shear stress or other stimuli, and to induce relaxation of the underlying vascular smooth muscle, is impaired in older individuals [3–5] and those with diabetes, hypertension, hypercholesterolemia, or tobacco exposure [6, 7].

An impairment of endothelial NOS not only reduces the ability of a blood vessel to relax, but also broadly disrupts vascular homeostasis. In addition to relaxing vascular smooth muscle, NO is a potent inhibitor of platelet adhesion and aggregation [8, 9]. In addition, NO suppresses vascular inflammation by reducing the expression of leukocyte adhesion molecules and inflammatory cytokines [10–13]. Consistent with these observations, in animal models, enhancement of NO synthesis (as with L-arginine administration or overexpression of eNOS protein) reduces the progression of atherosclerosis and myointimal hyperplasia [14–16]. The importance of NO in vascular homeostasis is supported by a large number of studies revealing that an impairment of endothelial vasodilator function is an independent risk factor for cardiovascular morbidity and mortality [17–19].

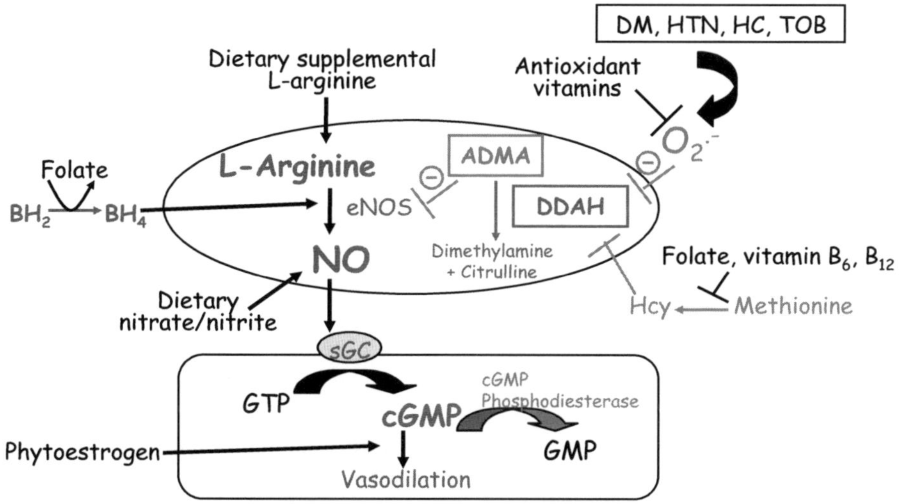

Fig. 7.1 The potential dietary interventions that may enhance the function of the nitric oxide synthase (NOS) pathway

A number of conditions associated with cardiovascular disease are also known to impair the NOS pathway. For example, diabetes mellitus is associated with mitochondrial dysfunction and oxidative stress [20] that can accelerate the degradation of NO [21]. Furthermore, diabetes mellitus favors the production of advanced glycosylation end products (AGEs) which can also disrupt eNOS activation, as with formation of *N*-acetylglucosamine (OGlcNAc) adducts with serine phosphorylation sites on eNOS [22, 23]. Aging also impairs the NOS pathway and increases the risk for major adverse cardiovascular events (MACE). Aging alters the phosphorylation and activation of eNOS in experimental animals [24]. Dyslipidemia is another major cause of endothelial vasodilator function. Hypercholesterolemia enhances the inhibitory interaction of caveolin-1 with eNOS; an effect that can be reversed by diet and exercise [25].

Components of Nutrition and Endothelial Function

As can be seen in Fig. 7.1, there are multiple dietary interventions that theoretically may enhance the function of the NOS pathway. Firstly, one might utilize dietary interventions to treat the cardiovascular risk factors that are known to impair the NOS pathway. Evidence indicates that optimal management of diabetes mellitus, hypertension, hypercholesterolemia, and hyperhomocysteinemia may improve endothelial vasodilator function [26–28]. Furthermore, it is possible that one might supplement portions of the pathway that are impaired. For example, one might provide exogenous dietary arginine to reverse the effects of asymmetric dimethylarginine

(ADMA; see below) or provide antioxidant vitamins to reduce the oxidative degradation of NO and/or the impairment of dimethylarginine dimethylaminohydrolase (DDAH). In what follows, we review the effect of components of the diet on endothelial function, and the evidence that dietary interventions might favorably influence endothelial vasodilator function.

Lipids

Saturated and trans-fatty acids. Evidence suggests that increased intake of saturated or trans-fatty acids impair endothelial vasodilator function within hours of the meal. For example, Plotnick et al. [29] showed that a single high-fat meal could impair endothelium-dependent vasodilation in human subjects, an effect that could be prevented by simultaneous consumption of antioxidant vitamins. Antioxidant vitamins may act to keep NOS cofactors such as tetrahydrobiopterin (BH4) from becoming oxidized, thereby increasing L-arginine conversion to NO; they may protect NO from oxidative degradation once produced and/or prevent oxidative impairment of DDAH, which is exquisitely oxidant sensitive [7]. Evidence for the latter comes from Fard et al. [30] who discovered that ADMA increases in the blood after eating a high-fat meal.

Evidence suggests that consumption of saturated fat may reduce cell membrane fluidity, vascular compliance, and the ability of the endothelium to respond appropriately to shear stress [31]. In addition, dietary saturated fat contributes to dyslipidemia [32], which can impair endothelial function [6]. Trans-fatty acids (e.g., partially hydrogenated vegetable oil) are found in processed foods, such as pretzels and potato chips. Trans-fatty acids also raise LDL cholesterol and impair endothelial vasodilator function [33, 34]. In a study conducted in the Netherlands, researchers found that men who consumed the most trans-fatty acids (6.4% of total calories) had twice the risk of developing heart disease than those who consumed the least amounts (2.4%) [35].

Mono and Polyunsaturated Fats

As a general rule, the amount of saturated fat in a plant-based diet tends to be lower than that in a diet rich in animal proteins. Furthermore, vegetables, nuts, and plant oils, such as olive or canola oil, are a source of monounsaturated fats (e.g., omega-9s or oleic acid). Monounsaturated fats reduce LDL cholesterol in the blood and have no effect on HDL cholesterol [33, 34]. In this regard, the Adventist Health study found that greater consumption of nuts reduces the risk of cardiovascular morbidity and mortality [36], possibly related to the monounsaturated fats in nuts.

Polyunsaturated fats include the omega-3 and omega-6 fatty acids. Both omega-3 and omega-6 fatty acids are considered essential fatty acids. Omega-6 fatty acids (linoleic acid) are polyunsaturated fats that are found in vegetable oils, and favorably influence the lipid profile. The marine omega-3 fatty acids, found in the oils of fish, seem to be vasoprotective. In 1980, Bang and Dyerberg reported that Greenland Eskimos have lower cardiovascular morbidity despite a high fat diet that includes whale blubber [37]. It is likely that the resistance of Eskimos to cardiovascular events is due to the high intake of marine lipids. About 40% of the fatty acids in fish are omega-3 polyunsaturated fatty acids (specifically eicosapentanoic acid (EPA) and docosohexanoic acid (DHA)). These fatty acids enhance membrane fluidity and vascular compliance [38]. Furthermore, omega-3s also improve endothelial function. They enhance the production of NO [39]. In addition, they substitute for arachidonic acid in the synthesis of an active form of prostacyclin (PGI3) and an inactive form of thromboxane (TxA3), thus improving blood vessel relaxation [40]. Accordingly, fish oil supplementation improves endothelium-dependent relaxation in patients with atherosclerosis [41]. Fish with the highest content of omega-3 tend to be fatty, cold-water fish such as mackerel, salmon, and trout (Table 7.1). Although the primary dietary source of omega-3 is from fish, some plants are also rich in omega-3 including flaxseed (about 50% of its oil is linolenic acid), some nuts, and dark leafy vegetables.

Protein

In general, vegetable protein (e.g., soy and beans) has greater cardiovascular benefit than animal protein, partly because vegetable protein contains little saturated fat. Furthermore, vegetable protein contains less methionine, which may be converted to homocysteine. Finally, vegetable proteins also have more phytonutrients and fiber than animal sources. These differences in the compositions of animal versus vegetable protein may explain the observation that oxidative stress in the blood goes up after eating a high-fat meal containing animal protein, while it goes down after a plant-based meal [42].

A complete protein source is a food that contains all nine essential amino acids in quantities that are similar to those needed by the body. Animal protein is generally complete. Proteins from plants are usually incomplete, with the exception of soy protein [43]. By combining them in a meal, vegetable proteins can become a complete source of the essential amino acids, e.g., beans and rice. Nonessential amino acids are those that can be synthesized endogenously. However in times of physiological stress, intense exercise, or trauma, some of these may become "conditionally" essential. We have suggested that L-arginine may be conditionally essential in patients with elevated levels of plasma ADMA (see section "Arginine"). In these subjects, we have shown that L-arginine supplementation may improve endothelium-dependent vasodilation, at least transiently [44].

Table 7.1 Omega-3 content of foods (g/100 g of food)

Foods	g/100 g
Mackerel	5.3
Herring	3.1
Salmon	2.0
Trout	1.6
Sardine, canned in oil	1.3
Halibut	0.9
Swordfish	0.8
Surimi	0.6
Shrimp	0.5
Catfish	0.4[a]
Fast food fish sandwich	0.4[a]
Fish sticks	0.4
Crab	0.3
Cod	0.3
Clams	0.3
Plaice	0.2
Haddock	0.2
Flounder	0.2
Tuna, canned in oil	0.2
Pike, northern	0.1
Flaxseed	22.8
Walnuts, English	6.8
Soybeans, roasted	1.5
Wheat germ	0.7
Beans, dry	0.6
Almonds	0.4
Avocado	0.1

[a]However, in these products, the fat is oxidized by frying so that there is minimal or no "good" fat

Sources: http://www.nal.usda.gov/fnic/foodcomp/Data/index.html

Carbohydrates

Fruits and vegetables are the best source of carbohydrates because, in addition to fiber, they are also loaded with micronutrients (such as Vitamins C and E, and β-carotene), enzyme cofactors (such as B_6, B_{12}, and folate), and minerals (such as potassium, zinc, and magnesium). By consuming at least seven servings of fruit and vegetables daily, most people can obtain adequate amounts of the majority of these micronutrients and vitamins.

The nutrient value of fresh fruits, nuts, vegetables, beans, and whole grains is generally higher than refined or commercially prepared food. This is because most of the micronutrients and fiber are lost in the refining process. For example, when wheat is refined, the bran (or outer shell) of the grain is removed. In an epidemiological study of about 34,000 Norwegians, those who ate the highest amounts of whole grains had a 23% reduction in cardiovascular mortality compared to those who ate less or no whole grains [45, 46].

Fiber is found only in plants. There are two different types of fiber: soluble and insoluble. Certain grains and plants, such as beans, legumes, and fruits, contain soluble fiber such as pectin and psyllium. Soluble fiber lowers blood cholesterol levels [47], thereby improving endothelial vasodilator function. Insoluble fiber is found in whole wheat goods, seeds, nuts, wheat bran, and the skin of fruits and vegetables. Insoluble fiber does not affect cholesterol levels but may help in weight management as it increases intestinal transit time.

Phytonutrients have antioxidant capacity, including carotenes (Vitamin A or β-carotene), tocopherols (Vitamin E compounds), and ascorbate (Vitamin C and polyphenols). Fruits and vegetables contain over 4,000 polyphenols which may have cardiovascular benefit. Polyphenols are often lost in processed foods. For example, the peel of many fruits may be enriched in polyphenols in comparison to the pulp (the peel also contains more fiber).

The most common groups of plant antioxidants are the flavonoids and the carotenoids. Plant flavonoids (e.g., from green tea or grapes) have been shown to reduce oxidation of LDL cholesterol, inhibit platelet aggregation, suppress inflammation, and enhance endothelium-dependent vasodilation [48, 49]. Chocolate contains polyphenols similar to those in tea, fruits, and vegetables. These antioxidants reduce the oxidation of cholesterol and protect and/or increase the production of NO [50]. However, because most sources of chocolate are high in calories and saturated fat, as well as the caffeine analog theobromine, moderation is advisable.

The French Paradox

How is it that the French eat large amounts of fatty foods, yet purportedly have fewer heart attacks than other Europeans? A possible factor could be the moderate consumption of red wine, which contains the antioxidant flavonoids, resveratrol and quercetin. One glass of red wine has the same antioxidant effect as 500 mg of Vitamin C and can blunt the adverse effect of a high-fat meal on the endothelium [51]. Fitzgerald et al. found that red wine (as opposed to white) relaxes vascular rings in vitro, an effect related to antioxidant activity in the skin of the grape (red wine is incubated with the skin of the grape, as opposed to most white wines). Teetotalers may take comfort in the observation that purple grape juice also improves the ability of the blood vessels to relax in people with impaired endothelial function [48].

That being said, several large population studies have suggested that any form of alcohol, when consumed in moderation, reduces the risk of cardiovascular events [52]. Moderate alcohol consumption may increase HDL cholesterol and have antiplatelet and anti-inflammatory effects [53, 54]. Epidemiological studies suggest that people with diabetes who consume one drink daily may have as much as an 80% reduction in risk of MACE by comparison to those who do not drink [55]. However, more than two or three drinks daily predisposes to hypertension, atrial arrhythmias, and cardiomyopathy.

The Western Diet and Endothelial Dysfunction

A single meal high in saturated fat transiently impairs endothelial vasodilator function [56]. The typical western diet is high in saturated fat and is associated with dyslipidemia and insulin resistance, each of which conditions are associated with impaired endothelial vasodilator function and elevated levels of ADMA [6, 30, 57, 58]. Even in young children with hypercholesterolemia, elevated ADMA and endothelial vasodilator dysfunction is observed [59]. In these metabolic disorders, endothelial vasodilator dysfunction is due in part to increased vascular oxidative stress, as an oral dose of Vitamin C can improve flow-mediated vasodilation [60, 61].

A diet high in animal protein and low in fruits and vegetables may provide insufficient dietary levels of B vitamins, namely, folate, B_6, and B_{12}. This relative vitamin B deficiency may cause hyperhomocysteinemia, which is further exacerbated by the higher levels of methionine in animal protein. Within hours, oral administration of methionine increases plasma homocysteine, with a temporally related increase in plasma ADMA and a decline in endothelial vasodilator function [62]. Homocysteine impairs the activity of DDAH [63] which may account for the correlation of plasma ADMA and homocysteine levels [62].

The Mediterranean Diet

The Lyon Diet Heart Study compared a Mediterranean-style diet to the standard low-fat diet recommended by the American Heart Association (AHA) at that time. The Lyon study was terminated after 27 months because of a marked (70%) reduction in MACE by comparison to the AHA diet. In a follow-up study, those assigned to the Mediterranean diet enjoyed a 50% reduction in risk of myocardial infarction. The Mediterranean diet of the Lyon Diet Heart Study consisted of less saturated fats, cholesterol and linoleic acid, and more oleic and α-linolenic acid [64]. The diet included more whole grain bread, more root and green vegetables, more fish and less red meat, more fruit, and replacement of butter and cream with canola or olive oil. The most notable distinction between the diets was the higher amount of polyunsaturated oils, namely, omega-3 fatty acid. This is significant in view of the prior Diet and Reinfarction Trial (DART). The DART study enrolled men who had sustained a myocardial infarction [65]. Those subjects that were instructed to increase their fish consumption enjoyed a 29% decline in all-cause mortality.

The benefit of the Mediterranean diet may be due in part to its effect to enhance endothelial function [66–68]. Emerging data also seem to suggest the benefits of a Mediterranean diet may be due to higher levels of nitrite and nitrate [69, 70].

The DASH Diet

The DASH (Dietary Approaches to Stop Hypertension) study indicated that diet may be as effective as any single antihypertensive agent in reducing blood pressure [71, 72]. This effect may be partly due to the high levels of nitrate consumed through such a diet [70, 73]. The DASH diet was enriched with vegetables, fruits, and low-fat dairy products, and was low in both fat and saturated fat. In a follow-up study, the diet also included techniques to reduce sodium intake (the use of spices, herbs, and fruit juices to season food; rinsing canned vegetables to remove excess salt; limiting cured meats and vegetables in brine solutions; and avoidance of canned ready-to-eat foods like soup). The findings of the DASH diet support the current AHA recommendations for at least seven servings of vegetables and fruit daily.

High Protein Diets

These diets (e.g., Atkins, Zone, and Stillman diets) are characterized by increased percentage of the diet as protein (30–60%) and fat (30–65%) with reductions in the percentage of diet as carbohydrate (5–40%). Some recent data suggest that high protein diets are somewhat more efficacious in reducing weight [74], but there is no evidence that they improve endothelial vasodilator function or reduce MACE. Because these diets are high in animal protein, they may theoretically impair endothelial function. Notably, epidemiological studies indicate that those subjects who consume more red meat are twice as likely to have a heart attack or stroke [75, 76].

Ultra-Low Fat Diets

Diets such as the Ornish or Kurzweil diets demand a reduction in fat to 10% of total calories. Ornish et al. have investigated the effects of a regimen that includes a vegetarian diet with less than 10% of calories from fat and minimal amounts of saturated fat (the "Reversal Diet"), together with moderate exercise, daily use of stress management techniques, group support and counseling, and a smoking cessation program. This program has been documented to reduce the progression of coronary artery disease [77]. However, potentially adverse effects of an ultra-low fat diet include a reduction in HDL cholesterol, as well as an increase in triglycerides and insulin resistance [78]. Nevertheless, for individuals with coronary artery disease and high LDL cholesterol, particular those that do not have low HDL cholesterol and high triglycerides, ultra-low fat vegetarian diets may be a reasonable alternative to the Mediterranean diet.

Dietary Supplements

About 40% of all Americans use some form of dietary supplement, such as vitamins and botanicals [79]. Dietary supplements are regulated by the Dietary Supplement Health and Education Act (DSHEA). However, evidence-based medicine is not required for the marketing of dietary supplements. Furthermore, quality control for dietary supplements is not as rigorous as that for drugs. For example, when the Good Housekeeping Institute analyzed six popular brands of St. John's Wort supplement, they found an almost 20-fold difference in supplement potency [80]. Accordingly, although some supplements described below have strong scientific evidence to support their utility in some indications, there may be heterogeneity in products within a particular class.

Omega-3 Fatty Acids

The omega-3 fatty acids DHA and EPA improve endothelial vasodilator function, perhaps due to their favorable effects on the lipid profile [81–84]. These agents also reduce oxidative stress and have anti-inflammatory effects [85–88]. Finally, there are data indicating that omega-3 fatty acids reduce myocardial dysrhythmia [89–92]. In the GISSI study, patients with cardiovascular disease that were given supplements of fish oil (1 g twice daily) were less likely to experience sudden death [93].

Supplemental DHA and EPA enhance the flow-induced vasodilation of forearm in healthy individuals [94], as well as hypercholesterolemic subjects [38, 95]. In diabetic patients, blood flow response to acetylcholine and NO release is enhanced after postprandial omega-3 fatty acid supplement [96]. Fish oil supplement increases endothelium-dependent vasodilation in patients with coronary artery disease [41, 97, 98]. Long-term consumption of fish or omega-3 fatty acids also lowers the blood pressure, lowers the incidence of coronary heart disease, preserves renal function, and reduces mortality [98, 99]. On the other hand, recent clinical trials do not indicate evidence of a strong protective effect of intake of omega-3 fatty acids from fish oil against ventricular arrhythmia in patients with implantable cardioverter-defibrillators [100, 101].

Unfortunately, unprocessed fish oil may contain traces of heavy metals (cadmium, copper, iron, magnesium, and lead) and/or pesticides [102–104]. Commercial processing of fish oil removes contaminants, but also removes natural antioxidants. Most commercial suppliers add antioxidants back to the fish oil, but this does not ensure stability. To avoid the problems with commercial processing of fish oils, some manufacturers are now obtaining EPA and DHA from algae [105, 106].

B Vitamins

Vitamin B_3 (Niacin). Food sources of vitamin B_3 include peanuts, brewer's yeast, fish, and meat, and (to a lesser extent) whole grains. Supplemental niacin in high doses (100–2,000 mg/day) can favorably alter the lipid profile, and thereby improve endothelium-dependent vasodilation [107]. Furthermore, evidence suggests that high-dose niacin can reduce the progression of atherosclerosis [108] and reduce MACE [109]. Niacin improves endothelial dysfunction in patients with coronary artery disease [110]. Niacin may inhibit vascular inflammation and protect against endothelial dysfunction independent of changes in plasma lipid levels [111]. However, even in amounts as low as 50–100 mg, individuals may experience flushing or headache. Higher doses (over 500 mg daily) of niacin should be taken only under the direct supervision of a physician, because of the concern for hepatic dysfunction or hyperuricemia.

Vitamins B_6, B_{12}, and folate. Supplementation with these B vitamins reduces homocysteine levels. In addition, low levels of the NOS cofactor tetrahydrobiopterin can be restored with B vitamins, in particular folate [112, 113]. The results of the Nurses' Health Study indicated that women who consume more B vitamins (either from food or vitamins) have half the risk of heart disease. However, large randomized clinical trials of B vitamins, to reduce plasma levels of homocysteine and thereby reduce MACE, have been disappointing [114, 115].

Antioxidants

Vitamins C and E. Oxidative stress plays a role in endothelial vasodilator dysfunction and in the initiation and progression of atherosclerosis. In experimental atherosclerosis, the endothelium produces excessive amounts of superoxide anion [116, 117]. Subsequently, oxidant-sensitive transcriptional pathways are activated, such as NFκβ-mediated activation of vascular cell adhesion molecule (VCAM-1) and monocyte chemoattractant peptide (MCP-1) [118]. Superoxide anion rapidly combines with NO to form peroxynitrite anion, which itself can interfere with cell signaling [119].

In preclinical studies, antioxidants enhanced NO bioactivity [120]. Furthermore, antioxidants preserve the activity of tetrahydropbiopterin [121], a necessary cofactor for NOS [122]. Finally, we have shown that antioxidants can protect the oxidant-sensitive enzyme DDAH [63, 123], and thereby reduce the accumulation of ADMA and N^G-methyl-l-arginine, the endogenous NOS inhibitors. Small clinical studies indicate that administration of Vitamin E or C to patients with hypercholesterolemia or diabetes mellitus can restore endothelial vasodilator function [124–126]. Epidemiological studies have shown that individuals who are in the upper quintile of antioxidant vitamin consumption are significantly less likely to incur MACE [127, 128].

These observations led to large randomized clinical trials of antioxidant vitamin therapy to reduce cardiovascular events. The Cambridge Heart Antioxidant Study (CHAOS) studied approximately 2,000 people with evidence of atherosclerosis. Participants who took 400–800 IU of Vitamin E daily enjoyed a reduced risk of heart attack and death from heart disease. However, larger randomized clinical trials have been compellingly negative [129]. In particular, the Heart Outcomes Prevention Evaluation (HOPE) study followed 10,000 patients for about 4.5 years. Subjects at high risk for cardiovascular events received 400 IU of Vitamin E or placebo daily, in a latin square design with ramipril 5 mg or placebo. Whereas ramipril reduced MACE by about 25%, vitamin E was not superior to placebo [130–132].

Subsequently, it was suggested that a combination of antioxidants would be more effective than high doses of Vitamin E or C alone. However, Greg Brown and colleagues found that a combination of antioxidant vitamins (β-carotene 12.5 mg BID, vitamin C 500 mg BID, vitamin E 400 IU BID, and selenium 50 μg BID) blunt the beneficial effects of statins to increase HDL cholesterol and induce regression of coronary artery disease [133]. Furthermore, the SU.VI.MAX Study (a primary prevention study in 13,000 subjects followed for over 7 years) showed no beneficial effect on MACE of combination antioxidant therapy that included Vitamin C and E, β-carotene, selenium, and zinc [134]. Furthermore, in a sub-study of over 1,000 subjects over 50 years in age, there was no beneficial effect of on vascular compliance or intima-media thickness [135]. Of concern, there was an increase in the number of carotid plaques in the antioxidant-supplemented group. Finally, there are some reports of increased bleeding in patients taking high doses of antioxidant vitamins with antiplatelet therapy [136, 137]. In light of these studies and others showing no benefit of antioxidant vitamins [138–142], it seems best to obtain a variety of antioxidants through generous and frequent servings of fruit and vegetables, rather than supplemental vitamins.

A number of other antioxidants have been shown to improve endothelial vasodilator function in human subjects, including coenzyme Q10, grape seed extract, and gingko biloba [143–145]. However, in view of the randomized clinical trials with antioxidant vitamins, the clinical relevance of these small clinical studies is questionable.

Soy Protein and Phytoestrogens

Epidemiological studies indicate that there is less prevalence of cardiovascular disease among individuals who consume more soy protein and other sources of phytoestrogens [146, 147]. This may be in part due to the effect of soy consumption to reduce cholesterol levels [148, 149]. Soy phytoestrogens, also called isoflavones, such as genistein and daidzein have antioxidant and anti-inflammatory effects. Furthermore, in postmenopausal women with high cholesterol, we found that soy isoflavones (genistein, daidzein, and glycitein) 50 mg once daily improve vasodilation, in the absence of an effect on the lipid profile [150]. However, enthusiasm for supplemental phytoestrogens should be muted in the absence of large randomized clinical trials. Furthermore, caution is advisable in view of the disappointing results of estrogen supplementation in women with cardiovascular disease [151, 152]. Nevertheless, because it contains fiber and a complex assortment of phytonutrients, moderate consumption of soy (i.e., to substitute for red meat) seems reasonable.

L-Carnitine

There have been several small, randomized, placebo-controlled clinical trials of L-carnitine in cardiovascular disease that suggest it might be useful, in part due to its beneficial effect on endothelial function. In patients with symptomatic CAD, L-carnitine (2–6 g daily) reduces symptoms and increases exercise tolerance [153, 154]. In a 12-month study of 472 patients postmyocardial infarction, L-carnitine 6 g daily reduced ventricular enlargement [155]. Propionyl L-carnitine may improve walking distance in patients with peripheral arterial disease (PAD) [156], an effect that is associated with increased serum levels of nitrogen oxides [157]. In healthy young adults, short-term administration of L-carnitine improved flow-mediated dilation and inflammatory response to a high-fat meal [158].

Arginine

We observed that supplemental L-arginine could improve endothelium-dependent vasodilation in hypercholesterolemic animals and humans [159–161]. Because NO could also inhibit processes that are involved in atherosclerosis, such as platelet aggregation [8] and vascular inflammation [162, 163], we reasoned that dietary arginine might have an antiatherogenic effect. Indeed, was the case in the hypercholesterolemic rabbit [14, 164]. Later, we showed that L-arginine supplementation could inhibit platelet aggregation and monocyte adhesion in hypercholesterolemic animals and humans [10, 165–168].

ADMA, the NOS antagonist. A likely explanation for the effect of supplemental L-arginine to enhance NO synthesis was provided by Vallance et al. [169], who demonstrated that endogenous methylarginines could antagonize the NOS pathway. The methylarginines include monomethylarginine (MMA), asymmetric dimethylarginine (ADMA), and symmetric dimethylarginine (SDMA), which are derived from the hydrolysis of proteins containing methylated arginine residues [170]. Both ADMA and MMA compete with L-arginine for binding to the NOS enzyme, whereas SDMA does not. Their antagonism of NOS is reversed by increasing the concentration of L-arginine [171].

Whereas 20% of the clearance of these compounds is through renal excretion, about 80% is due to the action of the cytoplasmic enzyme DDAH [172, 173], expressed ubiquitously as two isoforms [174]. This enzyme is inhibited by oxidative stress, possibly because it expresses a sulfhydryl group in its catalytic site [175]. We find that the enzyme activity is impaired in hypercholesterolemia, diabetes mellitus, and hyperhomocysteinemia [123, 172, 176, 177]. In addition, ADMA is a predictor of the severity of vascular diseases and is an independent predictor of MACE and mortality [178–181].

Reversing the effects of ADMA with l-*arginine supplementation.* Short-term administration of L-arginine improves endothelial vasodilator function and relieves symptoms in patients with coronary artery disease and congestive heart failure [182–184]. In a large randomized clinical trial, 792 patients with acute myocardial infarction were randomized to oral L-arginine (3 g TID for 30 days) or placebo [185]. MACE were reduced by about 10% ($p=0.06$).

However, in a subsequent study, patients with acute myocardial infarction were randomized to placebo or L-arginine supplementation for a longer term of therapy (6 months) following acute myocardial infarction. This trial was terminated early by the data and safety monitoring board when eight deaths occurred in the arginine-treated group [186].

We studied the potential benefit of L-arginine supplementation (3 g/day) versus placebo on endothelial vasodilator function and functional capacity in 220 patients with PAD [187]. After 6 months, those patients in the supplemented group had higher plasma L-arginine levels. However, measures of NO availability (including flow-mediated vasodilation, vascular compliance, plasma and urinary nitrogen oxides, and plasma citrulline formation) were reduced or not improved compared with placebo. Furthermore, the improvement in absolute claudication distance was significantly less in the arginine-treated group (28% vs. 12%; $p < 0.05$). Thus, in patients with PAD, long-term administration of L-arginine does not appear to be useful and may even induce a form of "arginine tolerance" or dysfunction of the NOS pathway.

Dietary Supplements and Interventions that Improve Cardiovascular Risk Factors

Hypolipidemic agents. By improving cardiovascular risk factors, some nutritional supplements may improve endothelial function. These include plant-based products containing sterols [188]. *Phytosterols* include sitosterol and campesterol which differ structurally from cholesterol by a methyl or ethyl group in their side chains. These structural differences cause them to be poorly absorbed. They associate with cholesterol and bile salts to increase cholesterol excretion. Clinical trials of medical foods that are enriched in plant sterols indicate they can reduce LDL cholesterol levels 10–20% [189, 190]. However, these products could reduce the absorption of fat-soluble vitamins. Accordingly, their use should probably be reserved for hypercholesterolemic adults in the secondary prevention of vascular events.

Red yeast rice contains HMG coA reductase inhibitors. In a small clinical trial, red yeast rice (2.4 g daily) for 8 weeks reduced total cholesterol by 17% [191]. However, like all dietary supplements, there is heterogeneity in product quality. A UCLA study showed that only one of nine red yeast rice supplements in health food stores contained all the monacolins that lower cholesterol [192]. Seven of nine brands contained a small amount of a toxic by-product of the fermentation process. Red yeast rice should not be combined with erythromycin, other statin drugs, fibrates, or high-dose niacin, except under medical supervision, as serious side effects (such as myositis or hepatic dysfunction) are theoretically possible. Currently, the FDA has imposed a ban on red yeast rice products in the United States.

There are a number of other dietary supplements that appear to improve the lipid profile and that may therefore enhance endothelial vasodilator function (Table 7.2). These include alliin (a component of garlic), guggulsterone, and policosanol [193–196]. It is most important to understand that the level of evidence for any of these dietary supplements is weak in comparison to FDA-approved hypolipidemic drugs. Large randomized clinical trials for primary and secondary prevention strongly support the use of FDA-approved cholesterol-lowering drugs such as statins [197].

For controlling blood sugar or blood pressure: *Chromium* has been theorized to enhance the action of insulin [198]. Severe chromium deficiency, as has been seen during total parenteral nutrition, is associated with elevated glucose, insulin, and lipid levels as well as CNS disturbance and peripheral neuropathy [199]. Milder chromium deficiency may be a cause of insulin resistance [200]. However, the results of randomized clinical trials have been mixed [201, 202]. Another supplement proposed to be useful in diabetes mellitus is *α-lipoic acid*, which in one clinical trial reduced the symptoms secondary to diabetic neuropathy [203]. Lipoic acid may also improve endothelial function and reduce markers of inflammation in the metabolic syndrome [204].

Table 7.2 Dietary supplements that *might* mitigate disorders associated with cardiovascular disease and *might* enhance endothelial function

Agents which may improve the lipid profile	
Red yeast rice	Garlic (alliin)
Guggulsterone	Phytosterol (sitosterol, campesterol, stigmasterol)
Policosanol	Omega-3 fatty acid
Vitamin B_3	Soy protein (phytoestrogen)
Agents which may be useful in diabetes mellitus	
Chromium	α-Lipoic acid
Agents which may reduce blood pressure	
Dietary salt substitutes (potassium, magnesium, and lysine)	
Garlic	

In general, large randomized clinical trials are not available to support these dietary interventions

Dietary salt substitutes (e.g., the combination of potassium, magnesium, and lysine) can reduce blood pressure by substituting for sodium chloride [205]. The use of these agents in hypertensive individuals should be under medical supervision of physician, particularly so if the individual is taking medication that can reduce potassium excretion (i.e., aldactone, angiotensin converting enzyme inhibitors, angiotensin receptor antagonists). In addition to its hypolipidemic effect, garlic preparations also have a mild antihypertensive effect [206], possibly due to enhancement of the NOS pathway [207].

Conclusion

There are a wide variety of dietary interventions and supplements that can improve the function of the NOS pathway. However, dietary supplements that improve human endothelial function in the short term do not necessarily have long-term benefit on vascular function and cardiovascular events. Indeed, some of these supplements may even have adverse effects when used long term. Furthermore, dietary supplements may have adverse interactions with pharmacotherapies (e.g., reports of increased bleeding in patients taking aspirin and fish oil supplementation; salt substitutes and ACEIs causing hyperkalemia). In addition, dietary supplements are often of heterogeneous quality. Accordingly, before enthusiastic adoption of a particular dietary supplement, the cautious physician will demand evidence for efficacy and safety from large randomized clinical trials, and will recommend supplements distributed by brand name vendors. With respect to various diets, the informed physician will recommend the Mediterranean diet for individuals with or at risk of cardiovascular disease, as this diet has the strongest support for a regimen that enhances endothelial vasodilator function and reduces cardiovascular events. The collective body of information provided in the coming chapters suggests that nitrite and nitrate content of certain foods in combination with antioxidants and systems promoting their reduction back to NO may need to be considered as other dietary factors of potential benefit.

References

1. Cooke JP, Rossitch Jr E, Andon NA, Loscalzo J, Dzau VJ. Flow activates an endothelial potassium channel to release an endogenous nitrovasodilator. J Clin Invest. 1991;88(5):1663–71.
2. Nishida K, Harrison DG, Navas JP, et al. Molecular cloning and characterization of the constitutive bovine aortic endothelial cell nitric oxide synthase. J Clin Invest. 1992;90(5):2092–6.
3. Taddei S, Virdis A, Ghiadoni L, et al. Age-related reduction of NO availability and oxidative stress in humans. Hypertension. 2001;38(2):274–9.
4. Gerhard M, Roddy MA, Creager SJ, Creager MA. Aging progressively impairs endothelium-dependent vasodilation in forearm resistance vessels of humans. Hypertension. 1996;27(4):849–53.
5. Egashira K, Inou T, Hirooka Y, et al. Effects of age on endothelium-dependent vasodilation of resistance coronary artery by acetylcholine in humans. Circulation. 1993;88(1):77–81.

6. Creager MA, Cooke JP, Mendelsohn ME, et al. Impaired vasodilation of forearm resistance vessels in hypercholesterolemic humans. J Clin Invest. 1990;86(1):228–34.
7. Cooke JP. Asymmetrical dimethylarginine: the Uber marker? Circulation. 2004;109(15):1813–8.
8. Stamler J, Mendelsohn ME, Amarante P, et al. N-acetylcysteine potentiates platelet inhibition by endothelium-derived relaxing factor. Circ Res. 1989;65(3):789–95.
9. Cooke JP, Stamler J, Andon N, Davies PF, McKinley G, Loscalzo J. Flow stimulates endothelial cells to release a nitrovasodilator that is potentiated by reduced thiol. Am J Physiol. 1990;259(3 Pt 2):H804–12.
10. Tsao PS, McEvoy LM, Drexler H, Butcher EC, Cooke JP. Enhanced endothelial adhesiveness in hypercholesterolemia is attenuated by L-arginine. Circulation. 1994;89(5):2176–82.
11. Tsao PS, Lewis NP, Alpert S, Cooke JP. Exposure to shear stress alters endothelial adhesiveness. Role of nitric oxide. Circulation. 1995;92(12):3513–9.
12. Tsao PS, Buitrago R, Chan JR, Cooke JP. Fluid flow inhibits endothelial adhesiveness. Nitric oxide and transcriptional regulation of VCAM-1. Circulation. 1996;94(7):1682–9.
13. Tsao PS, Wang B, Buitrago R, Shyy JY, Cooke JP. Nitric oxide regulates monocyte chemotactic protein-1. Circulation. 1997;96(3):934–40.
14. Cooke JP, Singer AH, Tsao P, Zera P, Rowan RA, Billingham ME. Antiatherogenic effects of L-arginine in the hypercholesterolemic rabbit. J Clin Invest. 1992;90(3):1168–72.
15. von der Leyen HE, Gibbons GH, Morishita R, et al. Gene therapy inhibiting neointimal vascular lesion: in vivo transfer of endothelial cell nitric oxide synthase gene. Proc Natl Acad Sci USA. 1995;92(4):1137–41.
16. Candipan RC, Wang BY, Buitrago R, Tsao PS, Cooke JP. Regression or progression. Dependency on vascular nitric oxide. Arterioscler Thromb Vasc Biol. 1996;16(1):44–50.
17. Schachinger V, Britten MB, Zeiher AM. Prognostic impact of coronary vasodilator dysfunction on adverse long-term outcome of coronary heart disease. Circulation. 2000;101(16): 1899–906.
18. Suwaidi JA, Hamasaki S, Higano ST, Nishimura RA, Holmes Jr DR, Lerman A. Long-term follow-up of patients with mild coronary artery disease and endothelial dysfunction. Circulation. 2000;101(9):948–54.
19. Gokce N, Keaney Jr JF, Hunter LM, et al. Predictive value of noninvasively determined endothelial dysfunction for long-term cardiovascular events in patients with peripheral vascular disease. J Am Coll Cardiol. 2003;41(10):1769–75.
20. Brownlee M. The pathobiology of diabetic complications: a unifying mechanism. Diabetes. 2005;54(6):1615–25.
21. Hink U, Li H, Mollnau H, et al. Mechanisms underlying endothelial dysfunction in diabetes mellitus. Circ Res. 2001;88(2):E14–22.
22. Musicki B, Kramer MF, Becker RE, Burnett AL. Inactivation of phosphorylated endothelial nitric oxide synthase (Ser-1177) by O-GlcNAc in diabetes-associated erectile dysfunction. Proc Natl Acad Sci U S A. 2005;102(33):11870–5.
23. Wells L, Vosseller K, Hart GW. Glycosylation of nucleocytoplasmic proteins: signal transduction and O-GlcNAc. Science. 2001;291(5512):2376–8.
24. Musicki B, Kramer MF, Becker RE, Burnett AL. Age-related changes in phosphorylation of endothelial nitric oxide synthase in the rat penis. J Sex Med. 2005;2(3):347–55.
25. Musicki B, Liu T, Strong T, et al. Low-fat diet and exercise preserve eNOS regulation and endothelial function in the penis of early atherosclerotic pigs: a molecular analysis. J Sex Med. 2008;5(3):552–61.
26. Ceriello A, Taboga C, Tonutti L, et al. Evidence for an independent and cumulative effect of postprandial hypertriglyceridemia and hyperglycemia on endothelial dysfunction and oxidative stress generation: effects of short- and long-term simvastatin treatment. Circulation. 2002;106(10):1211–8.
27. Jensen-Urstad KJ, Reichard PG, Rosfors JS, Lindblad LE, Jensen-Urstad MT. Early atherosclerosis is retarded by improved long-term blood glucose control in patients with IDDM. Diabetes. 1996;45(9):1253–8.
28. Ghiadoni L, Magagna A, Versari D, et al. Different effect of antihypertensive drugs on conduit artery endothelial function. Hypertension. 2003;41(6):1281–6.

29. Plotnick GD, Corretti MC, Vogel RA, Hesslink Jr R, Wise JA. Effect of supplemental phytonutrients on impairment of the flow-mediated brachial artery vasoactivity after a single high-fat meal. J Am Coll Cardiol. 2003;41(10):1744–9.
30. Fard A, Tuck CH, Donis JA, et al. Acute elevations of plasma asymmetric dimethylarginine and impaired endothelial function in response to a high-fat meal in patients with type 2 diabetes. Arterioscler Thromb Vasc Biol. 2000;20(9):2039–44.
31. Keogh JB, Grieger JA, Noakes M, Clifton PM. Flow-mediated dilatation is impaired by a high-saturated fat diet but not by a high-carbohydrate diet. Arterioscler Thromb Vasc Biol. 2005;25(6):1274–9.
32. Jenkins DJ, Wolever TM, Rao AV, et al. Effect on blood lipids of very high intakes of fiber in diets low in saturated fat and cholesterol. N Engl J Med. 1993;329(1):21–6.
33. de Roos NM, Bots ML, Katan MB. Replacement of dietary saturated fatty acids by trans fatty acids lowers serum HDL cholesterol and impairs endothelial function in healthy men and women. Arterioscler Thromb Vasc Biol. 2001;21(7):1233–7.
34. Mensink RP, Katan MB. Effect of dietary trans fatty acids on high-density and low-density lipoprotein cholesterol levels in healthy subjects. N Engl J Med. 1990;323(7):439–45.
35. Sebedio JL, Vermunt SH, Chardigny JM, et al. The effect of dietary trans alpha-linolenic acid on plasma lipids and platelet fatty acid composition: the TransLinE study. Eur J Clin Nutr. 2000;54(2):104–13.
36. Fraser GE, Sabate J, Beeson WL, Strahan TM. A possible protective effect of nut consumption on risk of coronary heart disease. The Adventist Health Study. Arch Intern Med. 1992;152(7):1416–24.
37. Dyerberg J, Bang HO, Stoffersen E, Moncada S, Vane JR. Eicosapentaenoic acid and prevention of thrombosis and atherosclerosis? Lancet. 1978;2(8081):117–9.
38. Mori TA, Watts GF, Burke V, Hilme E, Puddey IB, Beilin LJ. Differential effects of eicosapentaenoic acid and docosahexaenoic acid on vascular reactivity of the forearm microcirculation in hyperlipidemic, overweight men. Circulation. 2000;102(11):1264–9.
39. Abeywardena MY, Head RJ. Longchain n-3 polyunsaturated fatty acids and blood vessel function. Cardiovasc Res. 2001;52(3):361–71.
40. Funk CD, Powell WS. Release of prostaglandins and monohydroxy and trihydroxy metabolites of linoleic and arachidonic acids by adult and fetal aortae and ductus arteriosus. J Biol Chem. 1985;260(12):7481–8.
41. Fleischhauer FJ, Yan WD, Fischell TA. Fish oil improves endothelium-dependent coronary vasodilation in heart transplant recipients. J Am Coll Cardiol. 1993;21(4):982–9.
42. Tushuizen ME, Nieuwland R, Scheffer PG, Sturk A, Heine RJ, Diamant M. Two consecutive high-fat meals affect endothelial-dependent vasodilation, oxidative stress and cellular microparticles in healthy men. J Thromb Haemost. 2006;4(5):1003–10.
43. Food and Agriculture Organization of the United Nations. Protein quality evaluation. Joint FAO/WHO. FAO Food Nutr Pap. 1991;51:1–66.
44. Boger RH, Bode-Boger SM, Szuba A, et al. Asymmetric dimethylarginine (ADMA): a novel risk factor for endothelial dysfunction: its role in hypercholesterolemia. Circulation. 1998;98(18):1842–7.
45. Jacobs Jr DR, Meyer HE, Solvoll K. [Consumption of whole grain foods and chronic disease]. Tidsskr Nor Laegeforen. 2004;124(10):1399–401.
46. Lockheart MS, Steffen LM, Rebnord HM, et al. Dietary patterns, food groups and myocardial infarction: a case-control study. Br J Nutr. 2007;98(2):380–7.
47. Jenkins DJ, Kendall CW, Vuksan V, et al. Soluble fiber intake at a dose approved by the US Food and Drug Administration for a claim of health benefits: serum lipid risk factors for cardiovascular disease assessed in a randomized controlled crossover trial. Am J Clin Nutr. 2002;75(5):834–9.
48. Stein JH, Keevil JG, Wiebe DA, Aeschlimann S, Folts JD. Purple grape juice improves endothelial function and reduces the susceptibility of LDL cholesterol to oxidation in patients with coronary artery disease. Circulation. 1999;100(10):1050–5.
49. Widlansky ME, Hamburg NM, Anter E, et al. Acute EGCG supplementation reverses endothelial dysfunction in patients with coronary artery disease. J Am Coll Nutr. 2007;26(2):95–102.

50. Flammer AJ, Hermann F, Sudano I, et al. Dark chocolate improves coronary vasomotion and reduces platelet reactivity. Circulation. 2007;116(21):2376–82.
51. Djousse L, Ellison RC, McLennan CE, et al. Acute effects of a high-fat meal with and without red wine on endothelial function in healthy subjects. Am J Cardiol. 1999;84(6):660–4.
52. Cooper HA, Exner DV, Domanski MJ. Light-to-moderate alcohol consumption and prognosis in patients with left ventricular systolic dysfunction. J Am Coll Cardiol. 2000;35(7): 1753–9.
53. Ajani UA, Gaziano JM, Lotufo PA, et al. Alcohol consumption and risk of coronary heart disease by diabetes status. Circulation. 2000;102(5):500–5.
54. Imhof A, Woodward M, Doering A, et al. Overall alcohol intake, beer, wine, and systemic markers of inflammation in western Europe: results from three MONICA samples (Augsburg, Glasgow, Lille). Eur Heart J. 2004;25(23):2092–100.
55. Sacco RL, Elkind M, Boden-Albala B, et al. The protective effect of moderate alcohol consumption on ischemic stroke. JAMA. 1999;281(1):53–60.
56. Vogel RA, Corretti MC, Plotnick GD. Effect of a single high-fat meal on endothelial function in healthy subjects. Am J Cardiol. 1997;79(3):350–4.
57. Lundman P, Eriksson M, Schenck-Gustafsson K, Karpe F, Tornvall P. Transient triglyceridemia decreases vascular reactivity in young, healthy men without risk factors for coronary heart disease. Circulation. 1997;96(10):3266–8.
58. Yasuda S, Miyazaki S, Kanda M, et al. Intensive treatment of risk factors in patients with type-2 diabetes mellitus is associated with improvement of endothelial function coupled with a reduction in the levels of plasma asymmetric dimethylarginine and endogenous inhibitor of nitric oxide synthase. Eur Heart J. 2006;27(10):1159–65.
59. Engler MM, Engler MB, Malloy MJ, et al. Antioxidant vitamins C and E improve endothelial function in children with hyperlipidemia: Endothelial Assessment of Risk from Lipids in Youth (EARLY) Trial. Circulation. 2003;108(9):1059–63.
60. Plotnick GD, Corretti MC, Vogel RA. Effect of antioxidant vitamins on the transient impairment of endothelium-dependent brachial artery vasoactivity following a single high-fat meal. JAMA. 1997;278(20):1682–6.
61. Raitakari OT, Adams MR, McCredie RJ, Griffiths KA, Stocker R, Celermajer DS. Oral vitamin C and endothelial function in smokers: short-term improvement, but no sustained beneficial effect. J Am Coll Cardiol. 2000;35(6):1616–21.
62. Stuhlinger MC, Oka RK, Graf EE, et al. Endothelial dysfunction induced by hyperhomocyst(e) inemia: role of asymmetric dimethylarginine. Circulation. 2003;108(8):933–8.
63. Stuhlinger MC, Tsao PS, Her JH, Kimoto M, Balint RF, Cooke JP. Homocysteine impairs the nitric oxide synthase pathway: role of asymmetric dimethylarginine. Circulation. 2001;104(21):2569–75.
64. de Lorgeril M, Renaud S, Mamelle N, et al. Mediterranean alpha-linolenic acid-rich diet in secondary prevention of coronary heart disease. Lancet. 1994;343(8911):1454–9.
65. Burr ML, Fehily AM, Gilbert JF, et al. Effects of changes in fat, fish, and fibre intakes on death and myocardial reinfarction: diet and reinfarction trial (DART). Lancet. 1989;2(8666):757–61.
66. Vogel RA, Corretti MC, Plotnick GD. The postprandial effect of components of the Mediterranean diet on endothelial function. J Am Coll Cardiol. 2000;36(5):1455–60.
67. Perez-Jimenez F, Castro P, Lopez-Miranda J, et al. Circulating levels of endothelial function are modulated by dietary monounsaturated fat. Atherosclerosis. 1999;145(2):351–8.
68. Fuentes F, Lopez-Miranda J, Sanchez E, et al. Mediterranean and low-fat diets improve endothelial function in hypercholesterolemic men. Ann Intern Med. 2001;134(12):1115–9.
69. Lundberg JO, Feelisch M, Bjorne H, Jansson EA, Weitzberg E. Cardioprotective effects of vegetables: is nitrate the answer? Nitric Oxide. 2006;15(4):359–62.
70. Hord NG, Tang Y, Bryan NS. Food sources of nitrates and nitrites: the physiologic context for potential health benefits. Am J Clin Nutr. 2009;90(1):1–10.
71. Appel LJ, Champagne CM, Harsha DW, et al. Effects of comprehensive lifestyle modification on blood pressure control: main results of the PREMIER clinical trial. JAMA. 2003;289(16):2083–93.

72. Maruthur NM, Wang NY, Appel LJ. Lifestyle interventions reduce coronary heart disease risk: results from the PREMIER trial. Circulation. 2009;119(15):2026–31.
73. Larsen FJ, Ekblom B, Sahlin K, Lundberg JO, Weitzberg E. Effects of dietary nitrate on blood pressure in healthy volunteers. N Engl J Med. 2006;355(26):2792–3.
74. Weigle DS, Breen PA, Matthys CC, et al. A high-protein diet induces sustained reductions in appetite, ad libitum caloric intake, and body weight despite compensatory changes in diurnal plasma leptin and ghrelin concentrations. Am J Clin Nutr. 2005;82(1):41–8.
75. Fung TT, Stampfer MJ, Manson JE, Rexrode KM, Willett WC, Hu FB. Prospective study of major dietary patterns and stroke risk in women. Stroke. 2004;35(9):2014–9.
76. Iso H, Stampfer MJ, Manson JE, et al. Prospective study of fat and protein intake and risk of intraparenchymal hemorrhage in women. Circulation. 2001;103(6):856–63.
77. Ornish D, Scherwitz LW, Billings JH, et al. Intensive lifestyle changes for reversal of coronary heart disease. JAMA. 1998;280(23):2001–7.
78. Lefevre M, Champagne CM, Tulley RT, Rood JC, Most MM. Individual variability in cardiovascular disease risk factor responses to low-fat and low-saturated-fat diets in men: body mass index, adiposity, and insulin resistance predict changes in LDL cholesterol. Am J Clin Nutr. 2005;82(5):957–63; quiz 1145–6.
79. Archer SL, Stamler J, Moag-Stahlberg A, et al. Association of dietary supplement use with specific micronutrient intakes among middle-aged American men and women: the INTERMAP Study. J Am Diet Assoc. 2005;105(7):1106–14.
80. Good Housekeeping Institute. New Good Housekeeping Institute study finds drastic discrepancy in potencies of popular herbal supplement. News Release, Consumer Safety Symposium on Dietary Supplements and Herbals, New York City, 1998.
81. Radack KL, Deck CC, Huster GA. n-3 Fatty acid effects on lipids, lipoproteins, and apolipoproteins at very low doses: results of a randomized controlled trial in hypertriglyceridemic subjects. Am J Clin Nutr. 1990;51(4):599–605.
82. Connor WE, DeFrancesco CA, Connor SL. N-3 fatty acids from fish oil. Effects on plasma lipoproteins and hypertriglyceridemic patients. Ann N Y Acad Sci. 1993;683:16–34.
83. Sacks FM, Stone PH, Gibson CM, Silverman DI, Rosner B, Pasternak RC. Controlled trial of fish oil for regression of human coronary atherosclerosis. HARP Research Group. J Am Coll Cardiol. 1995;25(7):1492–8.
84. Rivellese AA, Maffettone A, Iovine C, et al. Long-term effects of fish oil on insulin resistance and plasma lipoproteins in NIDDM patients with hypertriglyceridemia. Diabetes Care. 1996;19(11):1207–13.
85. Taccone-Gallucci M, Manca-di-Villahermosa S, Battistini L, Stuffler RG, Tedesco M, Maccarrone M. N-3 PUFAs reduce oxidative stress in ESRD patients on maintenance HD by inhibiting 5-lipoxygenase activity. Kidney Int. 2006;69(8):1450–4.
86. An WS, Kim HJ, Cho KH, Vaziri ND. Omega-3 fatty acid supplementation attenuates oxidative stress, inflammation, and tubulointerstitial fibrosis in the remnant kidney. Am J Physiol Renal Physiol. 2009;297(4):F895–903.
87. Svegliati-Baroni G, Candelaresi C, Saccomanno S, et al. A model of insulin resistance and nonalcoholic steatohepatitis in rats: role of peroxisome proliferator-activated receptor-alpha and n-3 polyunsaturated fatty acid treatment on liver injury. Am J Pathol. 2006;169(3):846–60.
88. Casos K, Zaragoza MC, Zarkovic N, et al. A fish oil-rich diet reduces vascular oxidative stress in apoE(-/-) mice. Free Radic Res. 2010;44(7):821–9.
89. Virtanen JK, Mursu J, Voutilainen S, Tuomainen TP. Serum long-chain n-3 polyunsaturated fatty acids and risk of hospital diagnosis of atrial fibrillation in men. Circulation. 2009;120(23):2315–21.
90. Leaf A, Albert CM, Josephson M, et al. Prevention of fatal arrhythmias in high-risk subjects by fish oil n-3 fatty acid intake. Circulation. 2005;112(18):2762–8.
91. Djousse L, Rautaharju PM, Hopkins PN, et al. Dietary linolenic acid and adjusted QT and JT intervals in the National Heart, Lung, and Blood Institute Family Heart study. J Am Coll Cardiol. 2005;45(10):1716–22.

92. Mozaffarian D, Ascherio A, Hu FB, et al. Interplay between different polyunsaturated fatty acids and risk of coronary heart disease in men. Circulation. 2005;111(2):157–64.
93. Marchioli R, Barzi F, Bomba E, et al. Early protection against sudden death by n-3 polyunsaturated fatty acids after myocardial infarction: time-course analysis of the results of the Gruppo Italiano per lo Studio della Sopravvivenza nell'Infarto Miocardico (GISSI)-Prevenzione. Circulation. 2002;105(16):1897–903.
94. Armah CK, Jackson KG, Doman I, James L, Cheghani F, Minihane AM. Fish oil fatty acids improve postprandial vascular reactivity in healthy men. Clin Sci (Lond). 2008;114(11): 679–86.
95. Goode GK, Garcia S, Heagerty AM. Dietary supplementation with marine fish oil improves in vitro small artery endothelial function in hypercholesterolemic patients: a double-blind placebo-controlled study. Circulation. 1997;96(9):2802–7.
96. McVeigh GE, Brennan GM, Johnston GD, et al. Dietary fish oil augments nitric oxide production or release in patients with type 2 (non-insulin-dependent) diabetes mellitus. Diabetologia. 1993;36(1):33–8.
97. Morgan DR, Dixon LJ, Hanratty CG, et al. Effects of dietary omega-3 fatty acid supplementation on endothelium-dependent vasodilation in patients with chronic heart failure. Am J Cardiol. 2006;97(4):547–51.
98. Holm T, Andreassen AK, Aukrust P, et al. Omega-3 fatty acids improve blood pressure control and preserve renal function in hypertensive heart transplant recipients. Eur Heart J. 2001;22(5):428–36.
99. Hu FB, Cho E, Rexrode KM, Albert CM, Manson JE. Fish and long-chain omega-3 fatty acid intake and risk of coronary heart disease and total mortality in diabetic women. Circulation. 2003;107(14):1852–7.
100. Brouwer IA, Raitt MH, Dullemeijer C, et al. Effect of fish oil on ventricular tachyarrhythmia in three studies in patients with implantable cardioverter defibrillators. Eur Heart J. 2009;30(7):820–6.
101. Brouwer IA, Zock PL, Camm AJ, et al. Effect of fish oil on ventricular tachyarrhythmia and death in patients with implantable cardioverter defibrillators: the Study on Omega-3 Fatty Acids and Ventricular Arrhythmia (SOFA) randomized trial. JAMA. 2006;295(22):2613–9.
102. Foran SE, Flood JG, Lewandrowski KB. Measurement of mercury levels in concentrated over-the-counter fish oil preparations: is fish oil healthier than fish? Arch Pathol Lab Med. 2003;127(12):1603–5.
103. Hamilton MC, Hites RA, Schwager SJ, Foran JA, Knuth BA, Carpenter DO. Lipid composition and contaminants in farmed and wild salmon. Environ Sci Technol. 2005;39(22):8622–9.
104. Storelli MM, Storelli A, Marcotrigiano GO. Polychlorinated biphenyls, hexachlorobenzene, hexachlorocyclohexane isomers, and pesticide organochlorine residues in cod-liver oil dietary supplements. J Food Prot. 2004;67(8):1787–91.
105. van Beelen VA, Spenkelink B, Mooibroek H, et al. An n-3 PUFA-rich microalgal oil diet protects to a similar extent as a fish oil-rich diet against AOM-induced colonic aberrant crypt foci in F344 rats. Food Chem Toxicol. 2009;47(2):316–20.
106. El Abed MM, Marzouk B, Medhioub MN, Helal AN, Medhioub A. Microalgae: a potential source of polyunsaturated fatty acids. Nutr Health. 2008;19(3):221–6.
107. Lee JM, Robson MD, Yu LM, et al. Effects of high-dose modified-release nicotinic acid on atherosclerosis and vascular function: a randomized, placebo-controlled, magnetic resonance imaging study. J Am Coll Cardiol. 2009;54(19):1787–94.
108. Blankenhorn DH, Azen SP, Crawford DW, et al. Effects of colestipol-niacin therapy on human femoral atherosclerosis. Circulation. 1991;83(2):438–47.
109. Cashin-Hemphill L, Mack WJ, Pogoda JM, Sanmarco ME, Azen SP, Blankenhorn DH. Beneficial effects of colestipol-niacin on coronary atherosclerosis. A 4-year follow-up. JAMA. 1990;264(23):3013–7.
110. Warnholtz A, Wild P, Ostad MA, et al. Effects of oral niacin on endothelial dysfunction in patients with coronary artery disease: results of the randomized, double-blind, placebo-controlled INEF study. Atherosclerosis. 2009;204(1):216–21.

111. Wu BJ, Yan L, Charlton F, Witting P, Barter PJ, Rye KA. Evidence that niacin inhibits acute vascular inflammation and improves endothelial dysfunction independent of changes in plasma lipids. Arterioscler Thromb Vasc Biol. 2010;30(5):968–75.
112. Shirodaria C, Antoniades C, Lee J, et al. Global improvement of vascular function and redox state with low-dose folic acid: implications for folate therapy in patients with coronary artery disease. Circulation. 2007;115(17):2262–70.
113. Doshi SN, McDowell IF, Moat SJ, et al. Folate improves endothelial function in coronary artery disease: an effect mediated by reduction of intracellular superoxide? Arterioscler Thromb Vasc Biol. 2001;21(7):1196–202.
114. Stampfer MJ, Hu FB, Manson JE, Rimm EB, Willett WC. Primary prevention of coronary heart disease in women through diet and lifestyle. N Engl J Med. 2000;343(1):16–22.
115. Rimm EB, Willett WC, Hu FB, et al. Folate and vitamin B6 from diet and supplements in relation to risk of coronary heart disease among women. JAMA. 1998;279(5):359–64.
116. Bonthu S, Heistad DD, Chappell DA, Lamping KG, Faraci FM. Atherosclerosis, vascular remodeling, and impairment of endothelium-dependent relaxation in genetically altered hyperlipidemic mice. Arterioscler Thromb Vasc Biol. 1997;17(11):2333–40.
117. Laursen JB, Somers M, Kurz S, et al. Endothelial regulation of vasomotion in apoE-deficient mice: implications for interactions between peroxynitrite and tetrahydrobiopterin. Circulation. 2001;103(9):1282–8.
118. Aikawa M, Sugiyama S, Hill CC, et al. Lipid lowering reduces oxidative stress and endothelial cell activation in rabbit atheroma. Circulation. 2002;106(11):1390–6.
119. Beckman JS, Beckman TW, Chen J, Marshall PA, Freeman BA. Apparent hydroxyl radical production by peroxynitrite: implications for endothelial injury from nitric oxide and superoxide. Proc Natl Acad Sci U S A. 1990;87(4):1620–4.
120. Forstermann U. Oxidative stress in vascular disease: causes, defense mechanisms and potential therapies. Nat Clin Pract Cardiovasc Med. 2008;5(6):338–49.
121. Baker TA, Milstien S, Katusic ZS. Effect of vitamin C on the availability of tetrahydrobiopterin in human endothelial cells. J Cardiovasc Pharmacol. 2001;37(3):333–8.
122. Heitzer T, Brockhoff C, Mayer B, et al. Tetrahydrobiopterin improves endothelium-dependent vasodilation in chronic smokers: evidence for a dysfunctional nitric oxide synthase. Circ Res. 2000;86(2):E36–41.
123. Lin KY, Ito A, Asagami T, et al. Impaired nitric oxide synthase pathway in diabetes mellitus: role of asymmetric dimethylarginine and dimethylarginine dimethylaminohydrolase. Circulation. 2002;106(8):987–92.
124. Green D, O'Driscoll G, Rankin JM, Maiorana AJ, Taylor RR. Beneficial effect of vitamin E administration on nitric oxide function in subjects with hypercholesterolaemia. Clin Sci (Lond). 1998;95(3):361–7.
125. Hirashima O, Kawano H, Motoyama T, et al. Improvement of endothelial function and insulin sensitivity with vitamin C in patients with coronary spastic angina: possible role of reactive oxygen species. J Am Coll Cardiol. 2000;35(7):1860–6.
126. Schindler TH, Nitzsche EU, Munzel T, et al. Coronary vasoregulation in patients with various risk factors in response to cold pressor testing: contrasting myocardial blood flow responses to short- and long-term vitamin C administration. J Am Coll Cardiol. 2003;42(5):814–22.
127. Rallidis LS, Lekakis J, Kolomvotsou A, et al. Close adherence to a Mediterranean diet improves endothelial function in subjects with abdominal obesity. Am J Clin Nutr. 2009;90(2):263–8.
128. Matteucci E, Passerai S, Mariotti M, et al. Dietary habits and nutritional biomarkers in Italian type 1 diabetes families: evidence of unhealthy diet and combined-vitamin-deficient intakes. Eur J Clin Nutr. 2005;59(1):114–22.
129. Stephens NG, Parsons A, Schofield PM, Kelly F, Cheeseman K, Mitchinson MJ. Randomised controlled trial of vitamin E in patients with coronary disease: Cambridge Heart Antioxidant Study (CHAOS). Lancet. 1996;347(9004):781–6.
130. Heart Outcomes Prevention Evaluation Study Investigators. Effects of ramipril on cardiovascular and microvascular outcomes in people with diabetes mellitus: results of the HOPE study and MICRO-HOPE substudy. Lancet. 2000;355(9200):253–9.

131. Dagenais GR, Yusuf S, Bourassa MG, et al. Effects of ramipril on coronary events in high-risk persons: results of the Heart Outcomes Prevention Evaluation Study. Circulation. 2001;104(5):522–6.
132. Sleight P, Yusuf S, Pogue J, Tsuyuki R, Diaz R, Probstfield J. Blood-pressure reduction and cardiovascular risk in HOPE study. Lancet. 2001;358(9299):2130–1.
133. Cheung MC, Zhao XQ, Chait A, Albers JJ, Brown BG. Antioxidant supplements block the response of HDL to simvastatin-niacin therapy in patients with coronary artery disease and low HDL. Arterioscler Thromb Vasc Biol. 2001;21(8):1320–6.
134. Hercberg S, Galan P, Preziosi P, et al. The SU.VI.MAX Study: a randomized, placebo-controlled trial of the health effects of antioxidant vitamins and minerals. Arch Intern Med. 2004;164(21):2335–42.
135. Zureik M, Galan P, Bertrais S, et al. Effects of long-term daily low-dose supplementation with antioxidant vitamins and minerals on structure and function of large arteries. Arterioscler Thromb Vasc Biol. 2004;24(8):1485–91.
136. Booth SL, Golly I, Sacheck JM, et al. Effect of vitamin E supplementation on vitamin K status in adults with normal coagulation status. Am J Clin Nutr. 2004;80(1):143–8.
137. Horwitt MK. Critique of the requirement for vitamin E. Am J Clin Nutr. 2001;73(6): 1003–5.
138. GISSI-Prevenzione Investigators. Dietary supplementation with n-3 polyunsaturated fatty acids and vitamin E after myocardial infarction: results of the GISSI-Prevenzione trial. Gruppo Italiano per lo Studio della Sopravvivenza nell'Infarto miocardico. Lancet. 1999;354(9177):447–55.
139. The Alpha-Tocopherol, Beta Carotene Cancer Prevention Study Group. The effect of vitamin E and beta carotene on the incidence of lung cancer and other cancers in male smokers. N Engl J Med. 1994;330(15):1029–35.
140. Hennekens CH, Buring JE, Manson JE, et al. Lack of effect of long-term supplementation with beta carotene on the incidence of malignant neoplasms and cardiovascular disease. N Engl J Med. 1996;334(18):1145–9.
141. Heart Protection Study Collaborative Group. MRC/BHF Heart Protection Study of cholesterol lowering with simvastatin in 20,536 high-risk individuals: a randomised placebo-controlled trial. Lancet. 2002;360(9326):7–22.
142. de Gaetano G. Low-dose aspirin and vitamin E in people at cardiovascular risk: a randomised trial in general practice. Collaborative Group of the Primary Prevention Project. Lancet. 2001;357(9250):89–95.
143. Tiano L, Belardinelli R, Carnevali P, Principi F, Seddaiu G, Littarru GP. Effect of coenzyme Q10 administration on endothelial function and extracellular superoxide dismutase in patients with ischaemic heart disease: a double-blind, randomized controlled study. Eur Heart J. 2007;28(18):2249–55.
144. Edirisinghe I, Burton-Freeman B, Tissa Kappagoda C. Mechanism of the endothelium-dependent relaxation evoked by a grape seed extract. Clin Sci (Lond). 2008;114(4):331–7.
145. Campos-Toimil M, Lugnier C, Droy-Lefaix MT, Takeda K. Inhibition of type 4 phosphodiesterase by rolipram and *Ginkgo biloba* extract (EGb 761) decreases agonist-induced rises in internal calcium in human endothelial cells. Arterioscler Thromb Vasc Biol. 2000;20(9):E34–40.
146. Lissin LW, Cooke JP. Phytoestrogens and cardiovascular health. J Am Coll Cardiol. 2000;35(6):1403–10.
147. Anderson JW, Johnstone BM, Cook-Newell ME. Meta-analysis of the effects of soy protein intake on serum lipids. N Engl J Med. 1995;333(5):276–82.
148. Weggemans RM, Trautwein EA. Relation between soy-associated isoflavones and LDL and HDL cholesterol concentrations in humans: a meta-analysis. Eur J Clin Nutr. 2003;57(8):940–6.
149. Zhuo XG, Melby MK, Watanabe S. Soy isoflavone intake lowers serum LDL cholesterol: a meta-analysis of 8 randomized controlled trials in humans. J Nutr. 2004;134(9):2395–400.
150. Lissin LW, Oka R, Lakshmi S, Cooke JP. Isoflavones improve vascular reactivity in post-menopausal women with hypercholesterolemia. Vasc Med. 2004;9(1):26–30.

151. Tempfer CB, Froese G, Heinze G, Bentz EK, Hefler LA, Huber JC. Side effects of phytoestrogens: a meta-analysis of randomized trials. Am J Med. 2009;122(10):939–46.e9.
152. Vincent A, Fitzpatrick LA. Soy isoflavones: are they useful in menopause? Mayo Clin Proc. 2000;75(11):1174–84.
153. Cacciatore L, Cerio R, Ciarimboli M, et al. The therapeutic effect of L-carnitine in patients with exercise-induced stable angina: a controlled study. Drugs Exp Clin Res. 1991;17(4):225–35.
154. Davini P, Bigalli A, Lamanna F, Boem A. Controlled study on L-carnitine therapeutic efficacy in post-infarction. Drugs Exp Clin Res. 1992;18(8):355–65.
155. Iliceto S, Scrutinio D, Bruzzi P, et al. Effects of L-carnitine administration on left ventricular remodeling after acute anterior myocardial infarction: the L-Carnitine Ecocardiografia Digitalizzata Infarto Miocardico (CEDIM) Trial. J Am Coll Cardiol. 1995;26(2):380–7.
156. Brevetti G, Perna S, Sabba C, Martone VD, Condorelli M. Propionyl-L-carnitine in intermittent claudication: double-blind, placebo-controlled, dose titration, multicenter study. J Am Coll Cardiol. 1995;26(6):1411–6.
157. Loffredo L, Pignatelli P, Cangemi R, et al. Imbalance between nitric oxide generation and oxidative stress in patients with peripheral arterial disease: effect of an antioxidant treatment. J Vasc Surg. 2006;44(3):525–30.
158. Volek JS, Judelson DA, Silvestre R, et al. Effects of carnitine supplementation on flow-mediated dilation and vascular inflammatory responses to a high-fat meal in healthy young adults. Am J Cardiol. 2008;102(10):1413–7.
159. Cooke JP, Andon NA, Girerd XJ, Hirsch AT, Creager MA. Arginine restores cholinergic relaxation of hypercholesterolemic rabbit thoracic aorta. Circulation. 1991;83(3):1057–62.
160. Rossitch Jr E, Alexander III E, Black PM, Cooke JP. L-arginine normalizes endothelial function in cerebral vessels from hypercholesterolemic rabbits. J Clin Invest. 1991;87(4):1295–9.
161. Creager MA, Gallagher SJ, Girerd XJ, Coleman SM, Dzau VJ, Cooke JP. L-arginine improves endothelium-dependent vasodilation in hypercholesterolemic humans. J Clin Invest. 1992;90(4):1248–53.
162. Morawietz H, Erbs S, Holtz J, et al. Endothelial protection, AT1 blockade and cholesterol-dependent oxidative stress: the EPAS trial. Circulation. 2006;114(1 Suppl):I296–301.
163. Hayashi T, Matsui-Hirai H, Miyazaki-Akita A, et al. Endothelial cellular senescence is inhibited by nitric oxide: implications in atherosclerosis associated with menopause and diabetes. Proc Natl Acad Sci U S A. 2006;103(45):17018–23.
164. Wang BY, Singer AH, Tsao PS, Drexler H, Kosek J, Cooke JP. Dietary arginine prevents atherogenesis in the coronary artery of the hypercholesterolemic rabbit. J Am Coll Cardiol. 1994;23(2):452–8.
165. Theilmeier G, Chan JR, Zalpour C, et al. Adhesiveness of mononuclear cells in hypercholesterolemic humans is normalized by dietary L-arginine. Arterioscler Thromb Vasc Biol. 1997;17(12):3557–64.
166. Tsao PS, Theilmeier G, Singer AH, Leung LL, Cooke JP. L-arginine attenuates platelet reactivity in hypercholesterolemic rabbits. Arterioscler Thromb. 1994;14(10):1529–33.
167. Wang BY, Candipan RC, Arjomandi M, Hsiun PT, Tsao PS, Cooke JP. Arginine restores nitric oxide activity and inhibits monocyte accumulation after vascular injury in hypercholesterolemic rabbits. J Am Coll Cardiol. 1996;28(6):1573–9.
168. Wolf A, Zalpour C, Theilmeier G, et al. Dietary L-arginine supplementation normalizes platelet aggregation in hypercholesterolemic humans. J Am Coll Cardiol. 1997;29(3):479–85.
169. Vallance P, Leone A, Calver A, Collier J, Moncada S. Accumulation of an endogenous inhibitor of nitric oxide synthesis in chronic renal failure. Lancet. 1992;339(8793):572–5.
170. Leiper J, Murray-Rust J, McDonald N, Vallance P. S-nitrosylation of dimethylarginine dimethylaminohydrolase regulates enzyme activity: further interactions between nitric oxide synthase and dimethylarginine dimethylaminohydrolase. Proc Natl Acad Sci USA. 2002;99(21):13527–32.
171. Boger RH, Bode-Boger SM, Brandes RP, et al. Dietary L-arginine reduces the progression of atherosclerosis in cholesterol-fed rabbits: comparison with lovastatin. Circulation. 1997;96(4):1282–90.

172. Ito A, Tsao PS, Adimoolam S, Kimoto M, Ogawa T, Cooke JP. Novel mechanism for endothelial dysfunction: dysregulation of dimethylarginine dimethylaminohydrolase. Circulation. 1999;99(24):3092–5.
173. Leiper J, Nandi M, Torondel B, et al. Disruption of methylarginine metabolism impairs vascular homeostasis. Nat Med. 2007;13(2):198–203.
174. Leiper JM, Santa Maria J, Chubb A, et al. Identification of two human dimethylarginine dimethylaminohydrolases with distinct tissue distributions and homology with microbial arginine deiminases. Biochem J. 1999;343(Pt 1):209–14.
175. Hong L, Fast W. Inhibition of human dimethylarginine dimethylaminohydrolase-1 by S-nitroso-L-homocysteine and hydrogen peroxide. Analysis, quantification, and implications for hyperhomocysteinemia. J Biol Chem. 2007;282(48):34684–92.
176. Rodionov RN, Dayoub H, Lynch CM, et al. Overexpression of dimethylarginine dimethylaminohydrolase protects against cerebral vascular effects of hyperhomocysteinemia. Circ Res. 2010;106(3):551–8.
177. Sydow K, Mondon CE, Schrader J, Konishi H, Cooke JP. Dimethylarginine dimethylaminohydrolase overexpression enhances insulin sensitivity. Arterioscler Thromb Vasc Biol. 2008;28(4):692–7.
178. Siroen MP, van Leeuwen PA, Nijveldt RJ, Teerlink T, Wouters PJ, Van den Berghe G. Modulation of asymmetric dimethylarginine in critically ill patients receiving intensive insulin treatment: a possible explanation of reduced morbidity and mortality? Crit Care Med. 2005;33(3):504–10.
179. Nicholls SJ, Wang Z, Koeth R, et al. Metabolic profiling of arginine and nitric oxide pathways predicts hemodynamic abnormalities and mortality in patients with cardiogenic shock after acute myocardial infarction. Circulation. 2007;116(20):2315–24.
180. Zeller M, Korandji C, Guilland JC, et al. Impact of asymmetric dimethylarginine on mortality after acute myocardial infarction. Arterioscler Thromb Vasc Biol. 2008;28(5):954–60.
181. Furuki K, Adachi H, Enomoto M, et al. Plasma level of asymmetric dimethylarginine (ADMA) as a predictor of carotid intima-media thickness progression: six-year prospective study using carotid ultrasonography. Hypertens Res. 2008;31(6):1185–9.
182. Quyyumi AA, Dakak N, Diodati JG, Gilligan DM, Panza JA, Cannon III RO. Effect of L-arginine on human coronary endothelium-dependent and physiologic vasodilation. J Am Coll Cardiol. 1997;30(5):1220–7.
183. Chin-Dusting JP, Kaye DM, Lefkovits J, Wong J, Bergin P, Jennings GL. Dietary supplementation with L-arginine fails to restore endothelial function in forearm resistance arteries of patients with severe heart failure. J Am Coll Cardiol. 1996;27(5):1207–13.
184. Parnell MM, Holst DP, Kaye DM. Augmentation of endothelial function following exercise training is associated with increased L-arginine transport in human heart failure. Clin Sci (Lond). 2005;109(6):523–30.
185. Bednarz B, Jaxa-Chamiec T, Maciejewski P, et al. Efficacy and safety of oral l-arginine in acute myocardial infarction. Results of the multicenter, randomized, double-blind, placebo-controlled ARAMI pilot trial. Kardiol Pol. 2005;62(5):421–7.
186. Schulman SP, Becker LC, Kass DA, et al. L-arginine therapy in acute myocardial infarction: the Vascular Interaction With Age in Myocardial Infarction (VINTAGE MI) randomized clinical trial. JAMA. 2006;295(1):58–64.
187. Wilson AM, Harada R, Nair N, Balasubramanian N, Cooke JP. L-arginine supplementation in peripheral arterial disease: no benefit and possible harm. Circulation. 2007;116(2):188–95.
188. Weingartner O, Lutjohann D, Ji S, et al. Vascular effects of diet supplementation with plant sterols. J Am Coll Cardiol. 2008;51(16):1553–61.
189. Miettinen TA, Railo M, Lepantalo M, Gylling H. Plant sterols in serum and in atherosclerotic plaques of patients undergoing carotid endarterectomy. J Am Coll Cardiol. 2005;45(11):1794–801.
190. Blair SN, Capuzzi DM, Gottlieb SO, Nguyen T, Morgan JM, Cater NB. Incremental reduction of serum total cholesterol and low-density lipoprotein cholesterol with the addition of plant stanol ester-containing spread to statin therapy. Am J Cardiol. 2000;86(1):46–52.

191. Heber D, Yip I, Ashley JM, Elashoff DA, Elashoff RM, Go VL. Cholesterol-lowering effects of a proprietary Chinese red-yeast-rice dietary supplement. Am J Clin Nutr. 1999;69(2):231–6.
192. Heber D, Lembertas A, Lu QY, Bowerman S, Go VL. An analysis of nine proprietary Chinese red yeast rice dietary supplements: implications of variability in chemical profile and contents. J Altern Complement Med. 2001;7(2):133–9.
193. Turner B, Molgaard C, Marckmann P. Effect of garlic (*Allium sativum*) powder tablets on serum lipids, blood pressure and arterial stiffness in normo-lipidaemic volunteers: a randomised, double-blind, placebo-controlled trial. Br J Nutr. 2004;92(4):701–6.
194. Urizar NL, Liverman AB, Dodds DT, et al. A natural product that lowers cholesterol as an antagonist ligand for FXR. Science. 2002;296(5573):1703–6.
195. Torres O, Agramonte AJ, Illnait J, Mas Ferreiro R, Fernandez L, Fernandez JC. Treatment of hypercholesterolemia in NIDDM with policosanol. Diabetes Care. 1995;18(3):393–7.
196. Berthold HK, Unverdorben S, Degenhardt R, Bulitta M, Gouni-Berthold I. Effect of policosanol on lipid levels among patients with hypercholesterolemia or combined hyperlipidemia: a randomized controlled trial. JAMA. 2006;295(19):2262–9.
197. Nissen SE, Tuzcu EM, Schoenhagen P, et al. Effect of intensive compared with moderate lipid-lowering therapy on progression of coronary atherosclerosis: a randomized controlled trial. JAMA. 2004;291(9):1071–80.
198. Urberg M, Zemel MB. Evidence for synergism between chromium and nicotinic acid in the control of glucose tolerance in elderly humans. Metabolism. 1987;36(9):896–9.
199. Freund H, Atamian S, Fischer JE. Chromium deficiency during total parenteral nutrition. JAMA. 1979;241(5):496–8.
200. Stupar J, Vrtovec M, Dolinsek F. Longitudinal hair chromium profiles of elderly subjects with normal glucose tolerance and type 2 diabetes mellitus. Metabolism. 2007;56(1):94–104.
201. Jeejeebhoy KN, Chu RC, Marliss EB, Greenberg GR, Bruce-Robertson A. Chromium deficiency, glucose intolerance, and neuropathy reversed by chromium supplementation, in a patient receiving long-term total parenteral nutrition. Am J Clin Nutr. 1977;30(4):531–8.
202. Ali A, Ma Y, Reynolds J, Wise JP, Inzucchi SE, Katz DL. Chromium effects on glucose tolerance and insulin sensitivity in people at risk for diabetes. Endocr Pract. 2010;14:1–21.
203. Ziegler D, Gries FA. Alpha-lipoic acid in the treatment of diabetic peripheral and cardiac autonomic neuropathy. Diabetes. 1997;46 Suppl 2:S62–6.
204. Sola S, Mir MQ, Cheema FA, et al. Irbesartan and lipoic acid improve endothelial function and reduce markers of inflammation in the metabolic syndrome: results of the Irbesartan and Lipoic Acid in Endothelial Dysfunction (ISLAND) study. Circulation. 2005;111(3):343–8.
205. Costa FV. Non-pharmacological treatment of hypertension in women. J Hypertens Suppl. 2002;20(2):S57–61.
206. Sobenin IA, Andrianova IV, Fomchenkov IV, Gorchakova TV, Orekhov AN. Time-released garlic powder tablets lower systolic and diastolic blood pressure in men with mild and moderate arterial hypertension. Hypertens Res. 2009;32(6):433–7.
207. Al-Qattan KK, Thomson M, Al-Mutawa'a S, Al-Hajeri D, Drobiova H, Ali M. Nitric oxide mediates the blood-pressure lowering effect of garlic in the rat two-kidney, one-clip model of hypertension. J Nutr. 2006;136(3 Suppl):774S–6.

Chapter 8
Dietary Flavonoids as Modulators of NO Bioavailability in Acute and Chronic Cardiovascular Diseases

Matthias Totzeck, Malte Kelm, and Tienush Rassaf

Key points

- Coronary heart disease and acute myocardial infarction remain the leading causes of death worldwide.
- Diet compositions have gained much attention in prevention from chronic cardiovascular diseases.
- Flavonoids – found in high concentration in green tea, red wine and cocoa – are naturally occurring polyphenols.
- Improvement of diseases leading to arteriosclerosis (diabetes, hypertension) has been demonstrated in several randomized clinical trials under treatment with flavonoids.
- Cytoprotection by flavonoids has been largely related to modulation of nitric oxide bioavailability.
- Experimental evidence also suggests that flavonoids can alleviate myocardial damage in acute events, e.g., an acute myocardial infarction.
- Randomized clinical trials investigating the role of dietary interventions with flavonoids in acute myocardial infarctions have yet to be conducted.

Keywords Polyphenols • Wine • Cocoa • Chocolate • Coronary artery disease • Hypertension

T. Rassaf (✉)
Department of Medicine, The University of Duesseldorf, Duesseldorf, Germany
Klinik für Kardiologie, Pneumologie, Angiologie, Moorenstrasse 5, 40225 Düsseldorf, Germany
e-mail: Tienush.Rassaf@med.uni-duesseldorf.de

N.S. Bryan and J. Loscalzo (eds.), *Nitrite and Nitrate in Human Health and Disease*,
Nutrition and Health, DOI 10.1007/978-1-60761-616-0_8,

Introduction

The definition of a "healthy" diet composition has gained much attention in the prevention and treatment of chronic cardiovascular, malignant, and infectious diseases. Initial experimental data also suggest that certain nutrition components can exert protection in acute disease events, e.g., during an acute myocardial infarction, the leading cause of death worldwide; however, the exact composition of a "protective" diet and the mechanisms of protection remain unclear. Both European and United States government authorities have issued recommendations for daily nutrients that rely on vegetables and fruits to a very large extent. This is certainly due to their low amount of calories and high content of fiber. However, plant-derived foods may also contain high amounts of naturally occurring chemicals called polyphenols in addition to their nitrite and nitrate content. Among all polyphenols, the flavonoid sub-group has been avidly investigated in intervention trials, which demonstrated that the intake of high amounts of flavonoids is inversely related to endothelial dysfunction [1–5]. Endothelial dysfunction, an early stage of atherosclerosis and cardiovascular diseases, has been linked to impaired endothelium-derived nitric oxide (NO) bioavailability [6]. There is accumulating evidence that plant-derived flavonoids can reverse endothelial dysfunction by modulation of NO activity. This chapter focuses on the possible role of flavonoids as a dietary approach to prevent the onset and progression of chronic cardiovascular diseases, and to protect myocardium during acute events such as acute coronary syndromes.

Endothelial Dysfunction, Atherosclerosis and Myocardial Infarction: Role of NO

Protection via dietary flavonoids has been largely related to an improvement of NO bioavailability. An imbalanced NO homeostasis, in turn, is a key element in the development of chronic cardiovascular diseases. Additionally, experimental data suggest an important role for NO in protection from myocardial damage during acute cardiovascular events (e.g., an acute myocardial infarction).

NO in Chronic Cardiovascular Diseases

Endothelial dysfunction precedes chronic cardiovascular disease and is caused by several risk factors, including aging, hypercholesterolemia, hypertension, diabetes mellitus, smoking, obesity, chronic systemic inflammation, hyperhomocysteinemia, and a family history of premature atherosclerosis. An intact endothelium, however, is required for the maintenance of vascular tone and architecture, blood fluidity, and antithrombotic protection. NO signaling plays a major role in maintaining these functions by regulation of vascular tone and smooth muscle cell proliferation, white blood cell adhesion, and platelet function.

Endothelial NO synthases (eNOS) produce NO from the amino acid L-arginine and several co-factors [7]. Disturbance in the NOS-dependent NO pathway is a key element in the development of endothelial dysfunction, which leads to the onset and progression of atherosclerosis. In endothelial dysfunction, reduced NO bioavailability is caused by an inhibition of eNOS function rather than by lower eNOS protein expression [8]. One of the events leading to altered eNOS activity has been related to decreased levels of the cofactor, (6R)-5,6,7,8-tetrahydro-L-biopterin, and has been termed "eNOS uncoupling" [9]. Under these pathological conditions, eNOS produces potentially harmful reactive oxygen species (ROS, e.g., superoxide) rather than NO. The extent of endothelial dysfunction relates to the risk of patients for developing chronic cardiovascular diseases and acute ischemic events, e.g., myocardial infarction and stroke.

Another key element in the maintenance of endothelial function is endothelial progenitor cells (EPCs, also referred to as circulating angiogenic cells). EPCs participate in the repair of vessel injury and neo-vascularization [10–12]. NO modulates the recruitment of EPCs, and balanced NO homeostasis is, therefore, required for appropriate interaction between resident endothelial cells and circulating EPCs. Imbalanced NO levels may thus contribute to impaired vascular regenerative processes, for example, in patients with chronic ischemic heart disease [1, 13].

NO produced from eNOS can react with hemoglobin in the circulation to undergo oxidation to nitrate. Owing to the immense concentrations of hemoglobin in blood, signaling by NO has therefore been considered to occur only at the site of NO production. Indeed, evidence suggests that free hemoglobin is an effective NO scavenger that can attenuate NO bioavailability and NO-related cardiovascular functions; in patients with end-stage renal disease, impaired vascular function occurs, in part, due to hemodialysis-related release of free hemoglobin into the circulation [14]. Thus, the question of how NO bioavailability is preserved throughout the cardiovascular system is an important one for vascular homeostasis. In vivo, several biochemical mechanisms exist, allowing NO to be transported as NO itself, to react with plasma compounds to form nitrosospecies, and to be oxidized to bioactive nitrite [15–18]. Like endocrine hormones, these mechanisms ensure NO bioavailability and signaling throughout the body. Thus, the complexity of the circulating NO pool opens new, considerable pharmacological and therapeutical possibilities in the diagnosis and therapy of cardiovascular diseases.

In order to detect endothelial (dys)function and to evaluate the impact of dietary interventions on NO-mediated vascular functions, several invasive and noninvasive techniques have been utilized. The measurement of the so-called flow-mediated dilation (FMD) has been used in numerous interventional trials to assess endothelial function. In FMD the relative diameter change (in percent) of the brachial artery from pre-ischemia (baseline) to reactive hyperemia is measured via high-resolution ultrasound. Ischemia is induced via inflation of a blood pressure cuff around the forearm, distal to the part of the artery assessed by ultrasound. During the ischemic period, the forearm vasculature vasodilates, and the final pressure release from the cuff consequently leads to increased flow in the conduit brachial artery. This is accompanied by an enhanced vessel wall shear stress, which stimulates eNOS to produce NO, which in turn causes vascular smooth muscle relaxation and accompanying arterial dilation. FMD is, therefore, generally regarded to be the gold standard in the characterization of endothelial function in vivo. Notably, the extent of endothelial function detected via FMD also correlates with the function in the coronary conduit arteries.

NO in Acute Cardiovascular Events

Acute myocardial infarction caused by rupture of an atherosclerotic plaque and subsequent occlusion of a large coronary artery by thrombus is the leading cause of death worldwide. Upon the onset of an acute myocardial infarction, the immediate and successful recanalization of the occluded coronary artery is the optimal therapy, which leads to an infarct size reduction and improves the prognosis of the patient. The process of restoring blood flow to the ischemic myocardium, however, can induce injury itself. This phenomenon, termed ischemia/reperfusion (I/R) injury, reduces the beneficial effects of myocardial reperfusion [19]. Apart from acute impairment of left ventricular (LV) function, large myocardial infarctions lead to a pathologic remodeling of the LV, limiting prognosis.

Four phenomena, namely, myocardial stunning, the no-reflow phenomenon, arrhythmias, and lethal reperfusion injury, contribute to the extent and impact of the ultimate myocardial I/R injury [20–25]. The key mediators of these patho-biological mechanisms are ROS, deranged calcium levels (Ca^{2+}), pH, the mitochondrial permeability transition pore (mPTP), metabolic changes, and the immune responses. For a full summary on the mechanisms of myocardial I/R injury, the reader is referred to relevant, recent reviews [20, 26, 27].

As demonstrated in numerous experimental studies, myocardial I/R injury can be modulated. Two of the most powerful protective mechanisms in this context are ischemic preconditioning (IPC) and ischemic postconditioning (PostC), which are capable of reducing final infarct size by ~30–60% of the myocardium at risk [26–28]. However, although numerous animal model-based studies showed much-reduced I/R injury, the respective translational clinical trials failed to demonstrate the same benefit in humans. Several of these experimental studies point to a protective role of the NO pool during myocardial I/R [29–32]. Administration of nitrite in an attempt to modulate I/R injury leads to an elevation of the circulating NO pool and exerts protective tissue effects in a mouse model of I/R injury in vivo [29, 31, 33]. The protective effects derive from a hemeprotein-dependent reduction of nitrite to NO, as demonstrated in mice without myoglobin [31]; NO derived from nitrite reduction via cardiac myoglobin reversibly modulates mitochondrial electron transport, thus decreasing reperfusion-derived oxidative stress and inhibiting cellular apoptosis leading to a smaller final infarct size. Notably, translational approaches to test nitrite for its clinical relevance in patients are still under investigation.

In summary, disturbed NO homeostasis is a hallmark of the development of chronic cardiovascular disease, and plays a key role in acute ischemic events, such as an acute myocardial infarction. However, the involvement of NO in all of the mechanisms described above also points to the possibility of therapeutic interventions that modulate NO bioavailability, e.g., through diets enriched with flavonoids.

Biochemistry of Flavonoids

Flavonoids represent a sub-class of polyphenols, which are widely distributed in plants [34]. They are synthesized as micronutrients in the secondary metabolism of plants to protect against microbes, fungi, and insects, or to serve as pigments [35]. The term, "flavonoid," has been used to describe a group of compounds that can be divided into six sub-classes (Fig. 8.1). More precisely, flavonoids are ketone-containing compounds (flavanols, flavones, and isoflavones), whereas non-ketone polyphenols are referred to as flavanoids (e.g., flavan-3-ols) [36]. All compounds are formed in a metabolic pathway through a series of enzymatic modifications yielding numerous end

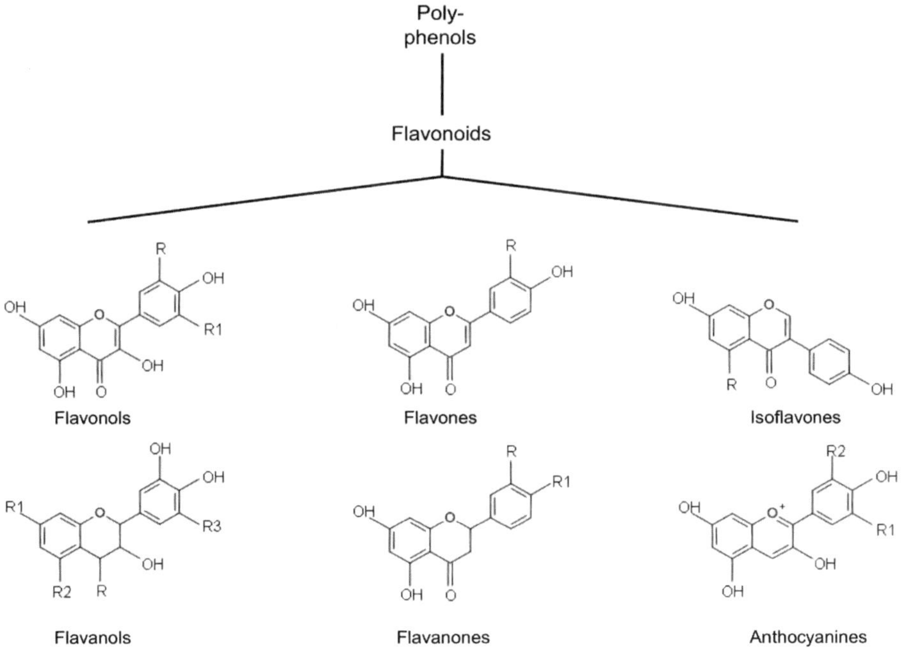

Fig. 8.1 Biochemistry of flavonoids. Metabolic processes in plants form flavonoid compounds that are classified into six groups according to their chemical structures

products. Key dietary products with high flavonoid content are green tea, red wine and products made from cocoa beans. The latter contains relatively high amounts of flavan-3-ols also known as flavanols [37]. Flavanols (e.g., catechin or (−)-epicatechin) or their oligomers, the so-called proanthocyanidines, are among the best-characterized substances in the field of micronutrients. As certain foods rich in flavonoids may contribute to cardiovascular health and protect from the onset of diseases, specific dietary recommendations are under intensive debate. Several steps will have to be taken before official guidelines can be issued. In a first step, certain foods will have to be evaluated in terms of their precise flavonoid content and the average population-based consumption. Environmental changes, as well as nutrition processing, may also influence total flavonoid levels. Moreover, methods for the determination of polyphenolic contents are avidly discussed currently, as recent studies have yielded divergent results. The US Department of Agriculture has, however, published a database with flavonoid contents of selected foods. On the basis of these data, the total actual daily intake of flavanols can be calculated. The next step for the development of recommendations is the assessment of flavonoid bio-metabolism. Despite their high concentrations in certain foods and beverages, several flavonoids show poor bioavailability due to chemical alterations in the digestive tract and liver metabolism. One example is the metabolism of catechin, which is already to large degree conducted in the small intestine [38, 39].

Flavonoids and Cardiovascular Diseases

Epidemiological data suggest an inverse relation between the intake of certain foods and beverages and the incidence of cardiovascular disease. Current literature databases account for more than 6,000 original publications and reviews examining this topic. In this context, epidemiological studies notoriously hold one major well-known flaw: the lack of demonstration of a cause and effect relationship. Moreover, research involving dietary interventions has to be conducted in a multiple step approach. International authorities in this field have, therefore, established the "International Conference on Polyphenols and Health." In its statutes criteria for both clinical trials and basic research with flavonoid intervention were defined in detail (for further information, please go to www.polyphenolsandhealth.org.uk/). Interventional trials are generally preferred; they need to be well controlled and conducted over a considerable length of time. Control substances should be well-balanced as, e.g., cocoa-derived products have a distinct taste and contain numerous other ingredients (e.g., theobromine and caffeine) [2].

Several interventional studies have revealed acute and chronic improvements of vascular function as measured by FMD, circulating NO pool, and EPC mobilization after the ingestion of flavanol-rich foods or beverages, such as red wine [40], purple grape juice [41], tea [42], and cocoa-based products (Table 8.1) [1, 3–5, 43, 44]. Even in optimally treated type 2 diabetic patients, beneficial health effects of flavanol-rich cocoa could be detected [3]. The latter study also identified that effects exerted by flavanols cannot only be found following prolonged ingestion

Table 8.1 Human interventional trials with flavanoid-rich food

Human interventional trials			
Disease	Treatment	Outcome	References
Healthy volunteers	Cocoa drink	PAT↑	Fisher et al. [76]
	Cocoa drink	FMD↑	Heiss et al. [43]
	Cocoa drink	FMD↑	Schroeter et al. [44]
	Chocolate	FMD↑	Grassi et al. [77]
	Chocolate	FMD↑	Engler et al. [78]
	Chocolate	FMD↑	Vlachopoulos et al. [79]
	Black tea	FMD↑	Grassi et al. [80]
	Red wine	FMD↑	Agewall et al. [40]
Cardiovascular risk factors	Cocoa drink	FMD↑	Heiss et al. [4, 5]
	Chocolate/cocoa drink	BP↓, FMD↑	Faridi et al. [81]
Coronary artery disease	Black tea	FMD↑	Duffy et al. [42]
	Grape juice	FMD↑	Stein et al. [41]
	Cocoa drink	FMD↑, EPC↑	Heiss et al. [1]
Hypertension	Chocolate	BP↓	Taubert et al. [82]
	Chocolate	BP↓	Taubert et al. [83]
	Chocolate	BP↓, FMD↑	Grassi et al. [84]
Hypercholesterolemia	Black tea	FMD↑	Hodgson et al. [85]
Diabetes	Cocoa drink	FMD↑	Balzer et al. [3]

PAT peripheral arterial tone measurement; *FMD* flow-mediated dilation measurement; *BP* blood pressure; *EPC* endothelial progenitor cells (mobilization)

(chronic FMD effects), as even an "acute-on-chronic" effect was described. The "positive" impact on FMD together with an enhanced circulating NO pool (nitroso species, nitrite) and a positive recruitment of EPCs provides strong evidence that the modulation of NO bioavailability is a key feature of diets rich in flavonoids.

Although the studies mentioned above have investigated the impact of a dietary intervention on chronic cardiovascular risk factors, the impact of flavonoids on acute ischemic events, such as I/R injury, can only be deduced from animal-model-based studies to date. Can we, however, easily translate diet-based experimental studies and assume a similar benefit for the corresponding patient cohort? Given that mice and rats in comparison to humans show several physiological differences in terms of eating habits, nutrition, digestion, metabolism, and immune system, the existing studies have to be interpreted very carefully in this regard.

In order to interpret data from rodent models of I/R, interspecies differences between mice, rats, and humans in absorption, metabolism, plasma concentration, and tissues distribution must be considered. The interspecies differences in the metabolism of *vitis vinifera* (red wine) derived resveratrol may serve as an example: while humans metabolize resveratrol almost entirely into the glucuronidated form, the most abundant metabolite found in rat plasma is the sulfated form [45]. Resveratrol treatment has been shown to produce cardioprotection from lethal I/R injury in rats [46–50]; this benefit, however, may not be observed in humans owing to the differences in resveratrol metabolism pointed out above.

The studies on flavonoids and their role in protecting the myocardium in the event of I/R injury make use of a variety of different compounds that are found, e.g., in red wine, green tea, cocoa beans, and vegetables. Even synthetic flavanol-based substances with questionable importance for dietary interventions were used in an attempt to modulate I/R injury induced by ROS [51, 52]. Notably, the food formulations used in the animal models were inconsistent; although some studies used the purified compound, others relied on extracts (green tea) or the whole fruit (flesh from grapes), the latter being closer to the flavonoids' natural distribution and formulation. In summary, a total of about 60 experimental studies using flavonoids to modulate I/R suggest that these compounds also exert protection in acute ischemic events, a hypothesis, which has yet to be proven in clinical intervention trials.

Protection-Mechanisms of Dietary Flavonoids

The in vivo effects of flavonoids strongly depend on the substances' pharmacodynamic and pharmacokinetic properties. Although numerous intervention trials using FMD strongly suggest involvement of NO, the exact mechanisms of protection by dietary flavonoids are still elusive. One of the key questions remaining in this context is whether flavonoids improve the bioavailability of NO via increased production or via inhibition of the metabolic inactivation of NO.

Flavonoids and the Nitrate–Nitrite–NO Pathway

In a recent in vitro study in human coronary artery endothelial cells, Ramirez-Sanchez et al. provided evidence that (−)-epicatechin, the most abundant flavonoid in cocoa, can activate eNOS together with the necessary involvement of phosphatidylinositol 3-kinase [53]. Additionally, Steffen et al. showed that the treatment of endothelial cells with (−)-epicatechin elevated intracellular levels of NO and cyclic guanosine monophosphate [54]. Further investigations revealed that (−)-epicatechin inhibited NADPH oxidase with a consequent reduction in superoxide anion production, which is known to scavenge NO. The inhibition of the NADPH oxidase, thus, increases bioavailability of NO rather than affecting NO synthesis. These results may explain the acute effects on endothelial function 1–2 h after the intake of a single flavanol-containing drink [4]. Prolonged intervention with flavanol-enriched drinks resulted in an increase in baseline levels of FMD [43]. In contrast to the acute impact, which can be related to the inhibition of NO metabolism, the chronic effects may be due to an increase in endothelial NOS expression in the vascular endothelium. Balanced NO homeostasis is required for a proper interaction between the endothelium and EPCs, which, in turn, participate in vascular regenerative processes. A very recent study in patients with coronary artery disease has shown that the treatment with flavanols over 30 days not only improves endothelial function as measured by FMD and the circulation NO pool [5, 15], but also significantly increased circulating EPC in these patients. The mobilization of EPC may serve as an additional explanation for the effects of a long-term dietary flavonoid intake on endothelial function and bio-repair.

The exact mechanisms of flavonoids on NO metabolism are currently unresolved, although, in light of the findings above, a strong interaction with eNOS activity can be assumed. Flavonoids might, however, interact with NO homeostasis in at least two other ways: (i) by increasing nitrite, which in turn exerts cytoprotection and (ii) by direct bioconversion of nitrite to NO. As stated above, clinical intervention strategies demonstrated an increase in nitrite following the consumption of drinks enriched with flavanols [3, 5, 44]. Plasma nitrite either derives from nutritional sources or via endogenous decomposition of NO. It is no longer regarded to be an inert oxidative product. Experimental data suggest that nitrite exerts cytoprotection in a wide variety of pathologies, first and foremost during I/R of the heart [29, 31]. These effects have been mainly related to nitrite bioconversion to NO. As they lead to increased nitrite levels, some of the protective effects from flavonoids may be related to this modulation in NO homeostasis. Nitrite reduction to bioactive NO can be mediated by deoxygenated hemoglobin [55] or myoglobin [32] in vivo. Notably, Gago et al. demonstrated that this might also occur directly via red wine polyphenols in the gut without the need of endogenous conversion mechanisms [56]. Following nitrite conversion, NO diffuses over the gut wall and induces muscle relaxation [57]. Taken together, the effects on the NO homeostasis maybe far reaching, ranging from increased NO production via eNOS and bioconversion of increased levels of nitrite and a modulated NO decomposition.

Cellular Targets of Flavonoids in I/R

Although the effects on endothelial and chronic cardiovascular function are mainly related to improved NO bioavailability, the mechanisms of protection from I/R injury by flavonoids remain unresolved. The processes during I/R injury are structurally complex and have been avidly investigated over the past two decades. As only a small number of studies on flavonoid protection from I/R injury have been published to date, a common mechanism or signaling cascade through flavonoid intervention remains unknown.

A considerable impact on oxidative stress is a common feature of most of the flavonoids in experimental trials on I/R injury. ROS formation during I/R may lead to cell damage, e.g., via interaction with lipids of the cell membrane. Possible direct ROS-reducing effects of flavonoids and especially flavanols have been discussed in detail, but the exact mechanisms remain unresolved. Likewise, the direct or indirect influence by dietary flavonoids on the formation of NO remains elusive. The tea polyphenol, epigallocatechin-3-gallate, applied via perfusate in a Langendorff heart model in guinea pigs decreased oxygen radicals, which was paralleled by an increase of NO [58]. Interestingly, eNOS inhibition with L-NAME completely abolished the positive effects of this green tea-derived flavonoid [59]. However, in vivo effects after oral administration of the substance may be considerably different due to the intestinal metabolism of epigallocatechin-3-gallate to form chemically altered and then absorbed substances – a process absent when using Langendorff heart preparations. Decreased ROS levels were also detected for procyanidin [60], osjain, pomiferin [61], proanthocyanidine [62], anthocyanin [63], and (–)-epicatechin [64]. Soy-isoflavone is the only substance that does not exert an ROS-reducing effect on the myocardium [65]. Increased NO levels were also detected following an intervention with proanthocyanidine in cell cultures of cardiomyocytes [66]. Controversial effects were published for quercetin and resveratrol. Rabbits treated with quercetin undergoing ischemia and reperfusion had less eNOS and inducible NOS mRNA and protein expression compared with control [67], while another group reported that cardioprotective effects by resveratrol were completely abolished in inducible NOS-deficient mice [68]. The metabolism of quercetin is complex. After oral intake of quercetin or quercetin glucosides, an extensive biotransformation takes place in the gut and small intestine before absorption of the active species. In the blood, orally ingested quercetin appears almost entirely in conjugated forms after phase-II metabolism [69]. These conjugates, and not purified and commercially available quercetin, may be responsible for its beneficial effects.

Only a few studies focussed on the molecular analysis of signaling proteins involved in I/R responses. Myocardial protection strategies induce a specific signaling pathway that ultimately leads to an inhibition of mPTP opening. In this context, Akt plays a central role; it can phosphorylate a number of proteins, such as eNOS, GSK-3β, and Bcl-2, and modify their (enzymatic) activities [26]. Protection from I/R by NO is closely linked to the role of Akt. Inhibition of either Akt or NOS decreased NO levels and led to higher cell death in proanthocyanidine-treated cardiomyocytes [66]. GSK-3β is phosphorylated by a number of proteins, including Akt or Erk. Phosphorylation of GSK-3β leads to its inactivation. High levels of phosphorylated GSK-3β appear to exert an anti-apoptotic effect [70]. Application of resveratrol led to an increase in phosphorylated GSK-3β. Notably, the soluble guanylyl cyclase inhibitor, 1-H [1,2,4] oxadiazolo[4,3-a]quinoxalin-1-one (ODQ), completely abolished the effect indicating an important involvement of the NO pathway. In this study, increased levels of GSK-3β were also found in the mitochondria of isolated cardiomyocytes along with a decrease in mitochondrial swelling. The authors deduced that, via phosphorylation of GSK-3β and its translocation, opening of mPTP is inhibited and cell damage decreased [71]. mPTP opening also induces Ca^{2+} overload, which was significantly decreased in epicatechin-3-gallate-treated hearts [58]. Signaling in apoptosis partly depends on signaling proteins Bax and Bcl-2. Although increased levels of Bax induce cell death, high levels of Bcl-2 appear to be anti-apoptotic [26]. Several authors have described an increase in Bcl-2 and a decrease in Bax and therefore anti-apoptotic effect for flavonoid-treated hearts [50, 52, 72–74]. The role of the signaling protein JNK is controversial. It plays a key role in phosphorylating numerous other downstream mediators. Treatments with extracts containing proanthocyanidine do not only decrease ROS and apoptosis but also show decreased JNK and c-jun levels [75].

Conclusion

The cardiovascular effects of flavonoids have been extensively investigated over the past decades. Epidemiological data in concert with results from intervention trials suggest that flavonoids might contribute to the prevention of the onset and progression of chronic cardiovascular diseases largely via modulation of NO bioavailability. Despite numerous trials showing beneficial effects for vascular function and cardiovascular risk factors, studies that examine primary clinical endpoints, such as survival or pain-free walking distance in peripheral arterial disease, have not been published to date.

Although a number of studies have examined their impact on chronic diseases, little is known about the effect of flavonoids on acute ischemic events, e.g., a myocardial I/R injury. These effects must be deduced from studies in animal models. This leads quite readily to difficulties in the interpretation in terms of the relevance for clinical practice: (i) the respective studies were carried out in rodents with considerably different metabolisms compared with humans; (ii) in most cases, they used purified compounds or even synthetic flavonoids; (iii) they lacked the proper balanced control substances, and (iv) they were conducted in some cases using isolated organ models, thus completely bypassing metabolism. Future animal studies with flavonoid intervention will, therefore, need to focus on the aspects of absorption/metabolism as necessary basis for translational studies in humans. In the past, many animal studies with substances tested positive for cardioprotection from I/R injury failed when translated into the clinical practice [20]. Nevertheless, the heterogeneous group of flavonoids has the capability to ameliorate myocardial I/R injury in animal-model-based experimental studies. If proven true in human studies, functional foods might serve as a powerful tool in primary and secondary cardiovascular prevention.

References

1. Heiss C, Jahn S, Taylor M, Real WM, Angeli FS, Wong ML, et al. Improvement of endothelial function with dietary flavanols is associated with mobilization of circulating angiogenic cells in patients with coronary artery disease. J Am Coll Cardiol. 2010;56(3):218–24.
2. Heiss C, Kelm M. Chocolate consumption, blood pressure, and cardiovascular risk. Eur Heart J. 2010;31(13):1554–6.
3. Balzer J, Rassaf T, Heiss C, Kleinbongard P, Lauer T, Merx MW, et al. Sustained benefits in vascular function through flavanol-containig cocoa in mediated diabetic patients: a double-masked, randomized, controlled trial. J Am Coll Cardiol. 2008;51:2141–9.

4. Heiss C, Kleinbongard P, Dejam A, Perre S, Schroeter H, Sies H, et al. Acute consumption of flavanol-rich cocoa and the reversal of endothelial dysfunction in smokers. J Am Coll Cardiol. 2005;46(7):1276–83.
5. Heiss C, Dejam A, Kleinbongard P, Schewe T, Sies H, Kelm M. Vascular effects of cocoa rich in flavan-3-ols. JAMA. 2003;290(8):1030–1.
6. Heiss C, Lauer T, Dejam A, Kleinbongard P, Hamada S, Rassaf T, et al. Plasma nitroso compounds are decreased in patients with endothelial dysfunction. J Am Coll Cardiol. 2006;47:573–9.
7. Moncada S, Higgs A. The L-arginine-nitric oxide pathway. N Engl J Med. 1993; 329(27):2002–12.
8. Li H, Wallerath T, Munzel T, Forstermann U. Regulation of endothelial-type NO synthase expression in pathophysiology and in response to drugs. Nitric Oxide. 2002;7(3):149–64.
9. Thum T, Fraccarollo D, Schultheiss M, Froese S, Galuppo P, Widder JD, et al. Endothelial nitric oxide synthase uncoupling impairs endothelial progenitor cell mobilization and function in diabetes. Diabetes. 2007;56(3):666–74.
10. Werner N, Kosiol S, Schiegl T, Ahlers P, Walenta K, Link A, et al. Circulating endothelial progenitor cells and cardiovascular outcomes. N Engl J Med. 2005;353(10):999–1007.
11. Dimmeler S, Zeiher AM. Vascular repair by circulating endothelial progenitor cells: the missing link in atherosclerosis? J Mol Med. 2004;82(10):671–7.
12. Shantsila E, Watson T, Lip GY. Endothelial progenitor cells in cardiovascular disorders. J Am Coll Cardiol. 2007;49(7):741–52.
13. Dimmeler S. Regulation of bone marrow-derived vascular progenitor cell mobilization and maintenance. Arterioscler Thromb Vasc Biol. 2010;30(6):1088–93.
14. Meyer C, Heiss C, Drexhage C, Kehmeier ES, Balzer J, Muhlfeld A, et al. Hemodialysis-induced release of hemoglobin limits nitric oxide bioavailability and impairs vascular function. J Am Coll Cardiol. 2010;55(5):454–9.
15. Rassaf T, Bryan NS, Kelm M, Feelisch M. Concomitant presence of N-nitroso and S-nitroso proteins in human plasma. Free Radic Biol Med. 2002;33:1590–6.
16. Rassaf T, Bryan NS, Maloney RE, Specian V, Kelm M, Kalyanaraman B, et al. NO adducts in mammalian red blood cells: too much or too little? Nat Med. 2003;9:481–2.
17. Rassaf T, Kleinbongard P, Preik M, Dejam A, Gharini P, Lauer T, et al. Plasma nitrosothiols contribute to the systemic vasodilator effects of intravenously applied NO: experimental and clinical Study on the fate of NO in human blood. Circ Res. 2002;91(6):470–7.
18. Elrod JW, Calvert JW, Gundewar S, Bryan NS, Lefer DJ. Nitric oxide promotes distant organ protection: evidence for an endocrine role of nitric oxide. Proc Natl Acad Sci USA. 2008;105(32):11430–5.
19. Piper HM, Garcia-Dorado D, Ovize M. A fresh look at reperfusion injury. Cardiovasc Res. 1998;38(2):291–300.
20. Yellon DM, Hausenloy DJ. Myocardial reperfusion injury. N Engl J Med. 2007;357(11):1121–35.
21. Braunwald E, Kloner RA. The stunned myocardium: prolonged, postischemic ventricular dysfunction. Circulation. 1982;66(6):1146–9.
22. Bolli R, Marban E. Molecular and cellular mechanisms of myocardial stunning. Physiol Rev. 1999;79(2):609–34.
23. Ito H. No-reflow phenomenon and prognosis in patients with acute myocardial infarction. Nat Clin Pract Cardiovasc Med. 2006;3(9):499–506.
24. Manning AS, Hearse DJ. Reperfusion-induced arrhythmias: mechanisms and prevention. J Mol Cell Cardiol. 1984;16(6):497–518.
25. Heusch G. Stunning–great paradigmatic, but little clinical importance. Basic Res Cardiol. 1998;93(3):164–6.
26. Murphy E, Steenbergen C. Mechanisms underlying acute protection from cardiac ischemia-reperfusion injury. Physiol Rev. 2008;88(2):581–609.
27. Skyschally A, Schulz R, Heusch G. Pathophysiology of myocardial infarction: protection by ischemic pre- and postconditioning. Herz. 2008;33(2):88–100.

28. Zhao ZQ, Corvera JS, Halkos ME, Kerendi F, Wang NP, Guyton RA, et al. Inhibition of myocardial injury by ischemic postconditioning during reperfusion: comparison with ischemic preconditioning. Am J Physiol Heart Circ Physiol. 2003;285(2):H579–88.
29. Duranski MR, Greer JJ, Dejam A, Jaganmohan S, Hogg N, Langston W, et al. Cytoprotective effects of nitrite during in vivo ischemia-reperfusion of the heart and liver. J Clin Invest. 2005;115(5):1232–40.
30. Hamid SA, Totzeck M, Drexhage C, Thompson I, Fowkes RC, Rassaf T, et al. Nitric oxide/cGMP signalling mediates the cardioprotective action of adrenomedullin in reperfused myocardium. Basic Res Cardiol. 2010;105:257–66.
31. Hendgen-Cotta UB, Merx MW, Shiva S, Schmitz J, Becher S, Klare JP, et al. Nitrite reductase activity of myoglobin regulates respiration and cellular viability in myocardial ischemia-reperfusion injury. Proc Natl Acad Sci USA. 2008;105(29):10256–61.
32. Rassaf T, Flogel U, Drexhage C, Hendgen-Cotta U, Kelm M, Schrader J. Nitrite reductase function of deoxymyoglobin: oxygen sensor and regulator of cardiac energetics and function. Circ Res. 2007;100(12):1749–54.
33. Bryan NS, Calvert JW, Elrod JW, Gundewar S, Ji SY, Lefer DJ. Dietary nitrite supplementation protects against myocardial ischemia-reperfusion injury. Proc Natl Acad Sci USA. 2007;104(48):19144–9.
34. Scalbert A, Williamson G. Dietary intake and bioavailability of polyphenols. J Nutr. 2000;130(8S Suppl):2073S–85.
35. Galeotti F, Barile E, Lanzotti V, Dolci M, Curir P. Quantification of major flavonoids in carnation tissues (*Dianthus caryophyllus*) as a tool for cultivar discrimination. Z Naturforsch C. 2008;63(3–4):161–8.
36. Ververidis F, Trantas E, Douglas C, Vollmer G, Kretzschmar G, Panopoulos N. Biotechnology of flavonoids and other phenylpropanoid-derived natural products. Part I: Chemical diversity, impacts on plant biology and human health. Biotechnol J. 2007;2(10):1214–34.
37. Williamson G, Manach C. Bioavailability and bioefficacy of polyphenols in humans. II. Review of 93 intervention studies. Am J Clin Nutr. 2005;81(1 Suppl):243S–55.
38. Donovan JL, Crespy V, Oliveira M, Cooper KA, Gibson BB, Williamson G. (+)-Catechin is more bioavailable than (−)-catechin: relevance to the bioavailability of catechin from cocoa. Free Radic Res. 2006;40(10):1029–34.
39. Donovan JL, Crespy V, Manach C, Morand C, Besson C, Scalbert A, et al. Catechin is metabolized by both the small intestine and liver of rats. J Nutr. 2001;131(6):1753–7.
40. Agewall S, Wright S, Doughty RN, Whalley GA, Duxbury M, Sharpe N. Does a glass of red wine improve endothelial function? Eur Heart J. 2000;21(1):74–8.
41. Stein JH, Keevil JG, Wiebe DA, Aeschlimann S, Folts JD. Purple grape juice improves endothelial function and reduces the susceptibility of LDL cholesterol to oxidation in patients with coronary artery disease. Circulation. 1999;100(10):1050–5.
42. Duffy SJ, Keaney Jr JF, Holbrook M, Gokce N, Swerdloff PL, Frei B, et al. Short- and long-term black tea consumption reverses endothelial dysfunction in patients with coronary artery disease. Circulation. 2001;104(2):151–6.
43. Heiss C, Finis D, Kleinbongard P, Hoffmann A, Rassaf T, Kelm M, et al. Sustained increase in flow-mediated dilation after daily intake of high-flavanol cocoa drink over 1 week. J Cardiovasc Pharmacol. 2007;49:74–80.
44. Schroeter H, Heiss C, Balzer J, Kleinbongard P, Keen CL, Hollenberg NK, et al. (−)-Epicatechin mediates beneficial effects of flavanol-rich cocoa on vascular function in humans. Proc Natl Acad Sci USA. 2006;103(4):1024–9.
45. Yu C, Shin YG, Chow A, Li Y, Kosmeder JW, Lee YS, et al. Human, rat, and mouse metabolism of resveratrol. Pharm Res. 2002;19(12):1907–14.
46. Hattori R, Otani H, Maulik N, Das DK. Pharmacological preconditioning with resveratrol: role of nitric oxide. Am J Physiol Heart Circ Physiol. 2002;282(6):H1988–95.
47. Gurusamy N, Lekli I, Mukherjee S, Ray D, Ahsan MK, Gherghiceanu M, et al. Cardioprotection by resveratrol: a novel mechanism via autophagy involving the mTORC2 pathway. Cardiovasc Res. 2010;86:103–12.

48. Lekli I, Szabo G, Juhasz B, Das S, Das M, Varga E, et al. Protective mechanisms of resveratrol against ischemia-reperfusion-induced damage in hearts obtained from Zucker obese rats: the role of GLUT-4 and endothelin. Am J Physiol Heart Circ Physiol. 2008;294(2):H859–66.
49. Mokni M, Limam F, Elkahoui S, Amri M, Aouani E. Strong cardioprotective effect of resveratrol, a red wine polyphenol, on isolated rat hearts after ischemia/reperfusion injury. Arch Biochem Biophys. 2007;457(1):1–6.
50. Das S, Cordis GA, Maulik N, Das DK. Pharmacological preconditioning with resveratrol: role of CREB-dependent Bcl-2 signaling via adenosine A3 receptor activation. Am J Physiol Heart Circ Physiol. 2005;288(1):H328–35.
51. Wang S, Dusting GJ, May CN, Woodman OL. 3′, 4′-Dihydroxyflavonol reduces infarct size and injury associated with myocardial ischaemia and reperfusion in sheep. Br J Pharmacol. 2004;142(3):443–52.
52. Wang S, Fei K, Xu YW, Wang LX, Chen YQ. Dihydroxyflavonol reduces post-infarction left ventricular remodeling by preventing myocyte apoptosis in the non-infarcted zone in goats. Chin Med J (Engl). 2009;122(1):61–7.
53. Ramirez-Sanchez I, Maya L, Ceballos G, Villarreal F. (−)-Epicatechin activation of endothelial cell endothelial nitric oxide synthase, nitric oxide, and related signaling pathways. Hypertension. 2010;55:1398–405.
54. Steffen Y, Schewe T, Sies H. (−)-Epicatechin elevates nitric oxide in endothelial cells via inhibition of NADPH oxidase. Biochem Biophys Res Commun. 2007;359(3):828–33.
55. Cosby K, Partovi KS, Crawford JH, Patel RP, Reiter CD, Martyr S, et al. Nitrite reduction to nitric oxide by deoxyhemoglobin vasodilates the human circulation. Nat Med. 2003;9(12):1498–505.
56. Gago B, Lundberg JO, Barbosa RM, Laranjinha J. Red wine-dependent reduction of nitrite to nitric oxide in the stomach. Free Radic Biol Med. 2007;43(9):1233–42.
57. Rocha BS, Gago B, Barbosa RM, Laranjinha J. Diffusion of nitric oxide through the gastric wall upon reduction of nitrite by red wine: physiological impact. Nitric Oxide. 2010;22(3):235–41.
58. Hirai M, Hotta Y, Ishikawa N, Wakida Y, Fukuzawa Y, Isobe F, et al. Protective effects of EGCg or GCg, a green tea catechin epimer, against postischemic myocardial dysfunction in guinea-pig hearts. Life Sci. 2007;80(11):1020–32.
59. Potenza MA, Marasciulo FL, Tarquinio M, Tiravanti E, Colantuono G, Federici A, et al. EGCG, a green tea polyphenol, improves endothelial function and insulin sensitivity, reduces blood pressure, and protects against myocardial I/R injury in SHR. Am J Physiol Endocrinol Metab. 2007;292(5):E1378–87.
60. Chang WT, Shao ZH, Vanden Hoek TL, McEntee E, Mehendale SR, Li J, et al. Cardioprotective effects of grape seed proanthocyanidins, baicalin and wogonin: comparison between acute and chronic treatments. Am J Chin Med. 2006;34(2):363–5.
61. Florian T, Necas J, Bartosikova L, Klusakova J, Suchy V, Naggara EB, et al. Effects of prenylated isoflavones osajin and pomiferin in premedication on heart ischemia-reperfusion. Biomed Pap Med Fac Univ Palacky Olomouc Czech Repub. 2006;150(1):93–100.
62. Pataki T, Bak I, Kovacs P, Bagchi D, Das DK, Tosaki A. Grape seed proanthocyanidins improved cardiac recovery during reperfusion after ischemia in isolated rat hearts. Am J Clin Nutr. 2002;75(5):894–9.
63. Toufektsian MC, de Lorgeril M, Nagy N, Salen P, Donati MB, Giordano L, et al. Chronic dietary intake of plant-derived anthocyanins protects the rat heart against ischemia-reperfusion injury. J Nutr. 2008;138(4):747–52.
64. Yamazaki KG, Romero-Perez D, Barraza-Hidalgo M, Cruz M, Rivas M, Cortez-Gomez B, et al. Short- and long-term effects of (−)-epicatechin on myocardial ischemia-reperfusion injury. Am J Physiol Heart Circ Physiol. 2008;295(2):H761–7.
65. Suparto IH, Williams JK, Fox JL, Yusuf JT, Sajuthi D. Effects of hormone therapy and dietary soy on myocardial ischemia/reperfusion injury in ovariectomized atherosclerotic monkeys. Menopause. 2008;15(2):256–63.
66. Shao ZH, Wojcik KR, Dossumbekova A, Hsu C, Mehendale SR, Li CQ, et al. Grape seed proanthocyanidins protect cardiomyocytes from ischemia and reperfusion injury via Akt-NOS signaling. J Cell Biochem. 2009;107(4):697–705.

67. Wan LL, Xia J, Ye D, Liu J, Chen J, Wang G. Effects of quercetin on gene and protein expression of NOX and NOS after myocardial ischemia and reperfusion in rabbit. Cardiovasc Ther. 2009;27(1):28–33.
68. Imamura G, Bertelli AA, Bertelli A, Otani H, Maulik N, Das DK. Pharmacological preconditioning with resveratrol: an insight with iNOS knockout mice. Am J Physiol Heart Circ Physiol. 2002;282(6):H1996–2003.
69. Graf BA, Ameho C, Dolnikowski GG, Milbury PE, Chen CY, Blumberg JB. Rat gastrointestinal tissues metabolize quercetin. J Nutr. 2006;136(1):39–44.
70. Tong H, Imahashi K, Steenbergen C, Murphy E. Phosphorylation of glycogen synthase kinase-3beta during preconditioning through a phosphatidylinositol-3-kinase-dependent pathway is cardioprotective. Circ Res. 2002;90(4):377–9.
71. Xi J, Wang H, Mueller RA, Norfleet EA, Xu Z. Mechanism for resveratrol-induced cardioprotection against reperfusion injury involves glycogen synthase kinase 3beta and mitochondrial permeability transition pore. Eur J Pharmacol. 2009;604(1–3):111–6.
72. Ling H, Lou Y. Total flavones from *Elsholtzia* blanda reduce infarct size during acute myocardial ischemia by inhibiting myocardial apoptosis in rats. J Ethnopharmacol. 2005;101(1–3):169–75.
73. Hao Y, Sun Y, Xu C, Jiang X, Sun H, Wu Q, et al. Improvement of contractile function in isolated cardiomyocytes from ischemia-reperfusion rats by ginkgolide B pretreatment. J Cardiovasc Pharmacol. 2009;54(1):3–9.
74. Ji ES, Yue H, Wu YM, He RR. Effects of phytoestrogen genistein on myocardial ischemia/reperfusion injury and apoptosis in rabbits. Acta Pharmacol Sin. 2004;25(3):306–12.
75. Sato M, Bagchi D, Tosaki A, Das DK. Grape seed proanthocyanidin reduces cardiomyocyte apoptosis by inhibiting ischemia/reperfusion-induced activation of JNK-1 and C-JUN. Free Radic Biol Med. 2001;31(6):729–37.
76. Fisher ND, Hughes M, Gerhard-Herman M, Hollenberg NK. Flavanol-rich cocoa induces nitric-oxide-dependent vasodilation in healthy humans. J Hypertens. 2003;21(12):2281–6.
77. Grassi D, Lippi C, Necozione S, Desideri G, Ferri C. Short-term administration of dark chocolate is followed by a significant increase in insulin sensitivity and a decrease in blood pressure in healthy persons. Am J Clin Nutr. 2005;81(3):611–4.
78. Engler MB, Engler MM, Chen CY, Malloy MJ, Browne A, Chiu EY, et al. Flavonoid-rich dark chocolate improves endothelial function and increases plasma epicatechin concentrations in healthy adults. J Am Coll Nutr. 2004;23(3):197–204.
79. Vlachopoulos C, Aznaouridis K, Alexopoulos N, Economou E, Andreadou I, Stefanadis C. Effect of dark chocolate on arterial function in healthy individuals. Am J Hypertens. 2005;18(6):785–91.
80. Grassi D, Mulder TP, Draijer R, Desideri G, Molhuizen HO, Ferri C. Black tea consumption dose-dependently improves flow-mediated dilation in healthy males. J Hypertens. 2009;27(4):774–81.
81. Faridi Z, Njike VY, Dutta S, Ali A, Katz DL. Acute dark chocolate and cocoa ingestion and endothelial function: a randomized controlled crossover trial. Am J Clin Nutr. 2008;88(1):58–63.
82. Taubert D, Berkels R, Roesen R, Klaus W. Chocolate and blood pressure in elderly individuals with isolated systolic hypertension. JAMA. 2003;290(8):1029–30.
83. Taubert D, Roesen R, Lehmann C, Jung N, Schomig E. Effects of low habitual cocoa intake on blood pressure and bioactive nitric oxide: a randomized controlled trial. JAMA. 2007;298(1):49–60.
84. Grassi D, Necozione S, Lippi C, Croce G, Valeri L, Pasqualetti P, et al. Cocoa reduces blood pressure and insulin resistance and improves endothelium-dependent vasodilation in hypertensives. Hypertension. 2005;46(2):398–405.
85. Hodgson JM, Puddey IB, Burke V, Watts GF, Beilin LJ. Regular ingestion of black tea improves brachial artery vasodilator function. Clin Sci (Lond). 2002;102(2):195–201.

Chapter 9
Nitrite and Nitrate in Human Breast Milk: Implications for Development

Pamela D. Berens and Nathan S. Bryan

Key points

- Breast milk is nature's most perfect food with essential nutrients for the health and development of babies.
- Breast milk is enriched in both nitrite and nitrate and the ratio of these anions change as the composition of milk changes.
- Exposure rates of infants consuming colostrum from breast milk reach nearly 1 mg/kg which exceeds the ADI for nitrite by 14 times.
- Breast milk contains higher concentrations of nitrite and nitrate than most commercially available formulas.
- The presence of nitrite and nitrate in breast milk suggests an essential physiological function for the growth and development of the neonate.

Keywords Nursing • Infants • Formula • Health disparity • Colostrum • Neonate

Introduction

Dietary sources of breast milk serve as primary sources of calories and nutrients in infants and children. Human breast milk is recommended as the exclusive food for the first 6 months of life, and continuing, along with safe, nutritious complementary foods, up through 2 years [1, 2]. Breast milk is nature's most perfect food. In fact, the US Centers for Disease Controls in 2010 noted that breast milk is widely acknowledged as the most complete form of nutrition for infants, with a range of benefits for infants' health, growth, immunity, and development." Breast milk is a

N.S. Bryan (✉)
Brown Foundation Institute of Molecular Medicine,
The University of Texas, Health Science Center at Houston,
1825 Pressler Street 530C, Houston, TX 77030, USA
e-mail: Nathan.Bryan@uth.tmc.edu

N.S. Bryan and J. Loscalzo (eds.), *Nitrite and Nitrate in Human Health and Disease*,
Nutrition and Health, DOI 10.1007/978-1-60761-616-0_9,

unique nutritional source that cannot adequately be replaced by any other food, including infant formula. It remains superior to infant formula from the perspective of the overall health of both mother and child. Infants are fragile and susceptible to disease, partly because their immune system is not fully developed early on and they are dependent upon nutrients supplied by the nursing mother. They must be treated with special care and given adequate nourishment. Infant formulas have progressed over the years and are able to mimic a few of the nutritional components of breast milk, but the formulas cannot be hoped to duplicate the vast and constantly changing array of essential nutrients and immunologic properties of human milk that vary over time. Furthermore, there may be many nutrients present in human breast milk that are still unrecognized and, therefore, lacking in infant formulas. Discovery of unrecognized nutrients may allow improvement in artificial formula developments and/or provide new insight into beneficial properties of certain molecules or nutrients. This chapter invokes a role for nitrite and nitrate as important nutrients in breast milk.

Human milk is known to confer significant nutritional and immunological benefits for the infant [3–5]. According to the CDC National Immunization Survey in 2006, 74% of infants were breastfed in the early postpartum period, 43% were at 6 months, and 23% at 12 months of age. Among infants born in 2006, only 33 and 14% were exclusively breastfed through 3 and 6 months of age, respectively (CDC [6]). These data also suggest that approximately 25% of breastfeeding infants will receive formula supplementation in the first 2 days of life. Nearly 38% of breastfeeding infants have supplement by 3 months and 44.7% have formula supplement before 6 months. These data suggest that, in addition to breast milk, other forms of milk or complementary foods may be significant nutrient sources for infants. Therefore, it is critical to understand the potential dietary exposures to nitrate and nitrite from these milk sources to assess potential health benefits and risks associated with their consumption as well as to identify differences in these nutrients.

The relevance of dietary sources of nitrite and nitrate for human health is in its ascendancy. Historically, health risks due to nitrate in ground water have been associated with a risk of methemoglobinemia (blue-baby syndrome) in children [7]. Infants less than 6 months of age may be exposed to excess nitrate in bacterially contaminated well water that reduces nitrate to nitrite [8]. Infants consuming excess nitrite experience blue baby syndrome due to nitrite-mediated oxidation of ferric (Fe^{2+}) iron in oxyhemoglobin that leads to hypoxia and cyanosis and the resulting blue color [9, 10]. Under physiological conditions, methemoglobin reduction is accomplished primarily by red-cell-reduced nicotinamide adenine dinucleotide reductase so efficiently that there is <1% of methemoglobin in the circulating blood of healthy adults. Infants younger than 6 months old are particularly susceptible to nitrate-induced methemoglobinemia because of their low stomach acid production, large numbers of nitrate-reducing bacteria, the relatively easy oxidation of fetal hemoglobin, and immaturity of the methemoglobin reductase system [11, 12]. As such, an American Academy of Pediatrics (AAP) consensus panel concluded that all prenatal and well-infant visits should include questions about the home water supply; if the water source is a private well, the water should be tested for nitrate [7].

The panel concluded that infants fed commercially prepared infant foods are generally not at risk of nitrate poisoning, but that home-prepared infant foods from vegetables (e.g., spinach, beets, green beans, squash, carrots) should be avoided until infants are 3 months or older. Breastfed infants are not at risk of excessive nitrate exposure from mothers who ingest water with high nitrate content (up to 100 ppm nitrate nitrogen) as nitrate concentration is thought not to increase significantly in the breast milk under these circumstances [7].

It is noteworthy that the few human nitrate and nitrite exposure studies, including in children and adults, have not produced methemoglobinemia. Infants exposed to 175–700 mg nitrate per day did not experience methemoglobin levels above 7.5%, suggesting that nitrate alone is not causative for methemoglobinemia [13]. A more recent randomized 3-way crossover study exposed healthy volunteer adults to single doses of sodium nitrite that ranged from 150–190 mg per volunteer to 290–380 mg per volunteer [14]. Observed methemoglobin concentrations were 12.2% for volunteers receiving the higher dose of nitrite and 4.5% for those receiving the lower dose. Recent nitrite infusion studies of up to 110 μg/kg/min for 5 min induced methemoglobin concentrations of only 3.2% in adults [15]. These data have led investigators to propose alternative explanations for the observed methemoglobinemia in infants, including gastroenteritis and associated iNOS-mediated production of NO induced by bacteria-contaminated water [16, 17]. Experts have questioned the veracity of the evidence supporting the hypothesis that nitrate and nitrite are toxic for healthy post-infant populations [9, 18, 19]. Although the previously held hypothesis of nitrite toxicity appears biologically plausible, conclusive scientific evidence supporting the hypothesis has not been documented [16, 17]. These findings highlight serious, but context-specific, risks associated with nitrite overexposure in infants. Potential health risks due to excessive nitrite and nitrate consumption in these specific population subgroups led to regulatory limits on permissible concentration of nitrate in drinking water (50 mg nitrate/L in the European Union, 44 mg/L in the United States) in accordance with World Health Organization (WHO) recommendations first established in 1970 and reaffirmed in 2004 [20]. The US Environmental Protection Agency limits human exposure to inorganic nitrate to 0.10 mg/L (or 10 ppm nitrate nitrogen) and nitrite to 1 ppm nitrite nitrogen [21]. As a result, it is recommended that baby formulas and foods be made with nitrite- and nitrate-free water in order to reduce the risk of blue baby syndrome. However, there may be a low level of these anions that provide benefit to the infant without causing harm.

Breast Milk Confers Benefits to Infants Early in Life

What is indisputable is the established health benefits of breast milk. Human breast milk is the optimal source of infant nutrition and contains all the proteins, lipids, carbohydrates, and trace elements required for infant development [22]. The WHO recommends exclusive breastfeeding for the first 6 months of life and, thereafter, the introduction of appropriate solid foods with continuation of breastfeeding through 2 years and beyond. The AAP recommends that mothers breastfeed for at least up to the first year of a child's life and continue until they both feel they are ready to stop. In the first 6 months, the baby should be nourished exclusively by breast milk. The AAP asserts that breast milk has the perfect balance of nutrients for the infant. It has enough sustenance along with vitamin D supplementation approximately for the first 6 months of life and should follow as the child's staple diet throughout the first year. Breast milk contains essential nutrients to foster growth and development and promote optimal health.

Studies have demonstrated a number of important health benefits to breastfeeding for both mother and infant. It provides a number of infant health advantages beginning at birth and continuing throughout a child's life. In fact, a large number of the health problems that today's children face might be decreased, or even prevented, by breastfeeding the infant exclusively for at least the first 6 months of life. The longer the mother breastfeeds, the more likely her child will acquire the health benefits of breastfeeding. Human breast milk acts to develop the immune system of infants such that those who are breastfed exclusively for 6 months have a more developed immune system than those who are not exclusively breastfed [23, 24]. Many studies show that breastfeeding strengthens the immune system. During nursing, the mother passes antibodies to the child, which helps the child resist diseases and helps improve the normal immune response to certain vaccines. Breastfeeding has also been shown to reduce the risks of the infant developing asthma and allergies [25] as well as childhood leukemia [26]. Cardiovascular disease risk is also reduced through the reduction of obesity, blood pressure, and cholesterol [5, 27].

Having been breastfed as an infant provides benefit later in life. Additional benefits may become apparent as the child grows older. Breast-fed children are less likely to contract a number of diseases later in life, including type I diabetes mellitus, multiple sclerosis, heart disease, and cancer before the age of 15. In fact breastfed babies have been shown to have a small reduction in blood pressure later in life [28]. Breastfeeding during infancy is also associated with a reduction in risk of ischemic cardiovascular disease later in life [29]. What is it about breast milk that allows the health-promoting properties? The current theory is that immunoglobulins passed from mother to baby are responsible for these effects, but many of these benefits extend beyond the immune system.

Changes in Intestinal Microflora of the Developing Infant

Prior to uncomplicated birth, the fetus is sterile, and the first encounter with the microbial world begins during delivery. This exposure will vary depending on whether it is a cesarean or a vaginal delivery. During vaginal birth, bacterial colonization of a previously germ-free gut begins with organisms that are derived from the vagina, skin, and rectum of the mother. Vaginal microflora are some of the first colonizers in the gastrointestinal tract of newborns [30]. Naturally delivered babies experience a period of 2–3 days in which bacteria invading and reproducing within infant's gut are predominately aerobic bacteria such as *Enterobacteriaceae*, streptococci, and staphylococci [31]. These potentially pathogenic bacteria may not appear to be beneficial to the health of babies, but research has shown that the metabolisms of these early bacteria are believed to be positive factors in preparing for colonization by beneficial enteric bacteria such as bifidobacteria, lactobacilli, and other anaerobic bacteria that appear to reach the gut after 2–3 days [32]. After bacterial transmission from the mother to the baby during delivery, the mother provides a continual source of bacteria from oral and skin microbes through suckling, kissing, and caressing. These bacteria have been shown to be able to create a reduced environment favorable to anaerobic bacteria, which colonize the gut at the end of week 1 taking over from the aerobic bacteria.

An additional source of bacteria for breastfed neonates is mother's milk which contains up to 10 [9] microbes/L in healthy mothers [33]. The most common bacteria colonized from this mode of transmission are staphylococci, streptococci, corynebacteria, micrococci, propionibacteria, and bifidobacteria, as well as lactobacilli [34]. After the first 2 weeks of life, stable microflora are established and maintained. This microflora environment is dependent upon breast vs. formula feeding. Supplementation of a breast fed baby with formula milk induces a rapid shift in the bacterial pattern of a breastfed baby [32]. This relationship between the microflora composition and the health of the baby is critical and thought to provide routes of metabolism and nutrients for the developing baby similar to vitamin K metabolism in adults. We present data later in the chapter showing this bacterial symbiosis in conversion of nitrate and nitrite to NO production.

Changes in Milk Characteristics over Time

The quality and characteristics of breast milk change over time. This change in composition is thought to be due partly to the changing gut microflora in the baby and the changing metabolic demands as the baby grows. Colostrum, the early milk from the mothers' breasts, is usually present after the fifth or sixth month of pregnancy. Colostrum has a yellow color, is thick in consistency, and is high in protein and low in fat and sugar. Colostrum is designed to meet a newborn's special needs. The protein content is three times higher than that of mature milk. It is rich in antibodies being passed from the mother. Breast milk will change and increase in quantity about 48–72 h after giving birth [35]. In addition to the macronutrients, vitamins, minerals, and water in human milk, specific components contribute to the immunocompetence of the infant, including secretory IgA, IgG and IgM, lactoferrin, lysozyme, cytokines, and exosome-derived microRNAs [36–39]. These antibodies protect the baby and act as a natural laxative, helping the baby pass the first stool, called meconium. Once the baby is born, colostrum is present in small but gradually increasing amounts for the first 3 days to match the small size of the baby's stomach. Most babies do not need additional nutrition during this time.

This precious colostrum should be given to the newborn as soon as possible after giving birth. It is recommended that mothers breastfeed at least 8–12 times per 24 h, so the baby receives this valuable milk. Milk will typically increase in quantity by 72 h after giving birth, although it may take longer depending on when breastfeeding began and how frequently breastfeeding occurs.

Once a more mature milk supply is established, there is variation in the milk that occurs during feedings. When the breastfeeding begins, the first milk the baby receives is called foremilk. It is thin and watery with a light blue tinge. Foremilk has a higher water content needed to satisfy the baby's thirst. Hind milk is released after several minutes of nursing. It is similar in texture to cream and has the highest concentration of fat. The hind milk has a relaxing effect on the baby. Hind milk helps the baby feel satisfied and gain weight. Interestingly, several research studies have demonstrated relatively high concentrations of nitrite and nitrate in human breast milk. Ohta et al. found high concentrations of nitrite and nitrate (166–1,246 μM) in the breast milk of Japanese mothers from days 1 to 8 [40]. We recently found that the ratio of nitrite and nitrate in breast milk changes as the composition of milk changes whereby colostrum contains the highest amount of nitrite (~10 μM) [41] (Fig. 9.1). Iizuka et al. report a range of 19.3–82.4 μM with average concentration of nitrite in colostrum as 43.6 ± 5.3 μM and nitrate concentrations ranging from 86.6 to 278.7 μM with an average of 180.6 ± 63.1 μM [42]. In this particular study it was shown that when breast milk was ingested there was a production of NO gas in the stomach that was absent in formula-fed babies, which have much lower concentrations of nitrite and nitrate [42]. The nitrite consumed by the infant enters the acidic environment of the stomach leading to the generation of NO in the gastric lumen. The effects of NO in adult stomachs are documented relative to gastric mucosal integrity and blood flow from the reduction of nitrate to nitrite to NO.

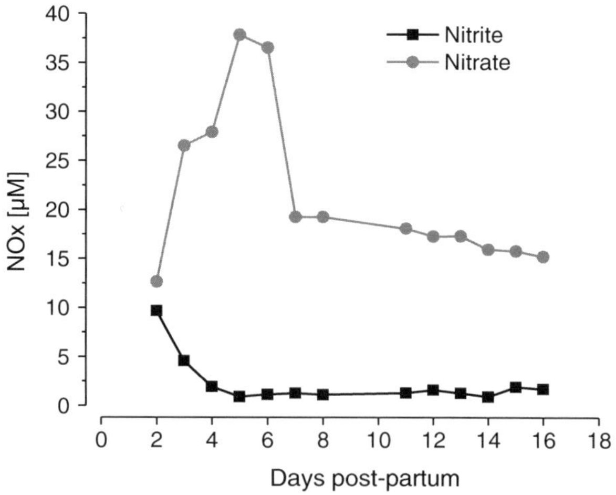

Fig. 9.1 A longitudinal study in the same nursing mothers from multiple collections over 16 days. Data are average of single donations from 2 nursing mothers for 16 days

Prior to the birthing process, the gastrointestinal tract of the infant is sterile, and it is rapidly colonized by bacteria originating from the mother and the environment, as discussed earlier [43, 44]. Reduction of nitrate to nitrite requires the commensal bacteria that normally reside in the body. However, in newborn infants, this pathway has not developed. Therefore, breast milk high in nitrite relative to nitrate overcomes the natural deficiency early in life. At later stages of development, nitrate becomes the predominant anion when a symbiosis exists with the colonized bacteria. This concept is supported by the data in Fig. 9.1 showing early post-partum colostrum enriched in nitrite. The nitrite concentration decreases concomitantly with nitrate increases during the time at which the gut becomes colonized by bacteria. There appears to be a complimentary system whereby the nitrite in breast milk can be reduced to NO. Colostrum is designed to meet a newborn's special needs. Our data suggest that the ratio of nitrite to nitrate also changes in order to meet the changing metabolic demands on the infant. Therefore, part of its role may be to supply a sufficient amount of nitrite in the early post-partum period until the gut microflora are then established to utilize nitrate as the primary substrate. Indeed, the pattern of nitrate and nitrite composition of human milk suggests that, as observed by others, these components also serve immunomodulatory and gastroprotective roles. In adults, the provision of nitrite, resulting from reduction by commensal lingual microbiota, is constant because of enterosalivary circulation of nitrate [45]. Interruption of the reduction of nitrate to nitrite has been shown to prevent a rise in plasma nitrite, hinder vasodilation with resulting decrease in blood pressure, and decreased platelet aggregation [46]. Therefore, it is reasonable to surmise that nitrite must be supplied through human milk to the newborn in order to derive the resulting vascular, immunologic, and gastroprotective benefits of NO to which it is metabolized.

Nitrite and Nitrate Anions in Breast Milk vs. Formula and Cows Milk

Currently there are stringent regulations on the amount of nitrite and nitrate allowed in our food supply and drinking water. These regulations may not be applicable to human milk. There have been several studies showing high concentrations of nitrite and nitrate in breast milk. Cekmen et al. found extremely high concentrations of nitrite in breast milk of healthy mothers that were reduced in pre-eclampsia patients [47]. Total nitrite levels were 56.09 ± 11.18 vs. 82.20 ± 12.01 μM, $P<0.05$, in colostrum of preeclamptics and controls, respectively. The level of total nitrite was 37.75 ± 12.10 vs. 53.28 ± 10.25 μM, $P<0.05$, in 30th-day milk of these same patients [47], consistent with our data showing the highest concentrations of nitrite in colostrum that declines over time (Fig. 9.1). Breast milk from certain mothers contained the highest nitrite concentration of any food or beverage product tested. Human breast milk contains high concentrations of nitrate and nitrite in the early postpartum period, and the relative concentrations of the two anions change from

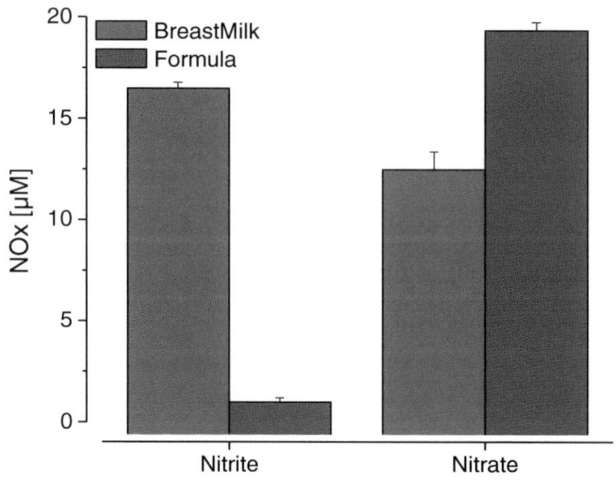

Fig. 9.2 Comparison of 3-day-old breast milk to Enfamil Baby Formula. Data are average ± SEM of $n=3$ samples of each

Table 9.1 Nitrate and nitrite concentrations from infant and pediatric formula milk products

Milk product	Nitrite (mg/100 mL)	Nitrate (mg/100 mL)
Colostrum	0.08	0.19
Boost Kid Essentials Lactose Free®	0.007	0.26
Bright Beginnings Soy Pediatric®	0.03	0.16
Similac®	0.00005	0.05
Pregestimil® – protein hydrolysate	0.00005	0.12
Enfamil EnfaCare®	0.00005	0.10
Nestle Carnation Instant Breakfast Plus® (lactose free)	0.00005	0.08
TwoCal HN® with FOS-Milk Based Hydrolasate	0.00005	0.18
Nutramigen® AA Elemental	0.00005	0.02
Market Pantry 2% Cow's Milk	0.0002	0.23
Silk® Soy Vanilla Milk	0.0036	3.48
Silk® Soy Egg Nog	0.013	0.34

colostrum to transition milk to mature milk. The concentrations are much higher than that found in commercial baby formulas. When comparing freshly expressed human breast milk at day 3 to Enfamil baby formula, human breast milk contained almost 20 times higher nitrite than that contained in Enfamil formula (Fig. 9.2). It is interesting to note that formula had much less nitrite but actually higher concentrations of nitrate than had the breast milk.

Table 9.1 contains nitrate and nitrite concentrations in a representative sample of colostrum, artificial formulas for infants and children, as well as samples of cow and soy milk. As shown below, colostrum contained the highest concentration of nitrite of any of the other milk products, and Silk Soy Vanilla milk contained the highest nitrate concentration. All of the pediatric formulas, with the exception of Bright Beginning Soy formula, had very low (barely detectable) levels of nitrite.

Source of Nitrite in Breast Milk

The origins of nitrate and nitrite in human milk are not known. Maternal nitrate and nitrite intakes are not reflected in nitrate and nitrite composition of human milk [7]. The data supporting this conclusion are sparse. One study published on this topic demonstrated that women who consumed water with a nitrate concentration <100 mg/L did not produce milk with elevated nitrate levels [48]. Physiological production of nitrate and nitrite in tissues is dynamic; it is dependent on the mammalian nitrate reductase activity of enzymes, such as xanthine oxidoreductase and a host of other redundant systems, [49–52] local conditions such as hypoxia, [53, 54]) and acidosis. Another hypothesis is that tissue production of nitrite and nitrate may be the result of bacterial nitrate reductase activity. Other work has demonstrated an important role for xanthine oxidoreductase (XO) in milk in the generation of NO_x. It is worth noting that XO activity of human milk, while generally very much lower than that of cow's milk [55], is exceptionally high in the first few weeks postpartum [56]. This is precisely the period when antibacterial activity is required in the neonatal gut, it coincides with particularly high levels of nitrite [42], and it correlates with the highest levels of XO-dependent NO generation found in human milk [57]. In the early postpartum stage, nitrite may exert critical antimicrobial activity. The presence of high levels of XO in breast milk, along with the high nitrite levels and low oxygen tension, allows for the generation of peroxynitrite, a potent bactericidal agent, from NO [58]. These proposed antimicrobial actions of nitrite and peroxynitrite may help fight infections in the peripartum period. It is also biologically plausible that the high concentration of nitrite in colostrum may be the byproduct of NO oxidation resulting from its production during the initiation of lactation by nitric oxide synthase (NOS) [59, 60]. NOS-derived NO, using L-arginine as a substrate, may assist in the "let down" reflex at the initiation of lactation [59, 61]. After initiation of lactation, there may be less NO required, and, hence, less nitrite produced by NO oxidation, to begin the flow of milk from the breast tissue. This may be one potential explanation as to why nitrite concentrations drop significantly in transitional and mature milk.

Levels of Exposure of Nitrite and Nitrate from Breast Milk

The published data on nitrite and nitrate content of human breast milk reveal that substantial exposure to the anions occurs in newborn infants. This becomes interesting in terms of levels of exposure based on nitrite ingestion relative to body weight in the infant (over 1 mg/kg). The Joint Food and Agricultural Organization/

World Health Organization has set the acceptable daily intake (ADI) for the nitrate ion at 3.7 mg/kg body weight and for the nitrite ion at 0.06 mg/kg body weight [62]. Normal daily infant breastmilk intake is 2–3 oz per pound per day or 200 mL/kg. That translates into 14–21 oz or 400–630 mL of milk for a typical 7 pound (3 kg) infant or more simply ~200 mL/kg. Taking an average of 50 μmoles/L concentration of nitrite or 2,300 μg/L, total daily nitrite exposure to nursing infants is roughly 1.2 mg/kg or 20 times higher than the ADI for nitrite. Among the milk samples analyzed from Table 9.1, nitrate intakes were highest from pediatric formulas (Boost Kids Essential® Lactose Free and Bright Beginnings Soy Pediatric®) and lowest from infant formulas (Similac® and Pregestimil®). The highest nitrate exposure of the milk samples tested, relative to the WHO ADI standard, is from Silk® Soy Vanilla, which could result in intakes of 104% of this standard. Boost Kids Essential® Lactose Free intake could result in 83% of the WHO ADI for nitrate. Regarding potential nitrite exposures, Bright Beginnings Soy Pediatric® intake could result in 383% and Boost Kids Essential® Lactose Free intake could result in 83% of the WHO ADI standard. Human milk colostrum samples had the highest nitrite concentration of any milk product tested. The small amount produced by the breast and, therefore, consumed by the infant still translates to 42% of WHO ADI at 100 mL intake.

The ability to exceed WHO ADI limits with usual intake levels of single foods, such as breast milk, spinach [63], or a desiccated vegetable supplement [64], indicates that these regulatory limits may not be applicable to all situations, such as food sources of nitrate and nitrite, for children and adults. This latter suggestion is warranted based on the consideration of intake estimates of nitrite and nitrate from other food sources. We have found that a DASH diet pattern with high-nitrate food choices exceeds the WHO ADI for nitrate by 550% for a hypothetical 60 kg adult. If this hypothetical adult were to consume, in addition to the high-nitrate DASH diet, three cups of the soy milk analyzed herein, the concentration of dietary nitrate from these foods (1,248 mg total or 20.8 mg/kg) would exceed WHO ADI levels by 562% for a 60 kg person.

Implications for Health

The nitrate and nitrite content of milk products demonstrate that these foods are significant contributors to total dietary nitrate and nitrite intakes. The data presented in this chapter suggests that human milk provides a dietary source for nitrite prior to the establishment of lingual and gastrointestinal microbiota. Once the microbiota are established, these commensal organisms are capable of reducing dietary nitrate, via enterosalivary circulation, to nitrite, and support gastrointestinal, immune, and cardiovascular health. The significant concentration of nitrate and nitrite in bovine milk demonstrated here indicates that these conclusions may be applicable across mammalian species. When we compare freshly express human breast milk at day 3 to Enfamil baby formula, human breast milk contains almost

20 times more nitrite. Studies have demonstrated a number of important health benefits to breastfeeding [5]. It provides a number of health advantages beginning at birth and continuing throughout a child's life. A large number of the health problems that today's children face might be decreased, or even prevented, by breastfeeding the infant exclusively for at least the first 6 months of life. The longer the mother breastfeeds, the more likely her child will get the health benefits of breastfeeding. Many studies show that breastfeeding strengthens the immune system. Infants who are breastfed exclusively for 6 months have a more developed immune system than those who are not exclusively breastfed [23, 24]. Breastfeeding has also been shown to reduce the risks of the infant developing asthma and allergies [25] as well as childhood leukemia [26]. Cardiovascular disease risk is also reduced through the reduction of obesity, blood pressure, and cholesterol [27]. Having been breastfed as an infant also provides benefit later in life. Breastfed children are less likely to contract a number of diseases later in life, including type 1 diabetes, multiple sclerosis, heart disease, and cancer before the age of 15 [65]. In fact breastfed babies have been shown to have a small reduction in blood pressure later in life [28]. Breastfeeding during infancy is also associated with a reduction in risk of ischemic cardiovascular disease later in life [29]. The early influence of nitrite and nitrate may explain, at least in part, some of that benefits.

In addition to the established cardiovascular and immunological benefits of NO-derived from breast milk nitrite, there is growing evidence related to its anti-cancer activity, as well. A review for the evidence of an association between breast-fed infants and certain cancers suggests that the children who are never breastfed or are breastfed for short term have a higher risk than those who are breastfed for ≥6 months of developing Hodgkin's disease (HD), but not non-Hodgkin's lymphoma or acute lymphoblastic leukemia [66]. Human milk contains an extensive array of anti-microbial activity and appears to stimulate early development of the infant immune system. Artificially fed infants negotiate exposure to infectious agents without the benefits of this immunologic armamentarium and do not do, as well as breast-fed infants, in resisting infection. Thus, human milk may make the breastfed infant better able to negotiate future carcinogenic insults by modulating the interaction between infectious agents and the developing infant immune system or by directly affecting the long-term development of the infant immune system. Part of this response may be mediated by the nitrite and nitrate content of breast milk.

Future research is required to establish definitively the role of nitrite and nitrate in formula fed vs. breastfed infants. The dietary and physiological determinants of milk nitrate and nitrite concentrations remain to be established. However, the presence of nitrate and nitrite in human milk provides indirect evidence for a physiological benefit of dietary nitrite for the protection of the gastrointestinal tract in the neonate prior to the establishment of commensal bacteria in the mouth and gut. The temporal relationship between the provision of nitrite in human milk and the development of commensal microbiota capable of reducing dietary nitrate to nitrite supports an hypothesis that milk nitrite may supply this component in the immediate term after birth. One conclusion that can be made from these exposure estimates is that humans are adapted to receive dietary nitrite and nitrate from birth and, therefore, they may not pose a significant risk at levels naturally found in certain foods. In fact, the absence of an essential nutrient, namely nitrite in baby formulas, may be involved in many of the health disparities in formula-fed babies, including necrotizing enterocolitis, infections, poor nutrient absorption, and even increased health risks later in life. If claims of nitrite and nitrate in promoting carcinogenesis were warranted, we would expect a higher incidence of cancer in breastfed infants compared with formula-fed infants, a result which has not been reported. In fact, there is a recognized and large disparity in the health of breastfed vs. formula-fed babies favoring breastfeeding. The current theory holds that this is due to immunoglobulins transferred from mother to baby which are responsible for the benefits of breast milk. A case can be made for nitrite as an essential molecule in the development and immune function of infants as well. Nature has devised a perfect system to nourish and foster the growth and development of nursing babies. Nitrite appears to be one of those indispensible nutrients.

References

1. Heinig MJ. The American Academy of Pediatrics recommendations on breastfeeding and the use of human milk. J Hum Lact. 1998;14(1):2–3.
2. Gartner LM, Morton J, Lawrence RA, Naylor AJ, O'Hare D, Schanter RJ, et al. Breastfeeding and the use of human milk. Pediatrics. 2005;115(2):496–506.
3. Hoddinott P, Tappin D, Wright C. Breast feeding. BMJ. 2008;336(7649):881–7.
4. James DC, Lessen R. Position of the American Dietetic Association: promoting and supporting breastfeeding. J Am Diet Assoc. 2009;109(11):1926–42.
5. Ip S, Chung M, Raman G, Chew P, Magula N, DeVine D, et al. Breastfeeding and maternal and infant health outcomes in developed countries. Evidence report/technology assessment. Rockville, MD: Tufts-New England Medical Center Evidence-based Practice center, under Contract NO. 290-02-0022; 2007.
6. Centre for Disease Control and Prevention (CDC). Breastfeeding trends and updated national health objectives for exclusive breastfeeding–United States, birth years 2000-2004. MMWR Morb Mortal Wkly Rep. 2007;56(30):760–3.
7. Greer FR, Shannon M. Infant methemoglobinemia: the role of dietary nitrate in food and water. Pediatrics. 2005;116(3):784–6.
8. Johnson CJ, Kross BC. Continuing importance of nitrate contamination of groundwater and wells in rural areas. Am J Ind Med. 1990;18(4):449–56.
9. McKnight GM, Duncan CW, Leifert C, Golden MH. Dietary nitrate in man: friend or foe? Br J Nutr. 1999;81(5):349–58.
10. Fan AM, Steinberg VE. Health implications of nitrate and nitrite in drinking water: an update on methemoglobinemia occurrence and reproductive and developmental toxicity. Regul Toxicol Pharmacol. 1996;23(1 Pt 1):35–43.
11. Dusdieker LB, Getchell JP, Liarakos TM, Hausler WJ, Dungy CI. Nitrate in baby foods. Adding to the nitrate mosaic. Arch Pediatr Adolesc Med. 1994;148(5):490–4.
12. Kross BC, Ayebo AD, Fuortes LJ. Methemoglobinemia: nitrate toxicity in rural America. Am Fam Physician. 1992;46(1):183–8.
13. Cornblath M, Hartmann AF. Methhemoglobinaemia in young infants. J Pediatr. 1948;33:421–5.

14. Kortboyer J, Olling M, Zeilmaker MJ. The oral bioavailability of sodium nitrite investigated in healthy adult volunteers. Bilthoven, tHE Netherlands: National Institute of Public Health and the Environment; 1997.
15. Dejam A, Hunter CJ, Tremonti C, Pluta RM, Han YY, Grimes G, et al. Nitrite infusion in humans and nonhuman primates: endocrine effects, pharmacokinetics, and tolerance formation. Circulation. 2007;116(16):1821–31.
16. L'Hirondel JL, Avery AA, Addiscott T. Dietary nitrate: where is the risk? Environ Health Perspect. 2006;114(8):A458–9; author reply A459–61.
17. Powlson DS, Addiscott TM, Benjamin N, Cassman KG, de kok TM, van Grinsven H, et al. When does nitrate become a risk for humans? J Environ Qual. 2008;37(2):291–5.
18. L'Hirondel JL. Nitrate and man: toxic, harmless or beneficial? Wallingford, UK: CABI; 2001.
19. Authority EFS. Nitrate in vegetables: scientific opinion of the panel on contaminants in the food chain. EFSA J. 2008;289:1–79.
20. World Health Organization. Recommendations; nitrate and nitrite. In: Guidelines for drinking water quality. 3rd ed. Geneva, Switzerland: WHO; 2004.
21. National Primary Drinking Water Regulations: Final Rule, 40. In: Agency USEP, editor Vol CFR parts 141–143. Federal Register; 1991:3526–97.
22. Newton ER. Breastmilk: the gold standard. Clin Obstet Gynecol. 2004;47(3):632–42.
23. Hawkes JS, Neumann MA, Gibson RA. The effect of breast feeding on lymphocyte subpopulations in healthy term infants at 6 months of age. Pediatr Res. 1999;45(5 Pt 1):648–51.
24. Hawkes JS, Gibson RA. Lymphocyte subpopulations in breast-fed and formula-fed infants at six months of age. Adv Exp Med Biol. 2001;501:497–504.
25. Fulhan J, Collier S, Duggan C. Update on pediatric nutrition: breastfeeding, infant nutrition, and growth. Curr Opin Pediatr. 2003;15(3):323–32.
26. Shu XO, Linet MS, Steinbuch M, Wen WQ, Buckley JD, Neglia JP, et al. Breast-feeding and risk of childhood acute leukemia. J Natl Cancer Inst. 1999;91(20):1765–72.
27. Singhal A. Early nutrition and long-term cardiovascular health. Nutr Rev. 2006;64(5 Pt 2): S44–49; discussion S72–91.
28. Martin RM, Gunnell D, Smith GD. Breastfeeding in infancy and blood pressure in later life: systematic review and meta-analysis. Am J Epidemiol. 2005;161(1):15–26.
29. Rich-Edwards JW, Stampfer MJ, Manson JE, Rosner B, Hu FB, Michels KB, et al. Breastfeeding during infancy and the risk of cardiovascular disease in adulthood. Epidemiology. 2004;15(5):550–6.
30. Brook I, Barrett CT, Brinkman III CR, Martin WJ, Finegold SM. Aerobic and anaerobic bacterial flora of the maternal cervix and newborn gastric fluid and conjunctiva: a prospective study. Pediatrics. 1979;63(3):451–5.
31. Morelli L. Postnatal development of intestinal microflora as influenced by infant nutrition. J Nutr. 2008;138(9):1791S–5.
32. Favier CF, Vaughan EE, De Vos WM, Akkermans AD. Molecular monitoring of succession of bacterial communities in human neonates. Appl Environ Microbiol. 2002;68(1):219–26.
33. West PA, Hewitt JH, Murphy OM. Influence of methods of collection and storage on the bacteriology of human milk. J Appl Bacteriol. 1979;46(2):269–77.
34. Martin R, Langa S, Reviriego C, Jiminez E, Marin ML, Xaus J, et al. Human milk is a source of lactic acid bacteria for the infant gut. J Pediatr. 2003;143(6):754–8.
35. Neville MC. Physiology of lactation. Clin Perinatol. 1999;26(2):251–79; v.
36. Hamosh M. Bioactive factors in human milk. Pediatr Clin North Am. 2001;48(1):69–86.
37. Garofalo R. Cytokines in human milk. J Pediatr. 2010;156(2 Suppl):S36–40.
38. Kosaka N, Izumi H, Sekine K, Ochiya T. microRNA as a new immune-regulatory agent in breast milk. Silence. 2010;1(1):7.
39. Admyre C, Johansson SM, Qazi KR, Filén JJ, Lahesmaa R, Norman M, et al. Exosomes with immune modulatory features are present in human breast milk. J Immunol. 2007;179(3):1969–78.
40. Ohta N, Tsukahara H, Ohshima Y, Nishii M, Ogawa Y, Sekine K, et al. Nitric oxide metabolites and adrenomedullin in human breast milk. Early Hum Dev. 2004;78(1):61–5.

41. Hord NG, Ghannam JS, Garg HK, Berens PD, Bryan NS. Nitrate and nitrite content of human, formula, bovine and soy milks: implications for dietary nitrite and nitrate recommendations. Breastfeed Med. 2010 Oct 19.[Epub ahead of print].
42. Iizuka T, Sasaki M, Oishi K, Uemura S, Koike M, Shinozaki M. Non-enzymatic nitric oxide generation in the stomachs of breastfed neonates. Acta Paediatr. 1999;88(10):1053–5.
43. Fujita K, Murono K. Nosocomial acquisition of *Escherichia coli* by infants delivered in hospitals. J Hosp Infect. 1996;32(4):277–81.
44. Mandar R, Mikelsaar M. Transmission of mother's microflora to the newborn at birth. Biol Neonate. 1996;69(1):30–5.
45. Petersson J, Carlstrom M, Schreiber O, Phillipson M, Christoffersson G, Jagare A, et al. Gastroprotective and blood pressure lowering effects of dietary nitrate are abolished by an antiseptic mouthwash. Free Radic Biol Med. 2009;46(8):1068–75.
46. Webb AJ, Patel N, Loukogeorgakis S, Okorie M, Aboud Z, Misra S, et al. Acute blood pressure lowering, vasoprotective, and antiplatelet properties of dietary nitrate via bioconversion to nitrite. Hypertension. 2008;51(3):784–90.
47. Cekmen MB, Balat A, Balat O, Aksoy F, Yurekli M, Erbagci AB, et al. Decreased adrenomedullin and total nitrite levels in breast milk of preeclamptic women. Clin Biochem. 2004;37(2):146–8.
48. Dusdieker LB, Stumbo PJ, Kross BC, Dungy CI. Does increased nitrate ingestion elevate nitrate levels in human milk? Arch Pediatr Adolesc Med. 1996;150(3):311–4.
49. Jansson EA, Huang L, Malkey R, Govoni M, Nihten C, Olssor A, et al. A mammalian functional nitrate reductase that regulates nitrite and nitric oxide homeostasis. Nat Chem Biol. 2008;4(7):411–7.
50. Dalsgaard T, Simonsen U, Fago A. Nitrite-dependent vasodilation is facilitated by hypoxia and is independent of known NO-generating nitrite reductase activities. Am J Physiol Heart Circ Physiol. 2007;292(6):H3072–8.
51. Li H, Cui H, Kundu TK, Alzawahra W, Zweier JL. Nitric oxide production from nitrite occurs primarily in tissues not in the blood: critical role of xanthine oxidase and aldehyde oxidase. J Biol Chem. 2008;283(26):17855–63.
52. Feelisch M, Fernandez BO, Bryan NS, Garcia-Saura MF, Bauer S, Whitlock DR, et al. Tissue processing of nitrite in hypoxia: an intricate interplay of nitric oxide-generating and -scavenging systems. J Biol Chem. 2008;283(49):33927–34.
53. Zweier JL, Wang P, Samouilov A, Kuppusamy P. Enzyme-independent formation of nitric oxide in biological tissues. Nat Med. 1995;1(8):804–9.
54. Bryan NS, Calvert JW, Elrod JW, Gundewar S, Ji SY, Lefer DJ. Dietary nitrite supplementation protects against myocardial ischemia-reperfusion injury. Proc Natl Acad Sci USA. 2007;104(48):19144–9.
55. Abadeh S, Killacky J, Benboubetra M, Harrison R. Purification and partial characterization of xanthine oxidase from human milk. Biochim Biophys Acta. 1992;1117(1):25–32.
56. Brown AM, Benboubetra M, Ellison M, Powell D, Reckless JD, Harrison R. Molecular activation-deactivation of xanthine oxidase in human milk. Biochim Biophys Acta. 1995;1245(2):248–54.
57. Stevens CR, Millar TM, Clinch JG, Kanczler JM, Bodamyali T, Blake DR. Antibacterial properties of xanthine oxidase in human milk. Lancet. 2000;356(9232):829–30.
58. Hancock JT, Salisbury V, Ovejero-Boglione MC, Cherry R, Hoare C, Eisenthal R, et al. Antimicrobial properties of milk: dependence on presence of xanthine oxidase and nitrite. Antimicrob Agents Chemother. 2002;46(10):3308–10.
59. Iizuka T, Sasaki M, Oishi K, Uemura S, Koike M, Minatogawa Y. Nitric oxide may trigger lactation in humans. J Pediatr. 1997;131(6):839–43.
60. Iizuka T, Sasaki M, Oishi K, Uemura S, Koike M. The presence of nitric oxide synthase in the mammary glands of lactating rats. Pediatr Res. 1998;44(2):197–200.

61. Lacasse P, Farr VC, Davis SR, Prosser CG. Local secretion of nitric oxide and the control of mammary blood flow. J Dairy Sci. 1996;79(8):1369–74.
62. Authority EFS. Nitrate in vegetables: scientific opinion of the panel on contaminants in the food chain. EFSA J. 2008;689:1–79.
63. Lundberg JO, Feelisch M, Bjorne H, Jansson EA, Weitzberg E. Cardioprotective effects of vegetables: is nitrate the answer? Nitric Oxide. 2006;15(4):359–62.
64. Hord NG, Tang Y, Bryan NS. Food sources of nitrates and nitrites: the physiologic context for potential health benefits. Am J Clin Nutr. 2009;90(1):1–10.
65. Schack-Nielsen L, Michaelsen KF. Breast feeding and future health. Curr Opin Clin Nutr Metab Care. 2006;9(3):289–96.
66. Davis MK. Review of the evidence for an association between infant feeding and childhood cancer. Int J Cancer Suppl. 1998;11:29–33.

Chapter 10
Regulation of Dietary Nitrate and Nitrite: Balancing Essential Physiological Roles with Potential Health Risks

Norman G. Hord

Key points

- US and European Union regulatory limits on nitrates in drinking water are necessary to limit environmental pollution known as eutrophication.
- Health concerns of excessive nitrate and nitrite consumption have driven regulatory actions due to perceived risk of methemoglobinemia in infants and gastrointestinal cancer risk in adults.
- The World Health Organization's Acceptable Daily Intake recommendations for nitrate can be exceeded by normal daily intakes of single foods and recommended dietary patterns, such as the DASH diet.
- Inconsistent positions on the health risks and benefits of foods containing nitrates and nitrites by the International Agency for Research on Cancer (IARC) and the European Food Safety Authority (EFSA) may contribute to confusion for consumers; regulators must take the opportunity to clarify and expand upon these positions in order to provide coherent dietary guidance.
- The established vasoprotective, blood pressure lowering, and antiplatelet aggregation properties of nitrite alone, or of nitrite originating from dietary nitrate, requires a new regulatory paradigm that incorporates the concepts of physiological deficiency, sufficiency and excess.
- There is a need to engage an independent panel of experts from academia, industry, and governmental and non-governmental sectors to undertake the first comprehensive, systematic review of the potential health risks and benefits of food sources of nitrates and nitrites.
- U.S. Institute of Medicine's Dietary Reference Intake paradigm may be a useful guide to the development of coherent dietary nitrate and nitrite intake recommendations.

N.G. Hord (✉)
Department of Food Science and Human Nutrition, Michigan State University, 208C G.M. Trout FSHN Building, East Lansing, MI 48824-1224, USA
e-mail: hord@msu.edu

N.S. Bryan and J. Loscalzo (eds.), *Nitrite and Nitrate in Human Health and Disease*, Nutrition and Health, DOI 10.1007/978-1-60761-616-0_10,

Keywords Accepted daily intake • Dietary recommendations • World Health Organization • Food safety • Risk benefit evaluation

Introduction

Current regulatory limits on nitrate concentrations in drinking water are necessary to reduce the risk of eutrophication; dietary intake limits for nitrate and nitrite based on potential risk of carcinogenicity and methemoglobinemia have more equivocal scientific support. The World Health Organization intake limits for nitrate are exceeded by normal daily intakes of single foods, such as soya milk and spinach, as well as recommended dietary patterns such as the Dietary Approaches to Stop Hypertension diet. Inconsistent regulatory positions on dietary exposures to nitrate and nitrite from processed meats and vegetables may be a source of public concern about the quality of dietary recommendations from health regulators. There is a great necessity to undertake a multidisciplinary, independent and systematic review of the potential health benefits and risks of dietary nitrates and nitrites. Regulatory bodies must consider all available data on the physiological roles of nitrate and nitrite in order to derive rational bases for dietary recommendations. The development of a more sophisticated regulatory paradigm that acknowledges physiological states of nitrate and nitrite deficiency, adequacy and excess, as exemplified by the Institute of Medicine's Dietary Reference Intake paradigm, may be needed. This paradigm should also consider the relevant metabolic interactions between food sources of nitrates and nitrites in specific physiologic and pathophysiologic contexts. These efforts could improve dietary guidance regarding nitrate and nitrite intakes that result in public health benefits.

> Everything should be made as simple as possible, but not simpler.
>
> ~Albert Einstein's comment on the philosophy of parsimony of William of Occam, English philosopher and Franciscan friar (c. 1285–1349).

Nitrite and Nitrate in Biology

Nitrate enters the food chain through plant foods via the action of lightning and soil bacteria. In addition to being a required nutrient for plants, it is an approved food additive [1]. Nitrate from vegetables is the major dietary source of this chemical while surface and ground water is a minor contributor. Nitrate intakes from vegetables are determined by the type of vegetable consumed, the levels of nitrate in the vegetables (including the nitrate content of fertilizer), the amount of vegetables consumed, and the level of nitrate in the water supply [2]. Nitrate concentration is the highest in the leaves whereas lower concentrations occur in seeds or tubers [3].

Nitrite exposure is mainly from endogenous nitrate conversion via a process known as enterosalivary circulation [4, 5]. Endogenous reduction of nitrate to nitrite is a source of nitric oxide (NO) and related metabolites (referred to as NO_x) in tissues and, in hypoxia, the vasculature [6, 7].

Regulatory Limits on Nitrate in Drinking Water

Environmental contamination with excess nitrate from fertilizer use is a persistent and growing problem. Nitrate and nitrite are naturally occurring ions that serve as nutrients for plants via fixation by soil bacteria. The enhancement of nitrogen fixation by the provision of nitrate-containing fertilizers has surpassed the amount that occurs naturally [8]. The resulting contamination of ground and surface water with excess nitrate by these agricultural practices is a global concern. Regulatory positions that address this concern are exemplified by the European Union's Nitrate Directive. The Nitrates Directive aims to protect water quality across Europe by preventing nitrates from agricultural sources polluting ground and surface waters and by promoting the use of good farming practices [9]. This type of pollution is referred to as eutrophication and is characterized by the excessive development of certain types of algae disturbs the aquatic ecosystems and becomes a health risk for animals and humans. The primary cause of eutrophication is an excessive concentration of plant nutrients, including nitrates, originating from agricultural practices or sewage treatment. While it is beyond the scope of this chapter to discuss these environmental issues, the contribution of contaminated drinking water sources to potential health effects of dietary nitrate and nitrite is highly relevant.

The permissible concentration of nitrate in drinking water is 50 mg nitrate/L in the European Union and 44 mg/L in the United States in accordance with World Health Organization recommendations first established in 1970 and reaffirmed in 2004 [10]. The harmony of US and European Union nitrate and nitrite regulations are betrayed by a growing body of evidence that ascribes essential functions in vascular and tissue homeostasis and immune function to these dietary constituents. Whereas accidental toxic exposures of nitrates and nitrites have occurred [11, 12], the health risks due to excessive nitrate and nitrite consumption, such as methemoglobinemia, appear only in specific subgroups of the population.

Experts have questioned the veracity of the evidence supporting the hypothesis that nitrates and nitrites are toxic for healthy adolescent and adult populations due to their ability to cause methemoglobinemia [13–16]. Many scientists now interpret the available data as evidence that the condition is caused by bacterial infection-induced enteritis rather than nitrate [15, 16]. Thus, it appears that the biologically plausible hypothesis of nitrite toxicity with regard to methemoglobinemia has essentially transformed a plausible hypothesis into sacrosanct dogma [13], despite the lack of proof [14, 16].

The excessive concentration of nitrate in drinking water, typically a minor contributor to dietary nitrate and nitrite concentrations, must be considered a serious health concern, particularly for infants [17]. Even so, the Society of Agricultural and Biological Engineers has, in parallel to those made this volume for nitrate and nitrite exposures from foods, called for a more rational approach to setting exposure limits on nitrogen-containing effluents in wastewater treatment [18]. To wit,

> Considering that definitive evidence of nitrate health risks is conspicuously lacking, a more rational approach to setting effluent limits for waste treatment systems is needed, one that considers costs/benefits and recognizes factors that act to limit nitrogen buildup in groundwater. Such factors include nitrogen removal by soil microorganisms, and aquifer hydrogeology [18].

Regulators must integrate data from potential exposures from drinking water with those of food sources of nitrate and nitrite. The balance of this chapter will address the potential health concerns regarding dietary nitrate and nitrite from the context of food sources and the physiological contexts of the actions of their metabolic products.

Regulatory Limits on Dietary Nitrate and Nitrite Intakes

The US Environmental Protection Agency limits human exposure to inorganic nitrates to 0.10 mg/L (or 10 ppm nitrate nitrogen) and nitrites to 1 ppm nitrite nitrogen [19]. The Joint Food and Agricultural Organization/World Health Organization has set the acceptable daily intake (ADI) for the nitrate ion at 3.7 mg/kg body wt and for the nitrite ion at 0.06 mg/kg body wt [1]. Likewise, Environmental Protection Agency has set a Reference Dose for nitrate of 1.6 mg nitrate nitrogen/kg body wt/day (equivalent to 7.0 mg nitrate ion/kg body wt/day).

Dietary Intake Estimates of Nitrate and Nitrite

The mean intake estimates for nitrate and nitrite from food in the United States and Europe vary from ~40–100 to 31–185 mg/day, respectively [20, 21]. Nitrite intakes vary from 0 to 20 mg/day. Nitrate intakes from sources other than vegetables, including drinking water and cured meats, have been estimated to average 35–44 mg/person/day for a 60-kg human [1]. On the basis of a conservative recommendation to consume 400 g of different fruits and vegetables per day at median nitrate concentrations, the dietary concentration of nitrate would be ~157 mg/day [1]. In the European Union, fruit consumption (average nitrate concentration: 10 mg/kg FW) constitutes more than half of the recommended intake of 400 g, nitrate intakes are estimated to be ~81–106 mg/day before additional nitrate losses from washing, peeling, and/or cooking are taken into consideration. In a convenience sample of foods analyzed for nitrate and nitrite content, two dietary patterns based upon the Dietary Approaches to Stop Hypertension that featured either low- or high-nitrate foods were found to contain between 174 mg nitrate/0.41 mg nitrite and 1,222 mg nitrate/0.35 mg nitrite [22]. It is noteworthy that the bioavailability of dietary nitrate is 100% [23].

Dietary Intakes in the Context of WHO ADI Levels

The appreciation of four facts regarding human exposures to nitrate and nitrite casts concern over current regulatory limits on nitrate and nitrite consumption. First, it is possible to approach or exceed WHO ADI limits with usual intake levels of single foods, such as colostrum (at 100 mL intake in a newborn infant delivering 42% of the WHO ADI intake limit), soya milk (750 mL intake for a hypothetical 6.8 kg infant yields 104% of the WHO ADI intake limit) [24], spinach [25], or a dessicated vegetable supplement [22]. Second, recommended dietary intakes of vegetables and fruits, such as a Dietary Approaches to Stop Hypertension pattern with high nitrate food choices, exceed the World Health Organization's Acceptable Daily Intake for nitrate by 550% for a 60-kg adult [22]. Third, for adults consuming the recommended intakes of vegetables and fruits, the origin of over 80% of dietary nitrate and nitrite, the concentration of nitrate in saliva, via enterosalivary circulation, can reach up to three times the concentration in most global regulatory limits for drinking water [26]. Fourth, provision of dietary nitrate, as beetroot juice [27], dietary nitrate [28], or in a traditional Japanese dietary pattern [29], is effective in lowering blood pressure in humans. These facts indicate that WHO intake limits may not reflect optimal nitrate and nitrite concentrations from foods that confer health benefits. If nitrates and nitrites act as nutrients, it is likely that they do so to bolster the reserve of nitrite-derived NO_x metabolites required for optimal functioning through periods of physiologic stress (e.g., hypoxia and acidosis) and diseases characterized by endothelial dysfunction [30–33].

Potential Negative Health Effects of Dietary Nitrate and Nitrite Exposures

Epidemiologic and clinical studies have associated nitrate and nitrite consumption with increased risk of gastrointestinal cancers, thyroid dysfunction and thyroid cancer [34], chronic obstructive pulmonary disease in women [35] as well as other conditions [36]. Nitrate and nitrite exposures have been associated with gastrointestinal cancer risk through the consumption of cured and processed meats [37, 38]. A recent meta-analysis has challenged this conclusion; these results conclude that available epidemiologic evidence is not sufficient to support a clear and unequivocal independent positive association between processed meat consumption and colorectal cancer risk [39]. Nitrates added to meats serve as antioxidants, develop flavor, and stabilize the red color in meats but must be converted to nitrite to exert these actions. Sodium nitrite is used as a colorant, flavor enhancer, and antimicrobial agent in cured and processed meats. Nitrate and nitrite use in meat products, including bacon, bologna, corned beef, hot dogs, luncheon meats, sausages, and canned and cured meat and hams is subject to limits put forth in Food and Drug Administration (FDA) and US Department of Agriculture (USDA) regulations. These regulations can be found in the Code of Federal Regulations (CFR) (21CFR 170.60, 172.170, and 172.175 for FDA and 9CFR 318.7 for USDA regulations, respectively). The use of nitrites in bacon must be accompanied by the use of either sodium erythorbate or sodium ascorbate (vitamin C), antioxidants that inhibit the nitrosation effect of nitrites on secondary amines [40]. The use of these antioxidants, along with lower nitrate and nitrite levels in processed meats, has lowered residual nitrite levels in cured meat products in the US by ~80% since the mid-1970s [41].

While there exists little evidence that consumption of nitrates or nitrites are associated with risk of gastrointestinal cancers [16], the chemistry of gastric digestion of meals containing nitrite, dietary fat and vitamin C may promote the formation of carcinogenic *N*-nitrosamines (NOCs) [42, 43]. The generation of nitrite-derived NO is maximal at the gastroesophageal junction where salivary nitrite first meets gastric acid. This chemistry is modeled using a physiologically relevant two phase (aqueous and lipid) in vitro model system. Acid-catalyzed nitrosation was found to be enhanced by vitamin C in the presence of lipid. Thus, the presence of lipid, in the microenvironment of the stomach, converts vitamin C from inhibiting to promoting acid nitrosation. In the presence of 10% lipid (a food matrix component for processed meats), the presence of vitamin C increased the formation of nitrosodimethylamine, nitrosodiethylamine, and *N*-nitrosopiperidine 8-, 60-, and 140-fold, respectively [42]. This effect is attributable to the ability of vitamin C to assist in the generation of NO in the aqueous phase, which enables the regeneration of nitrosating species by reacting with oxygen in the lipid phase. This concept has been supported by observations in a carcinogen-induced rat model of colorectal cancer in which processed meat consumption enhanced the number of preneoplastic lesions in the colon [44]. Taken together, these observations provide a biologically plausible mechanism for the observed association between processed meat consumption and gastrointestinal cancer risk [45]. It has been postulated that gastric formation of NOCs may be inhibited by nutrients and other components of vegetables and fruit [46]. In this two phase model system, it has been shown that certain polyphenols block the formation of NOCs [47]. Clearly, more research is needed to address the potential interactions among food constituents to affect cancer risk.

Potential Health Benefits of Dietary Nitrate and Nitrite

Based on data from the laboratories of Drs. Jon Lundberg, Mark Gladwin, Jay Zweier, Nathan Bryan and others have formed the modern basis for proposed health benefits of dietary nitrates and nitrites [48–50]. The nitrate–nitrite–NO pathway has been demonstrated to serve as a backup system to ensure NO supply in situations when the endogenous L-arginine/NO synthase pathway is dysfunctional [51]. This redundant system of NO production in tissues has important implications for cardiovascular, gastrointestinal and immune function related to the provision of dietary nitrate and nitrite. As nitrite-dependent NO generation has been shown to play critical physiological and pathological roles, and is controlled by oxygen tension, pH, reducing substrates and nitrite levels [52], it is necessary to balance these contexts in a modern regulatory framework that acknowledges a physiological requirement for nitrate and nitrite supplied by dietary means.

Addressing Context-Specific Health Effects of Food Sources of Nitrate and Nitrite

Two recent advisories have been issued regarding the potential health effects of food sources of dietary nitrates and nitrites. The International Agency for Research on Cancer (IARC) considered the carcinogenic risk associated with nitrate and nitrite consumption [53]. This thorough review concluded that the carcinogenicity from drinking water and food sources of nitrates and nitrites were not strongly supported by the epidemiologic evidence. This group considered a combination of positive and negative results from epidemiological and animal studies that supported a positive, coherent association with gastrointestinal cancer risk in conditions that favor endogenous formation of *N*-nitroso compounds. They reported the strongest associations were recorded in individuals with high nitrite and low vitamin C intake, a combination that promotes these nitrosation reactions. The presence of factors in foods, such as flavonoids in vegetables, may account for the lack of association with cancer risk for nitrate in food. The lack of association for nitrate in drinking water was attributed to low exposure in the available studies. They did not rule out the potential for nitrate in drinking water to result in endogenous nitrosation because water can be consumed if it was consumed in the absence of nitrosation inhibitors. Overall, the IARC group concluded that "ingested nitrate or nitrite under conditions that result in endogenous nitrosation is probably carcinogenic to humans (group 2A)" [53].

The European Food Safety Authority (EFSA) considered the potential health effects of dietary nitrate and nitrite consumption from vegetables because these foods are recommended for health benefits and are the primary dietary source of nitrate [1]. This group concluded that diets of a small proportion of certain EU Member States that consume only high amounts of leafy vegetables, the ADI consumption limits for nitrate could be exceeded. Like other expert groups, EFSA concluded that epidemiological studies do not suggest that nitrate intake from diet or drinking water is associated with increased cancer risk. Furthermore, this group considers the evidence that high intake of nitrite might be associated with increased cancer risk to be equivocal.

The EFSA panel concluded their analysis of the risk and benefits of exposure to nitrate from vegetables by stating:

> Overall, the estimated exposures to nitrate from vegetables are unlikely to result in appreciable health risks, therefore the recognised beneficial effects of consumption of vegetables prevail. The Panel recognised that there are occasional circumstances e.g. unfavourable local/home production conditions for vegetables which constitute a large part of the diet, or individuals with a diet high in vegetables such as rucola which need to be assessed on a case by case basis [1].

The fact that the IARC and EFSA recommendations were the result of a review process that considered similar bodies of evidence is noteworthy. More striking is the fact that, in spite of generally similar conclusions, the scope or purview of each group was limited. The IARC group did not consider other potential health effects of nitrate and nitrite and the EFSA group considered only those associated with nitrates in vegetables. While these excellent reports serve their intended functions, they highlight the need for a broader examination of the potential health benefits and risks of dietary nitrates and nitrites.

Determinants of Regulatory Paradigm Change

Beyond the ascription of risks and benefits of dietary nitrates and nitrites, a recent review concluded that these dietary constituents, based on their demonstrated physiological functions, be considered nutrients [22]. Analogous to *all* essential or indispensable nutrients, intake of excess nitrate and nitrite exposure is, in specific contexts, associated with an increased risk of negative health outcomes. A set of dietary reference intake (DRI) categories are set by the Food and Nutrition Board of the National Academy of Sciences for essential nutrients to clearly define, where possible, the contexts in which intakes are deficient, safe, or potentially excessive. These DRI categories include the recommended dietary allowance (RDA), adequate intake (AI), tolerable upper level intake (TUL), and estimated average requirement (EAR) [54]. The process of setting DRIs for nutrients considers a broad range of physiologic factors, including nutritional status and potential toxicities. Rational methodologies such as these, including the consideration of normal dietary consumption patterns of nitrate and nitrite-containing foods, have not been applied in setting exposure limits or in considering the potential health benefits of dietary nitrates and nitrites. Consideration of dietary nitrates and nitrites as nutrients, and the determination of physiological concentrations considered low, sufficient or excessive will require a consensus among researchers, health professionals and regulators.

Time to Relax Regulatory Limits on Dietary Nitrate and Nitrite?

While there are compelling indications that dietary nitrate or nitrite may reduce cardiovascular disease risk, it has been suggested that the data are insufficient to relax standards for nitrate in drinking water and foods [55]. Of course, the lack of awareness of the association between nitrate and nitrite and blood pressure means that the necessary tools employed by clinical researchers, epidemiologists and trialists to assess the association between nitrate/nitrite concentrations in foods and specific health biomarkers are not widely available. This means that in clinical trials, such as those completed to assess the efficacy of the DASH diet, nitrate and nitrite concentrations in foods were, because they were not considered causally related to blood pressure, not measured. Because these unmeasured factors likely contributed to the hypotensive effects of the DASH dietary pattern [22], dietary nitrate and nitrite would be considered confounding factors. This characterization is apropos for studies attributing cardiovascular benefits to vegetable intake [56], plant-based diets [57] and Mediterranean diet interventions for the secondary prevention of cardiovascular disease [58]. Dietary nitrates and nitrites may, in part, contribute to the salutary effects of vegetable and fruit consumption. It is hoped that dietary concentrations of these effect modifiers are measured or estimated and reported in epidemiologic and clinical studies of cardiovascular risk.

Lack of inclusion of food nitrate and nitrite concentrations in standard food databases (e.g., USDA National Nutrient Database for Standard Reference) is another obstacle to the development of a solid epidemiologic basis for quantifying cardiovascular and other health benefits of dietary nitrates and nitrites in human populations. As such, the development and availability of a database of food nitrate/nitrite concentrations would encourage more thorough investigations of hypotheses associating dietary nitrate/nitrite and specific health outcomes.

The compelling results of clinical studies demonstrate great potential for the treatment and prevention of cardiovascular diseases including ischemia-reperfusion (IR) injury and hypertension by dietary means. It is incumbent upon regulators to carry out a comprehensive, systematic and independent review of all available evidence of health effects of dietary nitrate and nitrite. An Institute of Medicine (IOM)-type process would support the scholarly nature of this effort in an independent institution. This effort would be synergistic with the recent IOM report "A Population-Based Policy and Systems Change Approach to Prevent and Control Hypertension" [59]. Among other public health strategies, this report highlights population-based strategies to reduce the prevalence of hypertension. These strategies include behavioral and lifestyle interventions such as reducing sodium intake, increasing consumption of fruits and vegetables, and increasing physical activity. Many scientists understand that consideration of diet-based therapies emphasizing nitrate and nitrite intakes could benefit individuals with hypertension.

After decades of being subject to regulatory limits on dietary nitrate and nitrite based upon the poor practice of causal inference, the public deserves cohesive regulations that reflect the physiologic necessity of nitrate and nitrite while accounting for contexts in which these dietary substances may produce health risks. The necessary work of bringing together experts from disparate scientific disciplines to craft meaningful dietary recommendations for nitrate and nitrite intakes will be a boon for public health.

References

1. European Food Safety Authority. Nitrate in vegetables: scientific opinion of the panel on contaminants in the food chain. EFSA J. 2008;689:1–79.
2. Pennington J. Dietary exposure models for nitrates and nitrites. Food Control. 1998;9:385–95.
3. Santamaria P. Nitrate in vegetables; content, toxicity and EC regulation. J Sci Food Agric. 2005;86(1):10–7.
4. Spiegelhalder B, Eisenbrand G, Preussmann R. Influence of dietary nitrate on nitrite content of human saliva: possible relevance to in vivo formation of N-nitroso compounds. Food Cosmet Toxicol. 1976;14(6):545–8.
5. Petersson J, Carlstrom M, Schreiber O, Phillipson M, Christoffersson G, Jagare A, et al. Gastroprotective and blood pressure lowering effects of dietary nitrate are abolished by an antiseptic mouthwash. Free Radic Biol Med. 2009;46(8):1068–75.
6. Jansson EA, Huang L, Malkey R, Govoni M, Nihlen C, Olsson A, et al. A mammalian functional nitrate reductase that regulates nitrite and nitric oxide homeostasis. Nat Chem Biol. 2008;4(7):411–7.
7. van Faassen EE, Bahrami S, Feelisch M, Hogg N, Kelm M, Kim-Shapiro DB, et al. Nitrite as regulator of hypoxic signaling in mammalian physiology. Med Res Rev. 2009;29(5):683–741.
8. Camargo JA, Alonso A. Ecological and toxicological effects of inorganic nitrogen pollution in aquatic ecosystems: a global assessment. Environ Int. 2006;32(6):831–49.
9. Monteny GJ. The EU Nitrates Directive: a European approach to combat water pollution from agriculture. ScientificWorldJournal. 2001;1 Suppl 2:927–35.
10. World Health Organization. Recommendations; nitrate and nitrite. In: Guidelines for drinking water quality. 3rd ed. Geneva, Switzerland: WHO; 2004. p. 417–20.
11. Center for Disease Control and Prevention (CDC). Methemoglobinemia following unintentional ingestion of sodium nitrite–New York, 2002. MMWR Morb Mortal Wkly Rep. 2002;51(29):639–42.
12. Butler AR, Feelisch M. Therapeutic uses of inorganic nitrite and nitrate: from the past to the future. Circulation. 2008;117(16):2151–9.
13. McKnight GM, Duncan CW, Leifert C, Golden MH. Dietary nitrate in man: friend or foe? Br J Nutr. 1999;81(5):349–58.
14. L'Hirondel JL, Avery AA, Addiscott T. Dietary nitrate: where is the risk? Environ Health Perspect. 2006;114(8):A458–9; author reply A459–61.
15. L'hirondel J, L'hirondel JL. Nitrate and man: toxic, harmless or beneficial? Wallingford, UK: CABI; 2001.
16. Powlson DS, Addiscott TM, Benjamin N, Cassman KG, de Kok TM, vanGrinsven H, et al. When does nitrate become a risk for humans? J Environ Qual. 2008;37(2):291–5.
17. Greer FR, Shannon M. Infant methemoglobinemia: the role of dietary nitrate in food and water. Pediatrics. 2005;116(3):784–6.
18. Churchill JJ. Re-examining "nitrate toxicity": a call for a more rational approach to effluent limits for nitrogen in decentralized wastewater treatment. In: Paper presented at Eleventh Individual and Small Community Sewage Systems Conference Proceedings, Warwick, RI; 2007.
19. National Primary Drinking Water Regulations: Final Rule, 40. Federal Register. 1991; CFR Parts 141–143(56 (20)):3526–97.
20. Mensinga TT, Speijers GJ, Meulenbelt J. Health implications of exposure to environmental nitrogenous compounds. Toxicol Rev. 2003;22(1):41–51.
21. Gangolli SD, van den Brandt PA, Feron VJ, Janzowsky C, Koeman JH, et al. Nitrate, nitrite and N-nitroso compounds. Eur J Pharmacol. 1994;292(1):1–38.
22. Hord NG, Tang Y, Bryan NS. Food sources of nitrates and nitrites: the physiologic context for potential health benefits. Am J Clin Nutr. 2009;90(1):1–10.
23. van Velzen AG, Sips AJ, Schothorst RC, Lambers AC, Meulenbelt J. The oral bioavailability of nitrate from nitrate-rich vegetables in humans. Toxicol Lett. 2008;181(3):177–81.

24. Hord NG, Ghannam J, Garg H, Tang YP, Bryan NS. Nitrate and nitrite content of human, formula, bovine and soy milks: implications for dietary nitrite and nitrate recommendations Breastfeed Med. 2010 Oct 19. [Epub ahead of print].
25. Lundberg JO, Feelisch M, Bjorne H, Jansson EA, Weitzberg E. Cardioprotective effects of vegetables: is nitrate the answer? Nitric Oxide. 2006;15(4):359–62.
26. Li H, Thompson I, Carter P, Whitely A, Bailey M, Leifert C, et al. Salivary nitrate–an ecological factor in reducing oral acidity. Oral Microbiol Immunol. 2007;22(1):67–71.
27. Webb AJ, Patel N, Loukogeorgakis S, Okorie M, Aboud Z, Misra S, et al. Acute blood pressure lowering, vasoprotective, and antiplatelet properties of dietary nitrate via bioconversion to nitrite. Hypertension. 2008;51(3):784–90.
28. Kapil V, Milsom AB, Okorie M, Miliki-Toyserkani S, Akram F, Rehman F, et al. Inorganic nitrate supplementation lowers blood pressure in humans: role for nitrite-derived NO. Hypertension. 2010;56(2):274–81.
29. Sobko T, Marcus C, Govoni M, Kamiya S. Dietary nitrate in Japanese traditional foods lowers diastolic blood pressure in healthy volunteers. Nitric Oxide. 2010;22(2):136–40.
30. Stokes KY, Dugas TR, Tang Y, Garg H, Guidry E, Bryan NS. Dietary nitrite prevents hypercholesterolemic microvascular inflammation and reverses endothelial dysfunction. Am J Physiol Heart Circ Physiol. 2009;296(5):H1281–8.
31. Garg HK, Bryan NS. Dietary sources of nitrite as a modulator of ischemia/reperfusion injury. Kidney Int. 2009;75(11):1140–4.
32. Elrod JW, Calvert JW, Gundewar S, Bryan NS, Lefer DJ. Nitric oxide promotes distant organ protection: evidence for an endocrine role of nitric oxide. Proc Natl Acad Sci USA. 2008;105(32):11430–5.
33. Lundberg JO, Gladwin MT, Ahluwalia A, Benjamin N, Bryan NS, Butler A, et al. Nitrate and nitrite in biology, nutrition and therapeutics. Nat Chem Biol. 2009;5(12):865–9.
34. Ward MH, Kilfoy BA, Weyer PJ, Anderson KE, Folsom AR, Cerhan JR. Nitrate intake and the risk of thyroid cancer and thyroid disease. Epidemiology. 2010;21(3):389–95.
35. Jiang R, Camargo Jr CA, Varraso R, Paik DC, Willett WC, Barr RG. Consumption of cured meats and prospective risk of chronic obstructive pulmonary disease in women. Am J Clin Nutr. 2008;87(4):1002–8.
36. Panesar NS. Downsides to the nitrate-nitrite-nitric oxide pathway in physiology and therapeutics? Nat Rev Drug Discov. 2008;7(8):710; author reply 710.
37. Santarelli RL, Pierre F, Corpet DE. Processed meat and colorectal cancer: a review of epidemiologic and experimental evidence. Nutr Cancer. 2008;60(2):131–44.
38. Norat T, Bingham S, Ferrari P, Slimani N, Jenab M, Mazuir M, et al. Meat, fish, and colorectal cancer risk: the European Prospective Investigation into cancer and nutrition. J Natl Cancer Inst. 2005;97(12):906–16.
39. Alexander DD, Miller AJ, Cushing CA, Lowe KA. Processed meat and colorectal cancer: a quantitative review of prospective epidemiologic studies. Eur J Cancer Prev. 2010;19(5):328–41.
40. Rao GS, Osborn JC, Adatia MR. Drug-nitrite interactions in human saliva: effects of food constituents on carcinogenic N-nitrosamine formation. J Dent Res. 1982;61(6):768–71.
41. Cassens RG. Residual nitrite in cured meat. Food Technol. 1997;51:53–5.
42. Combet E, Paterson S, Iijima K, Winter J, Mullen W, Crozier A, et al. Fat transforms ascorbic acid from inhibiting to promoting acid-catalysed N-nitrosation. Gut. 2007;56(12):1678–84.
43. Combet E, Preston T, McColl KE. Development of an in vitro system combining aqueous and lipid phases as a tool to understand gastric nitrosation. Rapid Commun Mass Spectrom. 2010;24(5):529–34.
44. Santarelli RL, Vendeuvre JL, Naud N, Tache S, Gueraud F, Viau M, et al. Meat processing and colon carcinogenesis: cooked, nitrite-treated, and oxidized high-heme cured meat promotes mucin-depleted foci in rats. Cancer Prev Res (Phila). 2010;3(7):852–64.
45. World Cancer Research Fund, American Institute of Cancer Research. Food, nutrition, physical activity, and the prevention of cancer: a global perspective. Second Expert Report; 2007.
46. de Kok TM, Engels LG, Moonen EJ, Kleinjans JC. Inflammatory bowel disease stimulates formation of carcinogenic N-nitroso compounds. Gut. 2005;54(5):731.

47. Combet E, El Mesmari A, Preston T, Crozier A, McColl KE. Dietary phenolic acids and ascorbic acid: influence on acid-catalyzed nitrosative chemistry in the presence and absence of lipids. Free Radic Biol Med. 2010;48(6):763–71.
48. Lundberg JO, Govoni M. Inorganic nitrate is a possible source for systemic generation of nitric oxide. Free Radic Biol Med. 2004;37(3):395–400.
49. Dejam A, Hunter CJ, Schechter AN, Gladwin MT. Emerging role of nitrite in human biology. Blood Cells Mol Dis. 2004;32(3):423–9.
50. Bryan NS, Fernandez BO, Bauer SM, Garcia-Saura MF, Milsom AB, Rassaf T, et al. Nitrite is a signaling molecule and regulator of gene expression in mammalian tissues. Nat Chem Biol. 2005;1(5):290–7.
51. Lundberg JO, Weitzberg E. NO generation from inorganic nitrate and nitrite: role in physiology, nutrition and therapeutics. Arch Pharm Res. 2009;32(8):1119–26.
52. Zweier JL, Li H, Samouilov A, Liu X. Mechanisms of nitrite reduction to nitric oxide in the heart and vessel wall. Nitric Oxide. 2010;22(2):83–90.
53. Grosse Y, Baan R, Straif K, Secretan B, El Ghissassi F, Cogliano V. Carcinogenicity of nitrate, nitrite, and cyanobacterial peptide toxins. Lancet Oncol. 2006;7(8):628–9.
54. Otten JJ, Hellwig J, Meyers, LD, editors. Dietary reference intakes: the essential guide to nutrient requirements. Washington, DC: National Academy Press, Food and Nutrition Board, Institute of Medicine, National Academy of Sciences; 2006.
55. Katan MB. Nitrate in foods: harmful or healthy? Am J Clin Nutr. 2009;90(1):11–2.
56. Joshipura KJ, Hung HC, Li TY, Hu FB, Rimm EB, Stampfer MJ, et al. Intakes of fruits, vegetables and carbohydrate and the risk of CVD. Public Health Nutr. 2009;12(1):115–21.
57. Jenkins DJ, Wong JM, Kendall CW, Esfahani A, Ng VW, Leong TC, et al. The effect of a plant-based low-carbohydrate ("Eco-Atkins") diet on body weight and blood lipid concentrations in hyperlipidemic subjects. Arch Intern Med. 2009;169(11):1046–54.
58. de Lorgeril M, Salen P, Martin JL, Monjaud I, Delaye J, Mamelle N. Mediterranean diet, traditional risk factors, and the rate of cardiovascular complications after myocardial infarction: final report of the Lyon Diet Heart Study. Circulations. 1999;99(6):779–85.
59. Institute of Medicine. A population-based policy and systems change approach to prevent and control hypertension. Washington, D.C.: The National Academy of Sciences; 2010.

Part III
Nitrite and Nitrate in Therapeutics and Disease

Chapter 11
Nitric Oxide Signaling in Health and Disease

Nathan S. Bryan and Jack R. Lancaster Jr.

Key points

- Nitric oxide is a key signaling molecule in the cardiovascular, immune, and nervous system.
- There are three enzymes that produce NO from L-arginine, two are constitutive and produce low amounts of NO for short periods of time, the other inducible and can produce high levels of NO for prolonged periods.
- Nitric oxide reacts primarily with transition metals and other free radicals.
- A number of diseases are associated with dysregulation of NO production.
- The nitric oxide pathway is a key target for development of new therapies.

Keywords EDRF • Nitric oxide synthase • L-arginine • Nitrosothiols • Inflammation • Free radical

Introduction

The discovery in the 1980s of the mammalian biosynthesis of nitric oxide ($^{\bullet}NO$, nitrogen monoxide) and its roles in the immune [1–5], cardiovascular [6–9], and nervous [10] systems established a startling new paradigm in the history of cellular signaling mechanisms. Prior to this discovery, it was essentially inconceivable that cells would intentionally produce a toxic molecule as a messenger; $^{\bullet}NO$ is a common air pollutant, a constituent of cigarette smoke, and a toxic gas, which appears in the exhaust of motor cars and jet airplanes, causes acid rain, and destroys the ozone layer.

N.S. Bryan (✉)
Brown Foundation Institute of Molecular Medicine,
The University of Texas, Health Science Center at Houston,
1825 Pressler Street 530C, Houston, TX 77030, USA
e-mail: Nathan.Bryan@uth.tmc.edu

N.S. Bryan and J. Loscalzo (eds.), *Nitrite and Nitrate in Human Health and Disease*,
Nutrition and Health, DOI 10.1007/978-1-60761-616-0_11,

Amazingly, despite this nasty reputation, it is now known that ˙NO is one of a family of reactive signaling molecules, which includes both reactive nitrogen and reactive oxygen species [11, 12]. This is, in fact, a hallmark example of the propensity of nature to seek out and exquisitely utilize the unique properties of unusual molecules. Its best recognized role, for which the 1998 Nobel Prize in Physiology or Medicine was awarded to Dr. Robert Furchgott, Dr. Louis Ignarro, and Dr. Ferid Murad, was for the importance of ˙NO as a signal molecule in the vasculature and specifically in the control of blood pressure. In addition to this role, ˙NO is one of the most important signaling molecules in the body, and is involved in virtually every organ system where it is responsible for modulating an astonishing variety of effects. Indeed, there have been over 100,000 publications on ˙NO, more than half of which have appeared in the last 8 years. ˙NO has been shown to be involved in and affect (just to list a few major examples) neurotransmission, memory, stroke, glaucoma and neural degeneration, pulmonary hypertension, penile erection, angiogenesis, wound healing, atherogenesis, inflammation such as arthritis, nephritis, colitis, autoimmune diseases (diabetes, inflammatory bowel disease), invading pathogens, tumors, asthma, tissue transplantation, septic shock, platelet aggregation and blood coagulation, sickle cell disease, gastrointestinal motility, hormone secretion, gene regulation, hemoglobin delivery of oxygen, stem cell proliferation and differentiation, and bronchodilation.

Here, we will focus on the physiology/pathophysiology of the three major organ systems originally documented for the importance of ˙NO as a signal, the cardiovascular, immune, and nervous systems, and understandings relevant to nitrite/nitrate as particularly related to the cardiovascular system. Other important pathologies where ˙NO also plays critical roles, including cancer [13], sepsis [14], diabetes [15], and lung [16] and kidney injury [17], are covered elsewhere.

Enzymatic ˙NO Synthesis

The first ˙NO synthase (NOS) to be isolated and the gene sequenced was the enzyme from rat cerebellum, by Bredt and Snyder in 1991 [18, 19]. Since then, the genes for the two other major NOS isoforms have been identified and cloned, revealing that all three enzymes perform the same basic enzymatic reaction. These heme- and flavin-containing enzymes utilize electrons from NADPH and produce ˙NO by the mixed-function oxidation of one of the two equivalent guanidine nitrogen atoms of the amino acid L-arginine (Fig. 11.1) [20, 21]. The three major isoforms are most commonly referred to as the neuronal ("nNOS" or NOS1), inducible ("iNOS" or NOS2), and endothelial ("eNOS" or NOS3), classified according to their initially identified location and function. nNOS and eNOS enzymes are constitutive, cytosolic, and Ca^{2+}/calmodulin-dependent, and release ˙NO for short time periods in response to receptor or physical stimulation. The ˙NO released by these enzymes acts as a signaling mechanism underlying several physiological responses depending on when and where it is produced and producing

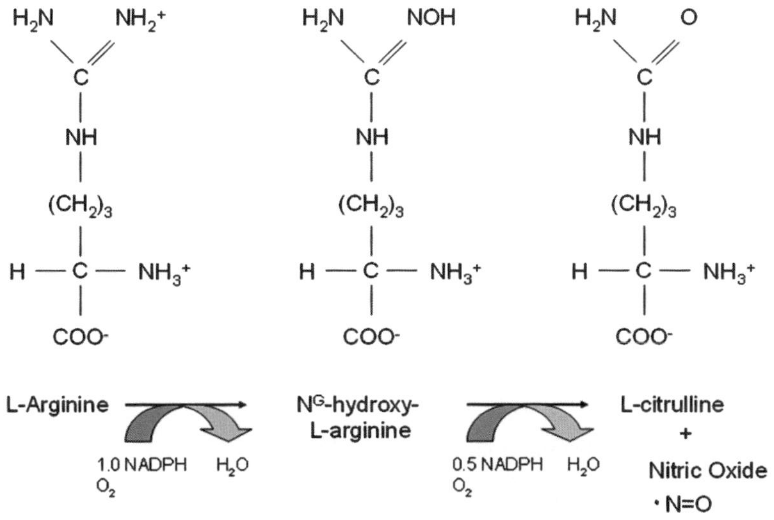

Fig. 11.1 Mechanism of •NO synthesis from L-arginine by NOS

relatively small amounts of •NO. The other enzyme type (iNOS) is induced after activation of macrophages, endothelial cells, and a number of other cells by endotoxin and pro-inflammatory cytokines, and once expressed, synthesizes NO for long periods of time and at higher levels than the constitutive enzymes. Furthermore, the inducible form is Ca^{2+} independent since calmodulin is already bound to the enzyme. It is now appreciated that eNOS is found in other cells and tissues beside the endothelium, nNOS is found in other cells than neurons, iNOS is found constitutively in some tissues, and there are inducible forms of both eNOS and nNOS, adding confusion to the nomenclature as it was first described. In addition, the activity and subcellular location of all three enzymes is regulated by a variety of different posttranslational modifications, especially eNOS [22]. Finally, there is evidence of NOS associated with erythrocytes [23] and also mitochondria [24, 25], although these possibilities are less well characterized.

Interconversions Between •NO and Nitrite/Nitrate

Formation of Nitrite/Nitrate from •NO

The major pathway for •NO metabolism is the stepwise oxidation to nitrite and nitrate [26]. Quantitatively the major mechanism for •NO metabolism in man in vivo is via reaction with oxyhemoglobin [27]. In plasma or other physiological fluids or buffers, •NO is oxidized almost completely to nitrite, where it remains stable for several hours. •NO and nitrite are rapidly oxidized to nitrate in whole blood.

The half life of nitrite (NO_2^-) in human blood is about 110 s. Nitrate (NO_3^-) on the other hand has a circulating half life of 5–8 h. For years, both nitrite and nitrate have been used as surrogate markers of •NO production in biological tissues. During fasting conditions with low intake of nitrite/nitrate, enzymatic •NO formation from NOS accounts for the majority of nitrite [28], and plasma nitrite reflects eNOS activity [29]. On the basis of these and other studies, it was believed that •NO is acutely terminated by oxidation to nitrite and nitrate; however, as described below it is now appreciated that nitrite or nitrate can recapitulate •NO physiology.

Formation of •NO from Nitrite/Nitrate

There are several reports of the biological production of •NO from nitrite which predate the discovery in mammals. The first identification of enzymatic synthesis of •NO was in the 1970s where it was shown that it is produced in plants by the enzymatic reduction of NO_2^- as an intermediate in nitrogen fixation [30, 31], although the resulting heme-nitrosyl most probably does not release free •NO. In 1983, formation of •NO from nitrite was proposed as the mechanism whereby nitrite prevents botulism [32]. •NO is a central intermediate in microbial denitrification [33].

It is now recognized that there are non-NOS sources of •NO [34]. The first description of such •NO formation was in 1994 when two independent groups demonstrated •NO detection from acidified nitrite in the stomach [35, 36], which is predictable based on well-established chemistry of acidified nitrite (but which may not apply to other, less acidic bodily compartments [37]). Nitrate per se is not metabolized by mammals, however, commensal facultative bacteria in the oral cavity efficiently convert nitrate to nitrite and is a significant source of circulating nitrite. As described in more detail below, there are multiple mechanisms for the reductive conversion of nitrite to •NO that may be therapeutically effective.

Biochemical Actions and Targets of •NO

Reactivity and Diffusivity of •NO

Although commonly characterized as a "highly reactive radical," the biochemistry of •NO is surprisingly simple since it is determined by one major principle: •NO undergoes chemistry that serves to stabilize its unpaired electron. There are major consequences of this chemistry. (1) •NO interacts almost exclusively with only two species, molecules that also contain an unpaired electron (thus allowing the electrons to pair up); and (2) transition metals to which •NO binds, thereby stabilizing the unpaired electron owing to special interactions with the electrons on metals in their *d* orbitals.

As a result of these two phenomena, in general •NO essentially does not react with molecules with covalent bonds (all electrons paired) that do not contain metal ions.

Being an extremely small, uncharged molecule with limited reactivity, •NO is highly diffusible in the biological milieu; at first glance, these properties would seem to violate the concept of specificity required for a biological signal and was initially a startling (and controversial) concept. Indeed, modeling studies predicted that with sustained •NO production (such as from iNOS), a gradient of •NO will extend great distances from the source, depending on the half-life of •NO [38, 39]. However, the actions of •NO in some systems are quite spatially confined, and, in fact, are highly localized to the immediate environment surrounding NOS [40]. The explanation(s) for this confinement of a highly diffusible molecule are unclear, but, could be explained by rapid reactivity at the source (e.g., with superoxide [41], see below) and also confinement due to pulsatile, as opposed to continuous, •NO production [42].

Intracellular •NO Targets: Transition Metals

The interaction between •NO and its primary protein target, the enzyme, soluble guanylyl cyclase (sGC), which mediates target cell responses such as vascular smooth muscle relaxation and platelet inhibition, has been well characterized although controversies remain [43, 44]. After entering the target cell, •NO binds to the heme moiety of sGC and activates the enzyme by inducing a conformational change that displaces iron out of the plane of the porphyrin ring [45]. sGC then catalyzes the production of cyclic GMP from GTP to elevate cyclic GMP. Cyclic GMP then triggers a cascade of intracellular events that culminate in a reduction in calcium-dependent vascular smooth muscle tone by inactivating myosin light chain kinase or MLCK [46]. MLCK normally phosphorylates the regulatory set of myosin light chains. This phosphorylation event activates cross-bridge cycling and initiates contraction. cGMP modulates MLCK activity by activating a cGMP-dependent protein kinase that phosphorylates MLCK. Phosphorylation of MLCK diminishes its affinity for calmodulin and, as a consequence, decreases the phosphorylation of myosin light chain, which, in turn, stabilizes the inactive form of myosin. In this manner, cGMP may induce vasorelaxation by indirectly decreasing myosin light chain-dependent myosin activation.

In addition to the heme iron of sGC, there are three other major intracellular iron targets of •NO. As mentioned previously, the reaction of •NO with intraerythrocytic hemoglobin (either oxygenated or deoxygenated) is quantitatively the most important in vivo reaction of •NO [27], the major product of which is nitrate. A very minor reaction comparatively is the formation of S-NO-hemoglobin, which is S-nitrosated at a specific cysteine residue (cysβ93), and which has been proposed as a major mechanism of blood flow and tissue oxygenation modulation in response to hypoxia [47, 48]. This controversial proposal is discussed in more detail below.

Another target of ${}^{\bullet}$NO is the mitochondrion, which dates back to 1979 [49]. At relatively low levels, relevant to ${}^{\bullet}$NO signaling, ${}^{\bullet}$NO binds to the same mitochondrial site as O_2, thus exhibiting competitive kinetics with O_2 [50]. This competition may have critically important consequences in intracellular signaling events, including the generation of reactive oxygen species (see below), as well as major influences on cellular and tissue distribution of ${}^{\bullet}$NO and of O_2, and possible contributions to tissue hypoxic vasodilation [39, 51].

Intracellular nonheme iron (NHI) is also a target for ${}^{\bullet}$NO. One effect of such ${}^{\bullet}$NO binding is to produce iron–${}^{\bullet}$NO complexes which are visible by electron paramagnetic resonance (EPR) spectroscopy. Observation of these signals was important in developing concepts of cellular actions of ${}^{\bullet}$NO in the early stages of the mammalian ${}^{\bullet}$NO field [52, 53]. It was initially believed that the origin of the NHI under these conditions is mitochondrial iron–sulfur centers (which are targets of ${}^{\bullet}$NO under conditions of high, inflammatory levels of ${}^{\bullet}$NO [54]), but quantitative and kinetic evidence now appears to support the so-called "chelatable iron pool" as the iron source [55].

Intracellular ${}^{\bullet}$NO Targets: Radicals

${}^{\bullet}$NO reacts at near-diffusion controlled rates with other radicals, and these reactions can have either damaging or protective effects due to three results of such reactions: (1) removal of ${}^{\bullet}$NO, (2) removal of the radical with which ${}^{\bullet}$NO reacts, and (3) formation of products that also may be very reactive [56]. Arguably the most important ${}^{\bullet}$NO/radical reaction (certainly the most discussed in the literature, especially with reference to the effects of ${}^{\bullet}$NO in disease conditions) is the extremely rapid reaction with the superoxide anion radical ($O_2^{\bullet-}$). $O_2^{\bullet-}$ is perhaps the "canonical" reactive oxygen species; the discovery in 1969 of the enzyme which detoxifies it (superoxide dismutase) by McCord and Fridovich being responsible for the birth of the entire field [57]. So it might seem that the ${}^{\bullet}NO + O_2^{\bullet-}$ reaction would only be protective, but in fact, the two other effects of the reaction are in principle damaging because it decreases the beneficial activities of ${}^{\bullet}$NO and also the product (peroxynitrite, $ONOO^-$) is an even more damaging ROS than $O_2^{\bullet-}$ [58, 59]. There are several well-established sources of $O_2^{\bullet-}$ that are relevant to the modulation of ${}^{\bullet}$NO signaling, as described below.

Current Aspects of ${}^{\bullet}$NO in Health and Disease: A Janus-Faced Molecule

The Cardiovascular System

The endothelium is the thin layer of cells that line the interior surface of blood vessels, forming an interface between circulating blood in the lumen and the rest of the vessel wall. Endothelial cells line the entire circulatory system, from the heart to the smallest capillary. Endothelial dysfunction is a loss of normal biochemical processes carried out by the endothelium and is a hallmark for a plethora of vascular diseases, and often leads to atherosclerosis. This is very common, for example, in patients with diabetes mellitus, hypertension, or other chronic pathophysiological conditions with a substantial cardiovascular component. Abnormal vasodilation as an important component of endothelial dysfunction arises from variations in blood flow observed in patients with atherosclerosis compared with healthy subjects [60]. In healthy subjects, activation of eNOS causes vasodilation in both muscular conduit vessels and resistance arterioles. In contrast, in subjects with atherosclerosis, similar stimulation yields attenuated vasodilation in peripheral vessels and causes paradoxical vasoconstriction in coronary arteries, thus indicating a decrease in the availability of ˙NO [61, 62]. Interestingly, endothelial dysfunction can be demonstrated in patients with risk factors for atherosclerosis in the absence of atherosclerosis itself [63, 64]. In addition, feeding healthy volunteers a high-fat meal leads to endothelial vasodilator dysfunction in a time span of just a few hours [65]. Experimental and clinical studies provide evidence that defects of endothelial ˙NO function is not only associated with all major cardiovascular risk factors, such as hyperlipidemia, diabetes, hypertension, smoking, and severity of atherosclerosis, but also has a profound predictive value for future atherosclerotic disease progression [60]. This concept is illustrated in Fig. 11.2. Therefore, the dysfunctional eNOS/˙NO pathway (including both ˙NO formation and also disappearance) provides an ideal target for therapeutic or preventive intervention once ˙NO homeostasis is better understood in atherosclerosis.

Potential mechanisms of endothelial dysfunction include increased vascular contraction to vasoconstrictors, such as endothelin-1, thromboxanes, and serotonin [66]; enhanced thrombus formation; and exacerbated smooth muscle cell (SMC) proliferation and migration [67]. Decreased ˙NO production may be the result of oxidized LDL-mediated displacement of eNOS from plasmalemmal caveolae, thereby inhibiting acetylcholine-induced activation of the enzyme [68]. Moreover, decreased ˙NO increases the tendency for lesion progression by enhancing vascular smooth muscle proliferation and migration, augmenting platelet activation and thrombosis, possibly participating in intravascular neovascularization, and favoring adverse lipid modification [69]. Once lesions have developed, endothelial dysfunction may exacerbate development of clinical events. Impaired endothelium may abnormally reduce vascular perfusion, produce factors that decrease plaque stability, and augment the thrombotic response to plaque rupture [70]. Augmentation of ˙NO or restoration of NOS function seems a logical means by which to inhibit atherosclerosis. However, overexpression of endothelial NOS accelerates lesion formation in apoE-deficient mice [71] demonstrating that enhanced NOS-derived ˙NO may not always be beneficial. Supplementation with tetrahydrobiopterin (BH_4) reduced the lesion size to those seen in Apo E knockout mice revealing the requirement of enzyme cofactors. Even with BH_4 supplementation, there was still no effect on lesion development.

The physiological effects of ˙NO extend well beyond the vascular endothelium. Radomski et al. [72] has shown that human platelets contain a NOS that is activated

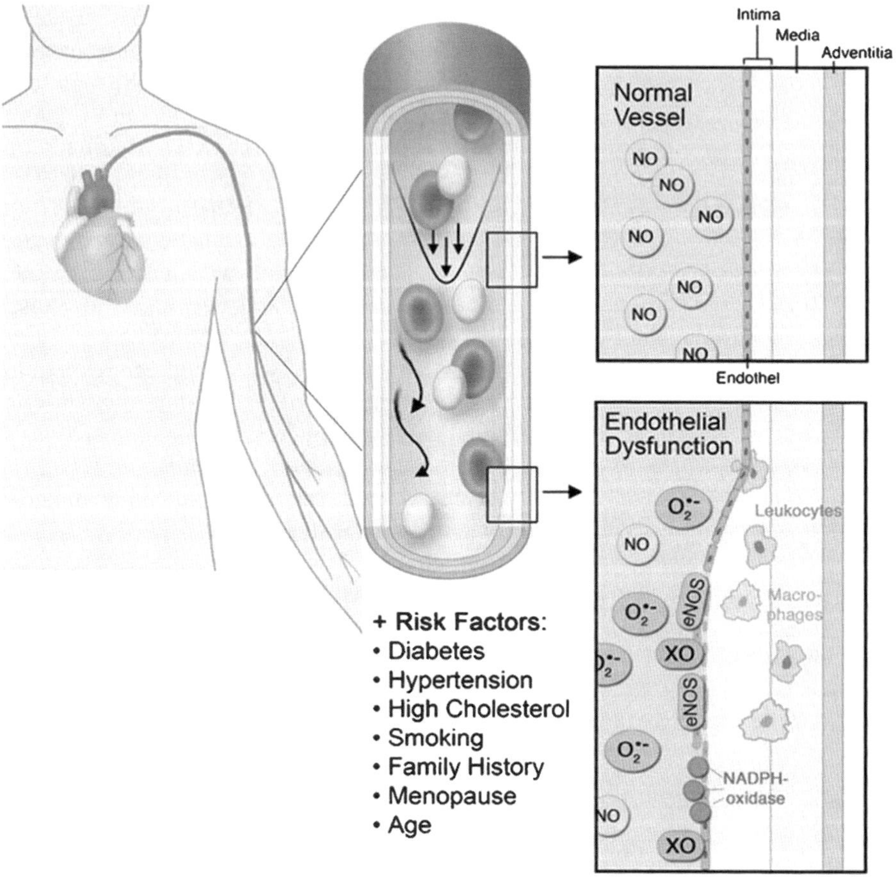

Fig. 11.2 The innermost lining of our blood vessels is single layer of endothelial cells. Healthy endothelial cells produce nitric oxide to combat disease and regulate blood pressure. Endothelial dysfunction is defined by the inability of the vessels to transduce ·NO activity, either by inability to produce from L-arginine or inactivation by superoxide ($O_2^{\bullet -}$)

when platelets are stimulated to aggregate. Thus, platelets themselves also have the enzymatic capacity to synthesize ·NO with both a constitutive and inducible form of NOS identified in human megakaryoblasts [73]. NOS activity increases with platelet activation, and this response appears to modulate platelet aggregation, thereby potentially limiting the self-amplification of platelet thrombus formation in vivo [74, 75]. It was also reported early on that human neutrophils inhibit platelet aggregation by releasing an ·NO-like factor [76]. These antithrombotic properties of the endothelium may be a consequence of the synergistic action of ·NO and prostacyclin. Radomski et al. [77] have shown the synergistic antiaggregatory effects of ·NO and prostacyclin on platelets. ·NO and prostacyclin may act in concert to oppose local vasospasm or thrombus formation at sites where platelets aggregate and the coagulation cascade is activated. It has also been proposed that the antiplatelet effects of endothelial-derived ·NO may prevent thromboembolic events during administration of potent prostacyclin inhibitors such as aspirin [78]. In this regard, •NO acts as an antiinflammatory molecule.

Endothelial dysfunction in the setting of cardiovascular risk factors has been shown to be, at least in part, dependent on the production of ROS, such as superoxide, and the subsequent decrease in vascular availability of nitric oxide. ROS production has been demonstrated to occur in the endothelial cell and also within the media and adventitia, all of which may impair •NO signaling within vascular tissue to endothelium-dependent but also endothelium-independent vasodilators [60]. More recent experimental but also clinical studies point to the pathophysiological importance of the xanthine oxidase, the vascular NADPH oxidase, mitochondria, and uncoupled endothelial nitric oxide synthase as significant enzymatic superoxide sources. These phenomena are described below.

The mitochondrial respiratory chain can be a major source of superoxide. During aerobic metabolism, the oxidation/reduction energy of mitochondrial electron transport is converted to the high-energy phosphate bond of ATP via a multicomponent electron transfer complex. Molecular oxygen serves as the final electron acceptor for cytochrome c oxidase (complex IV), the terminal component of the respiratory chain, and is ultimately reduced to water (H_2O). Up to 1–4% of O_2 even under normal physiological conditions may be incompletely reduced, resulting in $O_2^{\bullet-}$ formation, mainly at complex I (NADH coenzyme Q reductase) and complex III (ubiquinol Cyt c reductase) [79]. Increased mitochondrial $O_2^{\bullet-}$ generation can be enhanced in certain conditions such as conditions of metabolic perturbation, hypoxia-reoxygenation and ischemia-reperfusion, where the enhanced $O_2^{\bullet-}$ is at least partially responsible for an increase in endothelial permeability [80].

Xanthine oxidoreductase (XO) is a ubiquitous metalloflavoprotein found as one of two interconvertible yet functionally distinct forms, namely, xanthine dehydrogenase (XD), which is constitutively expressed in vivo, and XO, which is generated by the posttranslational modification of XD [81]. Functionally, both XD and XO catalyze oxidation of hypoxanthine to xanthine and xanthine to urate. However, whereas XD requires NAD^+ as an electron acceptor, XO instead requires the reduction of molecular O_2, thereby generating $O_2^{\bullet-}$. The conversion of XD to XO occurs either through reversible thiol oxidation of sulfhydryl residues on XD or via irreversible proteolytic cleavage of a segment of XD during hypoxia, ischemia, or in the presence of various proinflammatory mediators ((e.g., tumor necrosis factor-α (TNF-α)) [81]).

Dysfunctional or uncoupled NOSs are also a source of $O_2^{\bullet-}$. The essential NOS cofactor, tetrahydrobiopterin (BH_4), appears to have a key role in regulating NOS function by "coupling" the reduction of molecular O_2 to L-arginine oxidation as well as maintaining the stability of NOS dimers [82]. Thus, BH_4 availability may be a crucial factor in the balance between •NO and $O_2^{\bullet-}$ generation by eNOS. Furthermore, BH_4 itself is highly susceptible to oxidative degradation, and the initial oxidative loss of BH_4 in response to increased ROS production by NADPH oxidases (see below) has been shown to amplify oxidative stress through the resulting loss of •NO production and increased NOS-dependent $O_2^{\bullet-}$ generation [83]. In addition to increased catabolism or degradation, another reason for BH_4 depletion may be its reduced synthesis.

In recent years, it has become apparent that endothelial cells and other nonphagocytic cells constitutively express a $O_2^{\bullet-}$ -generating enzyme analogous to the phagocyte NADPH oxidase of neutrophils [84]. All the classical neutrophil oxidase components are expressed in endothelial cells, but the enzyme nevertheless exhibits several major differences from the neutrophil oxidase; for example, it continuously generates a low level of $O_2^{\bullet-}$ even in unstimulated cells, although its activity can be further increased by several agonists; and a substantial proportion of the O_2 generated by the enzyme is produced intracellularly, whereas neutrophil oxidase O_2 generation occurs mainly in the extracellular compartment.

Normal physiological processes are continuously generating superoxide and other oxygen radicals that can quickly and effectively inactivate ${}^{\bullet}$NO. Fortunately there are a number of antioxidant systems, both enzymatic and nonenzymatic, that help to limit the amount of ROS produced to preserve ${}^{\bullet}$NO activity. It appears that the production of ${}^{\bullet}$NO is a war of attrition. There are many circumstances in the production pathway that can diminish its output but there are also many physiological factors that can quickly inactivate ${}^{\bullet}$NO once it has been successfully produced.

ROS such as superoxide are produced in abundance in the dysfunctional endothelium and limitation of ROS generation increases the availability of ${}^{\bullet}$NO. For this reason antioxidant therapy with vitamin C and cholesterol-lowering therapy with statins (HMG-CoA reductase inhibitors) improve endothelial function [60]. An alternative approach to increase levels of bioactive ${}^{\bullet}$NO and to improve endothelial function is to increase the synthesis of ${}^{\bullet}$NO. Enhanced synthesis of ${}^{\bullet}$NO can be achieved by increased availability of agonists that stimulate release of ${}^{\bullet}$NO from the endothelial cells, like bradykinin, assuming a healthy endothelium and sufficient cofactor to supply the NOS enzymes. Another straightforward approach to increase ${}^{\bullet}$NO synthesis is to provide additional substrate (L-arginine) to the endothelial cell [85]. Furthermore, L-arginine takes part in protein production, endocrine functions, wound healing, and erectile function. It is not regarded as an essential amino acid as the adult human is able to synthesize L-arginine de novo via the urea cycle. In adults, the synthesis of L-arginine results in citrulline, a by-product of the glutamine metabolism in the gut and in the liver. Citrulline is excreted into the circulation and is reabsorbed in the kidney and converted to L-arginine. L-Citrulline is reformed if L-arginine is shunted through the NOS pathway. The dietetic application of L-arginine is the basic determinant of the L-arginine level in plasma, as the biosynthesis of L-arginine is not able to balance inadequate intake or deficiency. Providing supplementation substrate to individuals with inadequate ${}^{\bullet}$NO, therefore, has been suggested as a rational approach to increase ${}^{\bullet}$NO production by NOS [85].

Erythrocytes as Delivery Agents for ${}^{\bullet}$NO

The phenomenon of hypoxic vasodilation, whereby vessels dilate under decreased O_2 conditions, has been known for more than a century but the mechanism is not clear; in particular it is not clear as to what the "sensor" is that detects hypoxia and initiates a response. In 1995 Ellsworth et al. [86] proposed that the sensor is intraerythrocytic hemoglobin, which responds by increasing release of ATP to induce vasodilation. Shortly thereafter, Jia et al. [47] advanced a surprising alternative mechanism for the hemoglobin signal: rather than simply being an irreversible sink for ˙NO, hemoglobin within the red blood cell actually reacts with the ˙NO in the lung (where O_2 levels are high) forming S-nitrosohemoglobin (SNOHb) and releases it in vascular beds where O_2 is low (hypoxia). Important evidence for this was the report that the levels of SNOHb are higher in the arterial than the venous circulation, implying liberation of ˙NO upon transit through the hypoxic region (called the A/V transit). Gladwin et al. [87] subsequently found no change in SNOHb upon the A/V transit; however, there was significant consumption of plasma nitrite upon induction of hypoxia by exercise, and subsequent work suggested that deoxygenation of hemoglobin induces reaction with nitrite to produce ˙NO. These three mechanisms of hypoxic vasodilation (ATP, SNOHb, and nitrite) are highly controversial and may in fact be interrelated [88, 89].

Independent of its validity as the mechanism for hypoxic vasodilation, a series of recent studies have shown that nitrite administration is remarkably salutatory for a variety of clinically significant applications, including myocardial infarction, stroke, hypertension, angiogenesis, and organ transplantation [90]. In addition, a report by Kleinbongard et al. [91] demonstrates that plasma nitrite levels progressively decrease with increasing cardiovascular risk load indicating a reduction in nitric oxide produced. Risk factors considered include age, hypertension, smoking, and hypercholesterolemia, conditions all known for reduced availability of ˙NO. Although a correlation exists in the plasma, it is not known whether the situation is mirrored in the heart or other tissue at risk for ischemic injury or disease. If so, tissue nitrite may serve as an index of risk and restoring tissue nitrite may act as a first line of defense for protecting organs from ischemic and/or I/R injury. Since a substantial portion of steady-state nitrite concentrations in blood and tissue are derived from dietary sources, modulation of nitrite intake may provide a first line of defense for cardiovascular disease [92].

The Immune System

˙NO is also generated by macrophages and neutrophils as part of the human immune response. ˙NO is toxic to bacteria and other human pathogens. It is the inducible isoform of NOS that is responsible for macrophage ˙NO production. Inducible NOS has been found in many cell types, including macrophages [5], and is immunologically activated by exposure to bacterial endotoxin or pro-inflammatory cytokines such as interleukin-1, or interferon-gamma [93, 94], and tumor necrosis factor. The presence of iNOS message or protein can serve as a biomarker for inflammation in tissues. The iNOS protein may remain present for several days [95]. At these high concentrations and flux rates, ˙NO is cytotoxic and plays a key role in the immune response of macrophages to bacteria and other pathogens.

Antimicrobial activity and ˙NO production parallel tumor necrosis factor activity and a strong correlation exists between antimicrobial activity and production of l-Arg-derived ˙NO by cytokine-activated cells observed during in vitro studies [96]. The precise mechanism of ˙NO-mediated bactericidal and tumoricidal activity is unknown, but these observations suggest that macrophage ˙NO production contributes to nonspecific immunity. ˙NO from activated macrophages may be responsible for the profound loss of vascular tone seen in septic patients [97]. It is this relative overproduction of ˙NO and the subsequent vasodilation that are thought to mediate ˙NO's pathophysiologic role during sepsis and multiorgan failure during hypovolemia and hypoxia. Despite the rapid progress in our understanding of the complex physiological and pathophysiological processes involving ˙NO, uncertainties remain with regard to the critical cellular targets of ˙NO cytotoxicity, the relative importance of different ˙NO redox states and carrier molecules, and the importance of the ˙NO antimicrobial system in human phagocytes. Ultimately, the immunoregulatory and vasoregulatory activities of ˙NO may prove to be just as important as its antimicrobial properties during infection.

The Nervous System

In the central nervous system, ˙NO is a neurotransmitter that underpins several functions, including the formation of memory. Recurrent as in other organ systems, this ˙NO pathway may also play a role in the pathology of the central nervous system. The NOS isoform in the nervous system is activated by glutamate acting on N-methyl-D-aspartate receptors. In a matter of seconds the glutamate-induced increase in intracellular calcium concentration activates NOS via the calcium/calmodulin interaction as previously described. Under most circumstances, eNOS and nNOS are constitutive in the sense that their activation does not require new enzyme synthesis. However, both forms of NOS are inducible in that new enzyme synthesis occurs primarily under conditions of traumatic or pathological insult. The calcium influx that accompanies prolonged NMDA receptor activation is associated with degeneration of the neurons through a mechanism(s) that involves ˙NO, but is still not precisely clear [98]. The dichotomy of both the protective and deleterious actions of ˙NO is again revealed in the nervous system.

In the periphery, there is a widespread network of nerves, previously recognized as nonadrenergic and noncholinergic (NANC), that operate through a ˙NO-dependent mechanism to mediate some forms of neurogenic vasodilation and regulate various gastrointestinal, respiratory, and genitourinary tract functions, as well as autonomic innervation of smooth muscle in the gastrointestinal tract, the pelvic viscera, the airways, and other systems [99]. NOS has been detected in the gastric mucosa, and ˙NO appears to play a role in protecting the gastric mucosa during physiologic stress by acting as an endogenous vasodilator and thus supporting mucosal blood flow [100]. The exact mechanism of ˙NO's protective effect is unclear, but may relate to vasodilation, inhibition of platelet aggregation in the gastric microvasculature, or a protective effect on the epithelial cells themselves [101]. However, alternative means to enhance •NO in the periphery and stomach will certainly provide benefit to a number of conditions.

Conclusion

Nitric oxide research has expanded rapidly in the past 30 years, and the roles of •NO in physiology and pathology have been extensively studied. The pathways of •NO synthesis, signaling, and metabolism in vascular biological systems have been and continue to be a major area of research. As a gas (in the pure state and under standard temperature and pressure) and free radical with an unshared electron, •NO participates in various biological processes. Understanding its production, regulation, and molecular targets will be essential for the development of new therapies for various pathological conditions characterized by an un-balanced production and metabolism of •NO.

References

1. Wagner D, Tannenbaum SR. Enhancement of nitrate biosynthesis by *Escherichia coli* lipopolysaccharide. In: Magee PN, editor. Nitrosamines and human cancer. New York, NY: Cold Spring Harbor; 1982. p. 433–7.
2. Hegesh E, Shiloah J. Blood nitrates and infantile methemoglobinemia. Clin Chim Acta. 1982;125:107–15.
3. Stuehr DJ, Marletta MA. Mammalian nitrate biosynthesis: mouse macrophages produce nitrite and nitrate in response to *Escherichia coli* lipopolysaccharide. Proc Natl Acad Sci USA. 1985;82:7738–42.
4. Hibbs Jr JB, Taintor RR, Vavrin Z. Macrophage cytotoxicity: role for L-arginine deiminase and imino nitrogen oxidation to nitrite. Science. 1987;235:473–6.
5. Marletta MA, Yoon PS, Iyengar R, Leaf CD, Wishnok JS. Macrophage oxidation of L-arginine to nitrite and nitrate: nitric oxide is an intermediate. Biochemistry. 1988;27:8706–11.
6. Arnold WP, Mittal CK, Katsuki S, Murad F. Nitric oxide activates guanylate cyclase and increases guanosine 3′:5′-cyclic monophosphate levels in various tissue preparations. Proc Natl Acad Sci USA. 1977;74:3203–7.
7. Ignarro LJ, Byrns RE, Wood KS. Biochemical and Pharmacological Properties of Endothelium- Derived Relaxing Factor and its Similarity to Nitric Oxide Radical. In: Vanhoutte PM, editor. Vasodilatation: vascular smooth muscle, peptides, autonomic nerves, and endothelium. New York, NY: Raven; 1988. p. 427–35.
8. Furchgott RF. Studies on Relaxation of Rabbit Aorta by Sodium Nitrite: the Basis for the Proposal that the Acid-activatable Factor from Bovine Retractor Penis is Inorganic Nitrite and the Endothelium- Derived Relaxing Factor is Nitric Oxide. In: Vanhoutte PM, editor. Vasodilatation: vascular smooth muscle, peptides, autonomic nerves, and endothelium. New York, NY: Raven; 1988. p. 401–14.
9. Palmer RM, Ferrige AG, Moncada S. Nitric oxide release accounts for the biological activity of endothelium-derived relaxing factor. Nature. 1987;327:524–6.
10. Garthwaite J, Charles SL, Chess-Williams R. Endothelium-derived relaxing factor release on activation of NMDA receptors suggests role as intercellular messenger in the brain. Nature. 1988;336:385–8.

11. Forman HJ, Maiorino M, Ursini F. Signaling functions of reactive oxygen species. Biochemistry. 2010;49(5):835–42.
12. Hill BG, Dranka BP, Bailey SM, Lancaster Jr JR, Darley-Usmar VM. What part of NO don't you understand? Some answers to the cardinal questions in nitric oxide biology. J Biol Chem. 2010;285(26):19699–704.
13. Fukumura D, Kashiwagi S, Jain RK. The role of nitric oxide in tumour progression. Nat Rev Cancer. 2006;6(7):521–34.
14. Fernandes D, Assreuy J. Nitric oxide and vascular reactivity in sepsis. Shock. 2008;30 Suppl 1:10–3.
15. Capellini VK, Celotto AC, Baldo CF, Olivon VC, Viaro F, Rodrigues AJ, et al. Diabetes and vascular disease: basic concepts of nitric oxide physiology, endothelial dysfunction, oxidative stress and therapeutic possibilities. Curr Vasc Pharmacol. 2010;8(4):526–44.
16. Mehta S. The effects of nitric oxide in acute lung injury. Vascul Pharmacol. 2005;43(6):390–403.
17. Baylis C. Nitric oxide deficiency in chronic kidney disease. Am J Physiol Ren Physiol. 2008;294(1):F1–9.
18. Bredt DS, Snyder SH. Isolation of nitric oxide synthetase, a calmodulin-requiring enzyme. Proc Natl Acad Sci USA. 1990;87:682–5.
19. Bredt DS, Hwang PM, Glatt CE, Lowenstein C, Reed RR, Snyder SH. Cloned and expressed nitric oxide synthase structurally resembles cytochrome P-450 reductase. Nature. 1991;351:714–8.
20. Li H, Poulos TL. Structure-function studies on nitric oxide synthases. J Inorg Biochem. 2005;99(1):293–305.
21. Daff S. NO synthase: structures and mechanisms. Nitric Oxide. 2010;23(1):1–11.
22. Michel T, Vanhoutte PM. Cellular signaling and NO production. Pflugers Arch. 2010;459(6):807–16.
23. Ozuyaman B, Grau M, Kelm M, Merx MW, Kleinbongard P. RBC NOS: regulatory mechanisms and therapeutic aspects. Trends Mol Med. 2008;14(7):314–22.
24. Finocchietto PV, Franco MC, Holod S, Gonzalez AS, Converso DP, Arciuch VG, et al. Mitochondrial nitric oxide synthase: a masterpiece of metabolic adaptation, cell growth, transformation, and death. Exp Biol Med (Maywood). 2009;234(9):1020–8.
25. Lacza Z, Pankotai E, Busija DW. Mitochondrial nitric oxide synthase: current concepts and controversies. Front Biosci. 2009;14:4436–43.
26. Kelm M. Nitric oxide metabolism and breakdown. Biochim Biophys Acta. 1999;1411(2–3):273–89.
27. Wennmalm A, Benthin G, Edlund A, Kieler-Jensen N, Lundin S, Petersson AS, et al. Nitric oxide synthesis and metabolism in man. Ann NY Acad Sci. 1994;714:158–64.
28. Rhodes P, Leone AM, Francis PL, Struthers AD, Moncada S. The l-arginine:nitric oxide pathway is the major source of plasma nitrite in fasted humans. Biochem Biophys Res Commun. 1995;209:590–6.
29. Kleinbongard P, Dejam A, Lauer T, Rassaf T, Schindler A, Picker O, et al. Plasma nitrite reflects constitutive nitric oxide synthase activity in mammals. Free Radic Biol Med. 2003;35(7):790–6.
30. Aparicio PJ, Knaff DB, Malkin R. The role of an iron-sulfur center and siroheme in spinach nitrite reductase. Arch Biochem Biophys. 1975;169(1):102–7.
31. Lancaster Jr JR, Vega JM, Kamin H, Orme-Johnson NR, Orme-Johnson WH, Krueger RJ, et al. Identification of the iron-sulfur center of spinach ferredoxin- nitrite reductase as a tetranuclear center, and preliminary EPR studies of mechanism. J Biol Chem. 1979;254:1268–72.
32. Reddy D, Lancaster Jr JR, Cornforth DP. Nitrite inhibition of Clostridium botulinum: electron spin resonance detection of iron-nitric oxide complexes. Science. 1983;221:769–70.
33. Goretski J, Hollocher TC. Trapping of nitric oxide produced during denitrification by extracellular hemoglobin. J Biol Chem. 1988;263:2316–23.

34. Lundberg JO, Weitzberg E. NO-synthase independent NO generation in mammals. Biochem Biophys Res Commun. 2010;396(1):39–45.
35. Lundberg JO, Weitzberg E, Lundberg JM, Alving K. Intragastric nitric oxide production in humans: measurements in expelled air. Gut. 1994;35(11):1543–6.
36. Benjamin N, O'Driscoll F, Dougall H, Duncan C, Smith L, Golden M, et al. Stomach NO synthesis. Nature. 1994;368(6471):502.
37. Butler AR, Ridd JH. Formation of nitric oxide from nitrous acid in ischemic tissue and skin. Nitric Oxide. 2004;10(1):20–4.
38. Lancaster Jr JR. A tutorial on the diffusibility and reactivity of free nitric oxide. Nitric Oxide. 1997;1:18–30.
39. Thomas DD, Liu X, Kantrow SP, Lancaster Jr JR. The biological lifetime of nitric oxide: implications for the perivascular dynamics of NO and O_2. Proc Natl Acad Sci USA. 2001;98(1):355–60.
40. Hare JM. Nitric oxide and excitation-contraction coupling. J Mol Cell Cardiol. 2003;35(7):719–29.
41. Tziomalos K, Hare JM. Role of xanthine oxidoreductase in cardiac nitroso-redox imbalance. Front Biosci. 2009;14:237–62.
42. Salerno JC, Ghosh DK. Space, time and nitric oxide – neuronal nitric oxide synthase generates signal pulses. FEBS J. 2009;276(22):6677–88.
43. Derbyshire ER, Marletta MA. Biochemistry of soluble guanylate cyclase. *Handb Exp Pharmacol* 2009;(191):17–31.
44. Garthwaite J. New insight into the functioning of nitric oxide-receptive guanylyl cyclase: physiological and pharmacological implications. Mol Cell Biochem. 2010;334(1–2):221–32.
45. Ignarro LJ. Signal transduction mechanisms involving nitric oxide. Biochem Pharmacol. 1991;41:485–90.
46. Francis SH, Busch JL, Corbin JD, Sibley D. cGMP-dependent protein kinases and cGMP phosphodiesterases in nitric oxide and cGMP action. Pharmacol Rev. 2010;62(3):525–63.
47. Jia L, Bonaventura J, Stamler JS. S-nitrosohaemoglobin: a dynamic activity of blood involved in vascular control. Nature. 1996;380(6571):221–6.
48. Allen BW, Stamler JS, Piantadosi CA. Hemoglobin, nitric oxide and molecular mechanisms of hypoxic vasodilation. Trends Mol Med. 2009;15(10):452–60.
49. Stevens TH, Brudvig GW, Bocian DF, Chan SI. Structure of cytochrome a3-Cua3 couple in cytochrome c oxidase as revealed by nitric oxide binding studies. Proc Natl Acad Sci USA. 1979;76:3320–4.
50. Brunori M, Giuffre A, Sarti P. Cytochrome c oxidase, ligands and electrons. J Inorg Biochem. 2005;99(1):324–36.
51. Erusalimsky JD, Moncada S. Nitric oxide and mitochondrial signaling: from physiology to pathophysiology. Arterioscler Thromb Vasc Biol. 2007;27(12):2524–31.
52. Lancaster Jr JR, Hibbs Jr JB. EPR demonstration of iron-nitrosyl complex formation by cytotoxic activated macrophages. Proc Natl Acad Sci USA. 1990;87:1223–7.
53. Pellat C, Henry Y, Drapier JC. IFN-gamma-activated macrophages: detection by electron paramagnetic resonance of complexes between L-arginine-derived nitric oxide and non-heme iron proteins. Biochem Biophys Res Comm. 1990;166:119–25.
54. Hibbs JBJ, Taintor RR, Vavrin Z, Rachlin EM. Nitric oxide: a cytotoxic activated macrophage effector molecule [published erratum appears in Biochem Biophys Res Commun 1989;158(2):624]. Biochem Biophys Res Comm. 1988;157(1):87–94.
55. Toledo Jr JC, Bosworth CA, Hennon SW, Mahtani HA, Bergonia HA, Lancaster Jr JR. Nitric oxide-induced conversion of cellular chelatable iron into macromolecule-bound paramagnetic dinitrosyliron complexes. J Biol Chem. 2008;283(43):28926–33.
56. Lancaster Jr JR. Reactivity and Diffusivity of Nitrogen Oxides in Mammalian Biology. In: Forman HJ, Fukuto JM, Torres M, editors. Signal transduction by reactive oxygen and

nitrogen species: pathways and chemical principles. Dordecht: Kluwer Academic; 2003. p. 53–79.
57. McCord JM, Fridovich I. Superoxide dismutase. An enzymic function for erythrocuprein (hemocuprein). J Biol Chem. 1969;244(22):6049–55.
58. Beckman JS, Beckman TW, Chen J, Marshall PA, Freeman BA. Apparent hydroxyl radical production by peroxynitrite: implications for endothelial injury from nitric oxide and superoxide. Proc Natl Acad Sci USA. 1990;87:1620–4.
59. Pacher P, Beckman JS, Liaudet L. Nitric oxide and peroxynitrite in health and disease. Physiol Rev. 2007;87(1):315–424.
60. Forstermann U. Nitric oxide and oxidative stress in vascular disease. Pflugers Arch. 2010;459(6):923–39.
61. Lieberman EH, Gerhard MD, Uehata A, Selwyn AP, Ganz P, Yeung AC, et al. Flow-induced vasodilation of the human brachial artery is impaired in patients <40 years of age with coronary artery disease. Am J Cardiol. 1996;78(11):1210–4.
62. Ludmer PL, Selwyn AP, Shook TL, Wayne RR, Mudge GH, Alexander RW, et al. Paradoxical vasoconstriction induced by acetylcholine in atherosclerotic coronary arteries. N Engl J Med. 1986;315(17):1046–51.
63. Creager MA, Cooke JP, Mendelsohn ME, Gallagher SJ, Coleman SM, Loscalzo J, et al. Impaired vasodilation of forearm resistance vessels in hypercholesterolemic humans. J Clin Invest. 1990;86(1):228–34.
64. Celermajer DS, Sorensen KE, Georgakopoulos D, Bull C, Thomas O, Robinson J, et al. Cigarette smoking is associated with dose-related and potentially reversible impairment of endothelium-dependent dilation in healthy young adults. Circulation. 1993;88(5 Pt 1):2149–55.
65. Esposito K, Nappo F, Giugliano F, Giugliano G, Marfella R, Giugliano D. Effect of dietary antioxidants on postprandial endothelial dysfunction induced by a high-fat meal in healthy subjects. Am J Clin Nutr. 2003;77(1):139–43.
66. Lamping K, Faraci F. Enhanced vasoconstrictor responses in eNOS deficient mice. Nitric Oxide. 2003;8(4):207–13.
67. Huang PL. Endothelial nitric oxide synthase and endothelial dysfunction. Curr Hypertens Rep. 2003;5(6):473–80.
68. Blair A, Shaul PW, Yuhanna IS, Conrad PA, Smart EJ. Oxidized low density lipoprotein displaces endothelial nitric-oxide synthase (eNOS) from plasmalemmal caveolae and impairs eNOS activation. J Biol Chem. 1999;274(45):32512–9.
69. Loscalzo J. Nitric oxide insufficiency, platelet activation, and arterial thrombosis. Circ Res. 2001;88(8):756–62.
70. Faxon DP, Fuster V, Libby P, Beckman JA, Hiatt WR, Thompson RW, et al. Atherosclerotic vascular disease conference: writing group III: pathophysiology. Circulation. 2004;109(21):2617–25.
71. Ozaki M, Kawashima S, Yamashita T, Hirase T, Namiki M, Inoue N, et al. Overexpression of endothelial nitric oxide synthase accelerates atherosclerotic lesion formation in apoE-deficient mice. J Clin Invest. 2002;110(3):331–40.
72. Radomski MW, Palmer RM, Moncada S. An l-arginine/nitric oxide pathway present in human platelets regulates aggregation. Proc Natl Acad Sci USA. 1990;87:5193–7.
73. Sase K, Michel T. Expression of constitutive endothelial nitric oxide synthase in human blood platelets. Life Sci. 1995;57(22):2049–55.
74. Freedman JE, Loscalzo J, Barnard MR, Alpert C, Keaney JF, Michelson AD. Nitric oxide released from activated platelets inhibits platelet recruitment. J Clin Invest. 1997;100(2):350–6.
75. Freedman JE, Sauter R, Battinelli EM, Ault K, Knowles C, Huang PL, et al. Deficient platelet-derived nitric oxide and enhanced hemostasis in mice lacking the NOSIII gene. Circ Res. 1999;84(12):1416–21.
76. Salvemini D, de Nucci G, Gryglewski RJ, Vane JR. Human neutrophils and mononuclear cells inhibit platelet aggregation by releasing a nitric oxide-like factor. Proc Natl Acad Sci USA. 1989;86:6328–32.

77. Radomski MW, Palmer RM, Moncada S. Comparative pharmacology of endothelium-derived relaxing factor, nitric oxide and prostacyclin in platelets. Br J Pharmacol. 1987;92:181–7.
78. Bednar MM, Gross CE, Howard DB, Russell SR, Thomas GR. Nitric oxide reverses aspirin antagonism of t-PA thrombolysis in a rabbit model of thromboembolic stroke. Exp Neurol. 1997;146(2):513–7.
79. Lambert AJ, Brand MD. Reactive oxygen species production by mitochondria. Methods Mol Biol. 2009;554:165–81.
80. Pearlstein DP, Ali MH, Mungai PT, Hynes KL, Gewertz BL, Schumacker PT. Role of mitochondrial oxidant generation in endothelial cell responses to hypoxia. Arterioscler Thromb Vasc Biol. 2002;22(4):566–73.
81. Meneshian A, Bulkley GB. The physiology of endothelial xanthine oxidase: from urate catabolism to reperfusion injury to inflammatory signal transduction. Microcirculation. 2002;9(3):161–75.
82. Vasquez-Vivar J, Kalyanaraman B, Martasek P. The role of tetrahydrobiopterin in superoxide generation from eNOS: enzymology and physiological implications. Free Radic Res. 2003;37(2):121–7.
83. Landmesser U, Dikalov S, Price SR, McCann L, Fukai T, Holland SM, et al. Oxidation of tetrahydrobiopterin leads to uncoupling of endothelial cell nitric oxide synthase in hypertension. J Clin Invest. 2003;111(8):1201–9.
84. Lambeth JD. NOX enzymes and the biology of reactive oxygen. Nat Rev Immunol. 2004;4(3):181–9.
85. Boger RH. L-Arginine therapy in cardiovascular pathologies: beneficial or dangerous? Curr Opin Clin Nutr Metab Care. 2008;11(1):55–61.
86. Ellsworth ML, Forrester T, Ellis CG, Dietrich HH. The erythrocyte as a regulator of vascular tone. Am J Physiol. 1995;269(6 Pt 2):H2155–61.
87. Gladwin MT, Shelhamer JH, Schechter AN, Pease-Fye ME, Waclawiw MA, Panza JA, et al. Role of circulating nitrite and S-nitrosohemoglobin in the regulation of regional blood flow in humans. Proc Natl Acad Sci USA. 2000;97(21):11482–7.
88. Robinson JM, Lancaster Jr JR. Hemoglobin-mediated, hypoxia-induced vasodilation via nitric oxide: mechanism(s) and physiologic versus pathophysiologic relevance. Am J Respir Cell Mol Biol. 2005;32(4):257–61.
89. Jensen FB. The dual roles of red blood cells in tissue oxygen delivery: oxygen carriers and regulators of local blood flow. J Exp Biol. 2009;212(Pt 21):3387–93.
90. Calvert JW, Lefer DJ. Clinical translation of nitrite therapy for cardiovascular diseases. Nitric Oxide. 2010;22(2):91–7.
91. Kleinbongard P, Dejam A, Lauer T, Jax T, Kerber S, Gharini P, et al. Plasma nitrite concentrations reflect the degree of endothelial dysfunction in humans. Free Radic Biol Med. 2006;40(2):295–302.
92. Bryan NS. Nitrite in nitric oxide biology: cause or consequence? A systems-based review. Free Radic Biol Med. 2006;41(5):691–701.
93. Stuehr DJ, Marletta MA. Induction of nitrite/nitrate synthesis in murine macrophages by BCG infection, lymphokines, or interferon-gamma. J Immunol. 1987;139:518–25.
94. Ding AH, Nathan CF, Stuehr DJ. Release of reactive nitrogen intermediates and reactive oxygen intermediates from mouse peritoneal macrophages. Comparison of activating cytokines and evidence for independent production. J Immunol. 1988;141:2407–12.
95. Xie QW, Cho HJ, Calaycay J, Mumford RA, Swiderek KM, Lee TD, et al. Cloning and characterization of inducible nitric oxide synthase from mouse macrophages. Science. 1992;256:225–8.
96. Nathan CF, Hibbs Jr JB. Role of nitric oxide synthesis in macrophage antimicrobial activity. Curr Opin Immunol. 1991;3:65–70.
97. Cauwels A. Nitric oxide in shock. Kidney Int. 2007;72(5):557–65.
98. Garthwaite G, Garthwaite J. Differential dependence on Ca^{2+} of N-methyl-D-aspartate and quisqualate neurotoxicity in young rat hippocampal slices. Neurosci Lett. 1989;97(3):316–22.

99. Zhang J, Snyder SH. Nitric oxide in the nervous system. Annu Rev Pharmacol Toxicol. 1995;35:213–33.
100. Whittle BJ, Boughton-Smith NK, Moncada S. Biosynthesis and role of the endothelium-derived vasodilator, nitric oxide, in the gastric mucosa. Ann NY Acad Sci. 1992;664:126–39.
101. Calatayud S, Barrachina D, Esplugues JV. Nitric oxide: relation to integrity, injury, and healing of the gastric mucosa. Microsc Res Tech. 2001;53(5):325–35.

Chapter 12
Inhaled Nitric Oxide

Kenneth D. Bloch, Andrea U. Steinbicker, Lisa K. Lohmeyer, and Rajeev Malhotra

Key points

- Development of inhaled NO as a selective pulmonary vasodilator.
- Overview of how inhaled NO elicits its effects is provided, as well as alternate approaches to increasing pulmonary NO and cGMP concentrations.
- Consideration of the clinical applications of inhaled NO.
- Discussion of the methods required for the safe administration of NO gas.
- Potential applications of inhaled NO to the treatment of systemic vascular diseases.

Keywords Inhalative therapy • Pulmonary hypertension • Methemoglobinemia • Second messengers • cGMP

Introduction

In 1998, the Nobel Prize in Physiology or Medicine was awarded jointly to Robert Furchgott, Louis Ignarro, and Ferid Murad for their discoveries concerning "nitric oxide as a signaling molecule in the cardiovascular system." In 1977, Ferid Murad and colleagues proposed that nitroglycerin and related vasodilator compounds act by generating nitric oxide (NO). In 1980, Furchgott and Zawadski described endothelium-derived relaxing factor, which, ten years later, Ignarro and colleagues reported was equivalent to NO.

K.D. Bloch (✉)
Anesthesia Center for Critical Care Research, Department of Anesthesia, Critical Care, and Pain Medicine, Cardiovascular Research Center, Massachusetts General Hospital, Harvard Medical School, Boston, MA 02114, USA
e-mail: kdbloch@partners.org

N.S. Bryan and J. Loscalzo (eds.), *Nitrite and Nitrate in Human Health and Disease*, Nutrition and Health, DOI 10.1007/978-1-60761-616-0_12,

Nitric oxide is a free radical with a short half-life in biological fluids. It is not only rapidly oxidized to nitrite and nitrate, but can also react with several types of intracellular targets including superoxide leading to the production of the strong oxidant, peroxynitrite. Nitric oxide or NO metabolites can nitrosylate thiol-containing amino acids, thereby modulating the activities of a variety of enzymes, including caspases. Nitric oxide also reacts with heme-containing proteins including hemoglobin, mitochondrial cytochromes, and soluble guanylyl cyclase (sGC). The rapid reaction of NO with oxyhemoglobin (containing Fe^{2+}-heme) leads to the formation of methemoglobin (containing Fe^{3+}-heme) and nitrate. Nitric oxide inhibits electron transport by competing with oxygen for heme moieties in mitochondrial cytochrome enzymes.

Binding of NO to the heme moiety in sGC activates the enzyme leading to the synthesis of cyclic GMP (cGMP) from GTP. cGMP reacts with a variety of intracellular targets including phosphodiesterases (PDEs), cyclic nucleotide-gated ion channels, and cGMP-dependent protein kinases. cGMP activation of cGMP-dependent protein kinases modulates a broad range of biological processes including vascular relaxation, inhibition of platelet and leukocyte activation, and modulation of gastrointestinal motility and neurotransmission. The actions of cGMP are limited by PDEs, some of which are cGMP-specific (e.g., PDE5).

Nitric oxide-generating drugs, including nitroglycerin and sodium nitroprusside, have long been used by clinicians to treat patients with a broad spectrum of cardiovascular diseases including angina pectoris and systemic hypertension. The application of NO-donor compounds to treat patients with pulmonary hypertension is severely limited by the systemic vasodilator effects of these agents [1]. Moreover, in patients with lung injury and ventilation–perfusion mismatch, systemic administration of NO-donor compounds reduces hypoxic pulmonary vasoconstriction leading to a deterioration of oxygenation (due to enhanced perfusion of poor-ventilated lung regions).

A variety of novel drugs have been developed that target cGMP metabolism including PDE5 inhibitors used to treat male erectile dysfunction and pulmonary hypertension. More recently, agents which sensitize sGC to NO and/or directly activate the enzyme are being evaluated as treatments for pulmonary and systemic hypertension [2, 3].

In this chapter, we describe the development of inhaled NO as a therapeutic strategy to increase pulmonary vascular NO and cGMP concentrations. Inhaled NO induces pulmonary vasodilation without the side effects of systemic hypotension or hypoxemia. An overview of how inhaled NO elicits its effects is provided, and alternate approaches to increasing pulmonary NO and cGMP concentrations are considered. Several of the clinical applications of inhaled NO are presented, as is a discussion of the methods required for the safe administration of NO gas. Finally, we present experimental and clinical observations highlighting the potential applications of inhaled NO to the treatment of systemic vascular diseases.

Inhaled NO is a Selective Pulmonary Vasodilator

Pulmonary-selective therapies are useful in a variety of disease states. For example, inhaled β-adrenergic agents and glucocorticoids are highly efficacious for treatment of asthma, while avoiding the important side effects associated with these drugs when they are administered systemically. Could inhalation of NO gas or a NO-donor compound be used to treat pulmonary hypertension without causing systemic hypotension or impairing ventilation–perfusion matching?

Prior to 1990, NO gas was viewed, almost exclusively, as a poison. Nitric oxide is generated by combustion engines and in chemical synthesis reactions used in industry. Nitric oxide levels can reach 1,000 ppm in tobacco smoke [4]. Nitric oxide reacts with oxygen in the atmosphere to form nitrogen dioxide (NO_2), one of the greenhouse gases. NO_2, in turn, can react with water to form nitric acid, which causes lung injury. These considerations and others led the US Occupational Safety and Heath Administration to establish a permissible exposure limit for NO of 25 ppm averaged over an 8-h work shift.

Zapol and Frostell hypothesized that inhalation of low concentrations of NO gas would be sufficient to relax pulmonary vascular smooth muscle [5]. Moreover, upon reaching the bloodstream, NO would be rapidly inactivated by oxyhemoglobin avoiding the systemic vasodilation seen with NO-donor compounds (Fig. 12.1). The methemoglobin formed in the reaction of NO with oxyhemoglobin would be reduced to hemoglobin by methemoglobin reductase present in erythrocytes. These investigators studied conscious instrumented adult lambs in which pulmonary

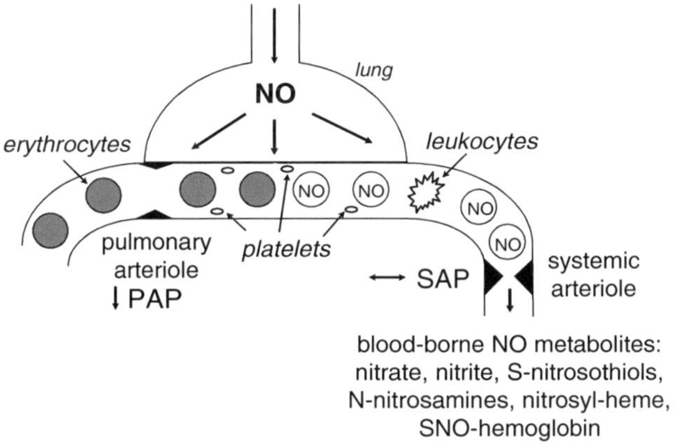

Fig. 12.1 Inhaled NO is a selective pulmonary vasodilator with actions on the systemic vasculature. A schematic of an alveolar-capillary unit is presented highlighting the ability of inhaled NO to dilate pulmonary arterioles and reduce pulmonary artery pressure (PAP). Inhaled NO does not dilate systemic arterioles or alter systemic arterial pressure (SAP) under normal conditions. Inhaled NO does have systemic effects, which are described in the text and which may be mediated by circulating cells exposed to NO in the lungs and blood-borne NO metabolites. Adapted with permission from Bloch et al. [103]

hypertension was induced by systemic administration of the thromboxane analog, U46619, or by reducing the oxygen concentration in the inhaled gas mixture. By minimizing the time during which NO was mixed with oxygen prior to delivery to the animal, levels of inhaled NO_2 were maintained below 5 ppm. Breathing NO rapidly reduced pulmonary artery pressure (PAP) in a dose-dependent manner; inhalation of 80 ppm nearly completely reversed the pulmonary hypertension. The pulmonary vasodilator effects of breathing NO were rapidly reversible. Interestingly in the absence of pulmonary vasoconstriction, breathing NO did not alter pulmonary vascular resistance (PVR). Most importantly, breathing up to 80 ppm NO did not alter systemic blood pressure and did not cause lung injury. Moreover, breathing 80 ppm NO for up to 3 h did not significantly increase methemoglobin levels.

Mechanisms Responsible for the Pulmonary Vasodilator Effects of Breathing NO

As noted above, NO signals via both cGMP-dependent and cGMP-independent mechanisms. Several lines of evidence suggest that inhaled NO induces pulmonary vasodilation via a cGMP-dependent mechanism. In early studies, it was reported that breathing NO led to an increase in plasma cGMP levels [6]. Differences in arterial and venous cGMP levels suggested that the lungs were net producers of cGMP during NO inhalation [7].

Strategies designed to decrease cGMP metabolism by inhibiting PDE5 augmented the pulmonary vasodilator effects of inhaled NO in animal models [8] and in patients with pulmonary hypertension [9, 10]. In addition, administration of an agent that sensitizes sGC to NO, BAY41-2272, augmented the pulmonary vasodilator effects of inhaled NO [11]. As noted above, pulmonary vasodilator effects of breathing NO do not persist after gas administration is discontinued. On the other hand, during BAY41-2272 infusion, the duration of pulmonary vasodilation after discontinuing NO was greater than the duration of the pulmonary vasodilation in the absence of BAY 41-2272. These findings suggest that it may be possible to develop strategies for pulmonary vasodilation that require only intermittent administration of NO thereby reducing the amount of gas needed and potentially improving patient convenience.

sGC is a heterodimer composed of an α subunit and a β subunit. Although the mammalian genome has two α subunits and two β subunits, only α1β1 and α2β1 appear to be active, and α1β1 is thought to be the predominant isoform in the heart and vasculature. Janssens and his team reported that inhaled NO cannot induce pulmonary vasodilation in mice deficient in the sGC α1β1 isoform [12].

Taken together, these results strongly support the concept that inhaled NO vasodilates the pulmonary vasculature via its ability to stimulate cGMP synthesis by sGC. The clinical applications of the combination of inhaled NO with PDE5 inhibitors or sGC activators remains relatively unexplored.

Alternative Strategies to Increase Pulmonary NO and/or cGMP Concentrations Using Inhaled Drugs

Inhalation of NO-donor compounds is effective for treatment of systemic and pulmonary vascular diseases. For example, inhaled nitroglycerin is used to treat chest pain in patients with angina pectoris. In animals with pulmonary hypertension, aerosol inhalation of nitroglycerin [13, 14] or sodium nitroprusside [15] reduced PAP, but higher doses also lowered systemic blood pressure. In small clinical series, inhalation of nitroprusside improved oxygenation in hypoxic neonates [16] and reduced PAP and PVR in children with congenital heart disease (CHD) [17].

Stamler and colleagues have explored the pulmonary vascular effects of inhaling an *S*-nitrosothiol, *O*-nitrosoethanol, which is resistant to reaction with oxygen or superoxide [18]. Inhalation of *O*-nitrosoethanol induced selective pulmonary vasodilation in hypoxic piglets. In a small non-randomized clinical study, these investigators reported that *O*-nitrosoethanol inhalation improved oxygenation and systemic hemodynamics in hypoxemic newborns [19].

Gladwin and colleagues have studied the therapeutic applications of nitrite as a NO reservoir [20]. Nitrite can be converted to NO by a variety of enzymes including hemoglobin and xanthine oxidoreductase. Hunter and colleagues reported that inhalation of high concentrations of nitrite reduced PAP in newborn lambs with hypoxia- and U46619-induced pulmonary vasoconstriction [21]. More recently, these investigators reported that intermittent administration of a low-dose nitrite aerosol could attenuate pulmonary vascular remodeling in rodent models of pulmonary arterial hypertension (PAH) [22].

A variety of other strategies using inhaled molecules have been developed to increase pulmonary cGMP concentrations. Studying awake lambs with U46619-induced pulmonary hypertension, Ichinose and colleagues demonstrated that inhalation of the PDE5 inhibitors, zaprinast or sildenafil, can selectively dilate the pulmonary vasculature and potentiate the effects of inhaled NO [23, 24]. More recently, Evgenov and colleagues studied the pulmonary vasodilator effects of inhaling sGC stimulators encapsulated in a biodegradable microparticle [25]. They observed that inhalation of microparticles containing BAY 41-8543 induced sustained pulmonary vasodilation in conscious sheep and could markedly potentiate the effects of inhaled NO. Of note, the ability of PDE5 inhibitors and sGC activators to potentiate the pulmonary vasodilator effects of NO appeared to be greater when the agents were delivered as an inhaled aerosol than when they were administered as an intravenous infusion.

Clinical Applications

Over the past 20 years, the application of inhaled NO to a wide variety of pulmonary and systemic disorders has been explored in experimental models and clinical trials. In this chapter, we highlight the application of inhaled NO to treat hypoxia in the newborn, to prevent the development of bronchopulmonary dysplasia (BPD) in premature infants, to improve matching of ventilation and perfusion, and to unload the right ventricle.

Persistent pulmonary hypertension of the newborn: Infants with PPHN typically present shortly after birth with respiratory distress and cyanosis, but with a structurally normal heart [26]. In these infants, the pulmonary vasculature fails to dilate normally with the first few breaths after birth, and high PVR leads to right-to-left shunting of deoxygenated blood via the ductus arterious and/or foramen ovale resulting in severe hypoxemia. The systemic hypoxemia in PPHN often fails to improve with conventional therapies (e.g., supplemental oxygen), and acidosis and further pulmonary vasoconstriction may develop [27]. Extracorporeal membrane oxygenation (ECMO) is commonly required to rescue these infants.

The incidence of PPHN is estimated at 0.2% of live-born term or near-term infants (≥34 weeks gestation). Persistent pulmonary hypertension of the newborn (PPHN) may be associated with normal lung parenchyma and excessively muscularized pulmonary vessels (idiopathic PPHN). In other cases, hypoplastic pulmonary vasculature, such as that associated with congenital diaphragmatic hernia, may cause pulmonary hypertension to persist after birth. In addition, PPHN is often associated with other conditions in which perinatal pulmonary vasodilation is inhibited including pulmonary diseases and sepsis.

Two pilot studies reported in the Lancet in 1992 demonstrated that breathing NO could increase oxygenation in severely hypoxemic newborns with PPHN without causing systemic hypotension [28, 29]. Multicenter randomized controlled studies confirmed that breathing NO can increase systemic oxygen levels in term and near-term babies with hypoxemia and pulmonary hypertension and reduce the need for ECMO [30, 31]. It is important to note that breathing NO did not improve survival in PPHN patients in these trials likely because ECMO was used as a rescue therapy for infants with persistent hypoxemia. Moreover, inhaled NO did not improve survival or reduce ECMO use in infants with hypoxemia associated with congenital diaphragmatic hernia [32]. Hypoxemia in term and near-term infants with pulmonary hypertension is the only indication for which treatment with inhaled NO is currently approved by the US Food and Drug Administration.

Bronchopulmonary dysplasia: BPD is a chronic respiratory disease observed in premature infants who experience lung injury due to oxygen therapy and mechanical ventilation. Pathologic findings are characterized by alveolar hypoplasia with reduced surface area available for gas exchange and abnormal pulmonary vascular development, which sometimes leads to pulmonary hypertension. In experimental animal models of BPD, breathing NO appeared to enhance lung growth [33–35] and decrease pulmonary vascular disease. An early report by Kinsella and colleagues suggested that breathing NO did not increase the survival of severely hypoxic premature newborns at risk for BPD [36]. However, in a single center randomized controlled trial, Schreiber and colleagues observed that breathing NO decreased the incidence of BPD and mortality in pre-term infants [37]. Multiple multicenter randomized controlled trials have followed up on these findings with mixed results. Comparison of these trials is challenging because of differences in the patient populations studied, as well as the dose, duration, and timing of NO administration. In two trials, early treatment of severely hypoxemic premature infants with inhaled NO did not reduce the incidence of BPD and tended to increase the frequency of severe intraventricular cerebral hemorrhage (IVH) [38, 39]. However, in subgroup analyses, it appeared that NO inhalation by larger infants (>1 kg birth weight) reduced the incidence of BPD and IVH. Moreover, in another multicenter clinical trial in which older infants (7–21 days of age) at increased risk of developing BPD were studied, breathing NO also reduced the incidence of BPD [40]. On the other hand, in a recently completed trial of premature infants with mild to moderate respiratory distress, inhaled NO did not prevent BPD and did not reduce brain injury [41]. Whether or not inhaled NO is beneficial in subgroups of premature infants remains to be determined.

Improving matching of ventilation and perfusion with inhaled NO: In acute respiratory distress syndrome (ARDS) and acute lung injury (ALI), perfusion of poorly-ventilated lung regions leads to hypoxemia that often is resistant to supplemental oxygen. By virtue of its method of administration, inhaled NO is selectively delivered to well-ventilated portions of the lung where it vasodilates the vasculature and improves matching of ventilation with perfusion and, thereby, enhances oxygenation. In 1993, Roissant and colleagues studied nine patients with severe ARDS using the multiple inert-gas-elimination technique [42]. They reported that inhaled NO improved oxygenation and reduced PAP in these patients associated with a reduction in intrapulmonary shunting. These investigators later reported that the ability of inhaled NO to improve systemic oxygenation in patients with ALI did not require a reduction in mean PAP [43]. Only very low doses of NO were required to improve oxygenation, and concentrations higher than 10 ppm could worsen oxygenation, possible by spilling over to poorly-ventilated lung regions. However, despite improving systemic oxygenation, inhaled NO did not improve outcomes in patients with ARDS in multiple randomized clinical trials [44]. Moreover, there was an increased risk of renal dysfunction in patients who were randomized to receive inhaled NO. Nonetheless, some clinicians use inhaled NO as a rescue therapy in patients with severe hypoxemia.

Unloading the right ventricle with inhaled NO: In adults and children undergoing cardiac surgery and cardiac transplantation, inhaled NO is frequently used to treat pulmonary hypertension and improve cardiac output in the perioperative period [45, 46]. Preoperative evaluation of pulmonary vasoreactivity is useful for identifying patients with pulmonary hypertension who are likely to respond to inhaled NO in the perioperative period (see "Evaluation of pulmonary vasoreactivity with inhaled NO" below).

Right ventricular failure in the perioperative period after placement of a left ventricular assist device (LVAD) is associated with increased mortality [47, 48]. Frequently, placement of a right ventricular assist device (RVAD) is required to maintain LVAD flow. Nitric oxide inhalation reduces PVR and can augment LVAD flow without causing systemic hypotension [49, 50]. In a small clinical trial of LVAD patients with elevated PVR, who were randomly assigned to treatment with and without inhaled NO, Argenziano and colleagues observed that inhaled NO reduced mean PAP and augmented LVAD flow [51]. In patients with RV failure after LVAD placement, clinicians routinely administer inhaled NO prior to implantation of an RVAD. However, a randomized clinical trial demonstrating that inhaled NO decreases the necessity for RVAD placement has yet to be reported.

In a subgroup of patients presenting with ST elevation myocardial infarction (MI) and occlusion of the right coronary artery, RV myocardial infarction (RVMI) can cause RV failure, resulting in cardiogenic shock without pulmonary edema. Prompt revascularization is the most effective means to treat cardiogenic shock due to RVMI [52]. However, in patients who present late in the course of their MI or in whom RV marginal coronary arteries cannot be fully revascularized, cardiogenic shock may require hemodynamic support with inotropic agents, intra-aortic balloon counterpulsation, or even RVAD placement. Several case reports [53–55] and one report of 13 patients [56] demonstrated that breathing NO can improve cardiac index in patients with cardiogenic shock due to RVMI. Moreover, in the subset of RVMI patients with hypoxemia due to right-to-left shunting via a patent foramen ovale, breathing NO reduces shunting and improves oxygenation [54, 56].

Evaluation of pulmonary vasoreactivity with inhaled NO: Acute pulmonary vasodilator testing (APVT) is performed in clinical practice for three primary indications. Most commonly, APVT is used in patients with World Health Organization (WHO) Group I PAH to identify those patients with greatest potential for sustained benefit to oral calcium-channel blocker (CCB) therapy and to evaluate clinical prognosis [57]. WHO Group I PAH patients have a resting mean pulmonary arterial pressure (mPAP) ≥25 mmHg at rest with a pulmonary capillary wedge pressure of ≤15 mmHg (to exclude patients with pulmonary venous hypertension) and a PVR ≥3 Wood units. This group includes patients with idiopathic PAH (IPAH) and familial PAH (FPAH), as well as patients with associated forms of PAH (APAH; e.g., pulmonary hypertension in conjunction with connective tissue disease, portal hypertension, HIV, drugs and toxins, etc.). Inhaled NO is the preferred agent of choice during APVT for PAH, since it is relatively well-tolerated, is selective for the pulmonary circulation, and has rapid onset and offset.

A positive response to APVT in PAH is currently defined as a decrease in mean PAP by at least 10 mmHg to an absolute level less than 40 mmHg without a decrease in cardiac output. The use of CCBs as a vasodilator in APVT should be avoided since a fraction of non-responders will develop severe adverse effects including shock and severe systemic hypotension [58]. On the other hand, the degree of vasoreactive responsiveness to inhaled NO during APVT is predictive of long-term benefit to CCB therapy in PAH [58, 59]. Although alternative agents such as intravenous epoprostenol or intravenous adenosine can be used during APVT, the acute administration of these agents has been associated with systemic hypotension and other adverse effects [60, 61].

APVT with inhaled NO has also been used in patients with CHD and pulmonary hypertension undergoing preoperative evaluation for either corrective surgery or cardiac transplantation [62]. The presence of fixed PH is a predictor of perioperative morbidity and mortality in CHD. Although exact criteria have not been prospectively validated, many centers use a PVR less than 10–14 Wood units, a pulmonary:systemic vascular resistance (PVR:SVR) ratio less than or equal to 0.66, and a PVR:SVR ratio less than or equal to 0.33 during vasodilator administration as thresholds for determining operability for surgical correction and identifying patients with improved surgical prognosis. Preoperative APVT with a combination of inhaled NO and oxygen is more sensitive in identifying appropriate operative candidates than the use of oxygen alone in vasodilator testing [63].

APVT is also used to assess perioperative risk in patients with acquired heart disease who are undergoing evaluation for heart transplantation. Elevated resting PVR and transpulmonary gradient are independent predictors of increased post-transplant mortality [64–66]. Patients with fixed PH have a poorer postoperative prognosis than do patients who exhibit reversibility with APVT [67]. Although no thresholds have been prospectively validated, typical values used to determine candidacy for heart transplantation include a PVR less than or equal to 6 Wood units at rest or less than 3 Wood units with maximal vasodilation [68]. Patients exceeding these threshold values can be considered for combined heart–lung transplantation. Inhaled NO has been used for APVT in the preoperative evaluation for cardiac transplantation [69–71], but operators must proceed with caution because NO inhalation can increase LV filling pressures in the presence of LV dysfunction and result in pulmonary edema [69, 72, 73].

Methods Required for Safe Administration of Inhaled NO

NO reacts with oxygen to form the pollutant nitrogen dioxide (NO_2). The amount of NO_2 formed depends on the concentrations of NO and oxygen and for how long the two gases are mixed. Exposure of the lung and respiratory epithelium to NO_2 can cause pulmonary injury. Moreover, because NO is unstable when mixed with oxygen, it must be shipped and stored in nitrogen.

During the clinical development of inhaled NO, new technologies and equipment were devised to enable the reliable and continuous delivery of low concentrations of NO gas and the mixture of NO with oxygen (at concentrations dictated by the needs of the patient) via endotracheal tube, tight-fitting face mask, or nasal canula. Additional considerations included the requirement that the NO delivery system be portable to enable transport of patients breathing NO. Commercially-available equipment minimizes the time during which oxygen and NO are mixed prior to administration to the patient and monitors the levels of NO, oxygen, and NO_2 administered to the patient in real-time. Clinicians who administer NO gas to patients need to be aware of several "side effects" associated with breathing NO including methemoglobinemia, inhibition of platelet function, and increases in pulmonary capillary wedge pressure (in the setting of LV dysfunction; see "Evaluation of pulmonary vasoreactivity with inhaled NO" above). Clinicians also need to be cognizant of the rebound pulmonary hypertension that occurs if inhaled NO is abruptly discontinued.

Methemoglobinemia: When NO reacts with oxyhemoglobin to form nitrate, methemoglobin is formed. Methemoglobin does not bind oxygen, impairing the ability of the blood to deliver oxygen from the lungs to the periphery. Methemoglobin is rapidly converted to ferrous hemoglobin by methemoglobin reductase in erythrocytes. Methemoglobin levels greater than 5% are typically seen only in patients who breathe 80 ppm NO [74] but are also rarely seen in infants breathing lesser concentrations [75].

Bleeding risk: Nitric oxide is well known to reduce the ability of platelets to adhere and become activated. Inhaled NO was reported to increase bleeding time in rabbits and people [76] and to reduce thrombosis after thrombolysis in dogs [77]. The risk of bleeding has been a particular concern in studies of the use of inhaled NO in premature infants who are already at increased risk of intracranial bleeding. However, recent studies of premature infants treated with inhaled NO have not shown an increased risk of IVH [41].

Rebound pulmonary hypertension: In the mid-1990s, two groups reported that pulmonary hypertension could develop as patients were weaned from breathing NO [78, 79]. Using an experimental lamb model, Black and Fineman and their colleagues reported that prolonged NO inhalation led to increased endothelin 1 levels and reduced NO synthase activity, both of which contributed to the pulmonary vascular rebound observed when inhaled NO was discontinued [80]. Administration of PDE5 inhibitors was found to attenuate the rebound seen when withdrawing inhaled NO [81, 82]. In general, however, pulmonary vascular rebound can be prevented by slowly weaning the concentration of inhaled NO.

Extrapulmonary Effects of Breathing NO

Initial studies by Zapol and colleagues suggested that the effects of breathing NO were limited to the lung, as reflected by the absence of systemic vasodilation [5]. However, in 1993, Hogman et al. reported that breathing NO could inhibit platelet function as reflected by a prolongation of the bleeding time [76]. These observations suggest that at least some of the inhaled NO reaching the bloodstream escaped inactivation by hemoglobin.

Although the impact of inhaled NO on platelet function remains controversial [83–85], Hogman's report led other research teams to identify additional systemic effects of breathing NO. For example, Lee and colleagues reported that breathing NO (80 ppm) for 2 weeks could reduce neointima formation in balloon-injured rat carotid arteries [86]. Semigran and colleagues demonstrated that breathing 80 ppm NO, but not 20 ppm, enhanced coronary artery patency in a canine model of coronary thrombosis after thrombolysis [77]. These investigators went on to show that PDE5 inhibitors markedly augment the ability of inhaled NO to prevent thrombosis [87].

Fox-Robichaud and colleagues explored the impact of inhaled NO on leukocyte function [88]. Studying a feline model of intestinal ischemia–reperfusion injury, these investigators reported that breathing 80 ppm NO, but not 20 ppm, could decrease leukocyte adherence and activation and maintain mesenteric blood flow.

Hataishi and colleagues extended these observations by studying the impact of breathing NO on cardiac ischemia–reperfusion injury [89]. Studying mice subjected to transient coronary artery ligation and reperfusion, they reported that breathing NO for 24 h (beginning during ischemia) could decrease MI size. Similar to the findings of Fox-Robichaud et al., Hataishi and his team reported that breathing 40 or 80 ppm NO was cardioprotective, whereas breathing 20 ppm NO was not. Liu and colleagues explored the impact of breathing NO on cardiac ischemia–reperfusion in a large animal model [90]. They observed that breathing 80 ppm NO reduced MI size in the hearts of pigs subjected to transient coronary artery occlusion, whereas intravenous administration of nitroglycerin did not. Of note, in both the mouse and pig models, breathing NO during cardiac ischemia and reperfusion reduced accumulation of leukocytes in the heart and importantly did not cause systemic hypotension. Taken together, these observations suggest that concentrations of NO required to elicit effects in the systemic circulation are greater than those required to modulate pulmonary vascular tone.

Promising results using inhaled NO in experimental models of ischemia–reperfusion injury has led to several small-scale clinical trials. Lang and colleagues reported that breathing 80 ppm NO reduced muscle inflammation in the lower extremities of patients undergoing surgical limb ischemia (for knee surgery requiring a tourniquet for a bloodless surgical field) and reperfusion [91]. These investigators went on to report that breathing NO reduced ischemia–reperfusion injury in liver transplantation associated with a reduction in hepatocyte apoptosis and improved synthetic function [92]. Although both of these studies were encouraging, they were small (~10 patients in each experimental group). Additional multicenter randomized controlled trials will be necessary to determine whether inhaled NO can improve outcomes in patients experiencing systemic organ ischemia–reperfusion injury.

Fox-Robichaud and colleagues reported that breathing NO elicited its greatest effects under conditions characterized by reduced NO levels, such as those caused by inhibiting NO synthase or induced by ischemia and reperfusion [88]. Studies performed by Gladwin and colleagues supported this hypothesis by demonstrating that breathing NO could increase forearm blood flow during regional blockade of NO synthesis [93, 94]. In contrast, Hataishi et al. failed to detect systemic vasodilation in response to inhaled NO in mice deficient in the endothelial NO synthase isoform (NOS3) [95].

In another model characterized by a systemic deficiency of NO, Gladwin and colleagues found that hemolysis induced systemic vasoconstriction via the ability of cell-free hemoglobin to scavenge NO generated by the endothelium. They reported that breathing 80 ppm NO dramatically reduced systemic blood pressure in dogs with systemic vasoconstriction induced by hemolysis associated with nearly complete oxidation of cell-free hemoglobin to methemoglobin [96]. These investigators hypothesized that oxidation of extracellular hemoglobin is required for the vasodilator effects of inhaled NO in hemolysis. This hypothesis was challenged by the findings of Yu et al. [97], who reported that pretreatment of mice with inhaled NO could prevent the vasoconstriction induced by the subsequent administration of cell-free hemoglobin without oxidizing the extracellular hemoglobin to methemoglobin. The observations of Yu and colleagues suggest that pretreatment with inhaled NO may enable administration of hemoglobin-based oxygen carriers as blood substitutes [98].

How does inhaled NO elicit its extrapulmonary effects? Because of the exceedingly short half-life of NO in biological fluids, it seems unlikely that NO reaching the bloodstream during NO inhalation will be transported to the periphery in an unmodified form. Several hypotheses have been proposed to account for the ability of inhaled NO to elicit systemic effects (Fig. 12.1). One hypothesis is that exposure of platelets and leukocytes to NO, as they transit the pulmonary circulation, would inhibit their function in the periphery. Fox-Robichaud et al. reported that breathing NO did not alter the activation of leukocytes exposed to an artificial surface suggesting that inhaled NO does not directly inactivate leukocytes [88].

An alternate hypothesis is that inhaled NO that reaches the circulation can react with plasma and/or cellular constituents to form stable NO metabolites such as nitrite and *S*-nitrosothiols. These stable NO metabolites, in turn, regenerate NO in the periphery. A variety of NO metabolites have been detected in animals and human beings breathing NO. NO can react with heme in hemoglobin forming nitrosyl-hemoglobin and can nitrosylate the cysteine[93] in the hemoglobin β unit to form SNO-hemoglobin. SNO-hemoglobin formation depends on oxygen concentrations: NO reacts with cysteine [93] in the presence of high oxygen concentrations in the lung, and SNO-hemoglobin releases NO under the low oxygen concentrations present in the periphery [99].

Breathing NO leads to a modest increase in the blood levels of nitrite and a more marked increase in nitrate levels [93]. Recent studies have demonstrated that administration of nitrite can reduce cardiac ischemia–reperfusion injury [100, 101] leading to the speculation that the cardioprotective effects of inhaled NO are attributable to nitrite.

Nitric oxide reacts with a broad spectrum of plasma components to produce *S*-nitrosothiols (RSNO) and *N*-nitrosamines (RNNO). Recently, Nagasaka and Fernandes and their colleagues performed a quantitative analysis of the NO metabolites formed in blood and tissues during NO breathing in mice [102]. They reported that NO inhalation led to the rapid accumulation of nitrate and nitrite in plasma and erythrocytes, as well as a dramatic increase in RSNO, RNNO, and nitrosyl-heme concentrations in erythrocytes (Fig. 12.2). These findings suggest that both blood compartments could contribute to the transport of NO metabolites

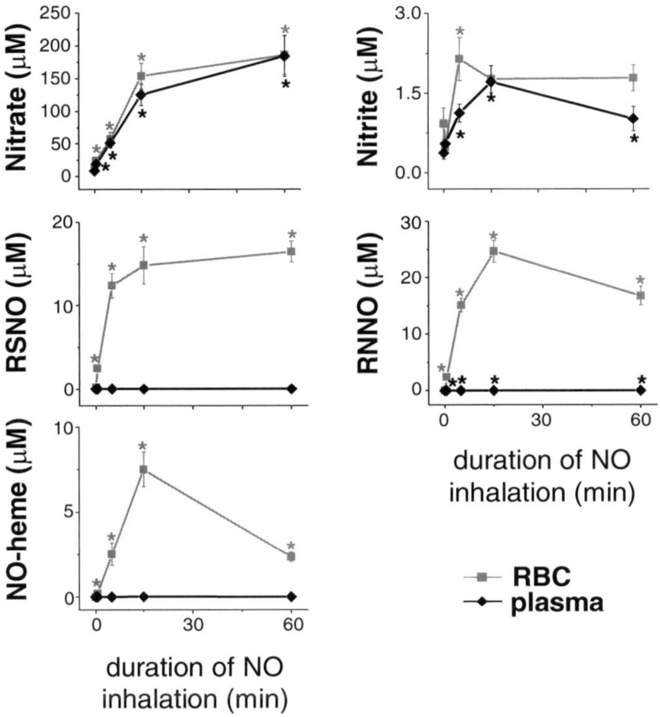

Fig. 12.2 Distribution and kinetics of accumulation of NO-metabolites in mice breathing nitric oxide (plasma and erythrocytes). Concentrations of NO-metabolites were measured in blood of mice breathing air supplemented with NO (80 ppm) for 0, 0.5, 5, 15, and 60 min (n=5–7). *RBC* erythrocytes; *NO-heme* nitrosyl-heme species; *RNNO* *N*-nitrosamines; *RSNO* *S*-nitrosothiols. *P<0.05 versus mice not breathing nitric oxide. Adapted with permission from reference Nagasaka et al. [102]

to the periphery. During NO inhalation, NO metabolites accumulated in all of the tissues sampled. However, the distribution and kinetics of NO metabolite accumulation during NO inhalation was organ-specific. Taken together, these observations support the hypothesis that inhaled NO elicits its systemic effects, at least in part, via transport of NO metabolites to the periphery.

Conclusions

Over 20 years ago, Frostell and Zapol learned that breathing low concentrations of a "poisonous" gas, NO, could reduce pulmonary vascular tone in a uniquely selective manner. Extensive basic and clinical research led to the rapid translation of the remarkably safe pulmonary vasodilator effects of breathing NO to benefit neonates with hypoxemia and pulmonary hypertension. On the other hand, randomized clinical trials have demonstrated that inhaled NO does not improve outcomes in most patients with ARDS, and the impact of inhaled NO on BPD is at best uncertain. The observations in animal models and small clinical trials that breathing NO can load the body with NO metabolites, such as nitrite and nitrate, sufficient to decrease ischemia–reperfusion injury without causing systemic hypotension are exciting and merit additional investigation.

Acknowledgements and Disclosures The authors acknowledge research funding support from the Deutsche Forschungsgemeinschaft DFG (AUS), NHLBI 5T32HL007208 (RM), and a sponsored research agreement with Ikaria, Inc. (KDB).

The authors acknowledge the mentorship and support of Warren M. Zapol MD. The authors thank Drs. Fumito Ichinose and Jesse D. Roberts Jr. for their critical review of this manuscript.

The MGH has obtained patents relating to the use of inhaled nitric oxide and has licensed them to Ikaria, Inc. and Linde Gas Therapeutics AB.

References

1. Packer M, Halperin JL, Brooks KM, Rothlauf EB, Lee WH. Nitroglycerin therapy in the management of pulmonary hypertensive disorders. Am J Med. 1984;76(6A):67–75.
2. Stasch JP, Hobbs AJ. NO-independent, haem-dependent soluble guanylate cyclase stimulators. Handb Exp Pharmacol. 2009;191:277–308.
3. Schmidt HH, Schmidt PM, Stasch JP. NO- and haem-independent soluble guanylate cyclase activators. Handb Exp Pharmacol. 2009;191:309–39.
4. Norman V, Keith CH. Nitrogen oxides in tobacco smoke. Nature. 1965;205:915–6.
5. Frostell C, Fratacci MD, Wain JC, Jones R, Zapol WM. Inhaled nitric oxide. A selective pulmonary vasodilator reversing hypoxic pulmonary vasoconstriction. Circulation. 1991;83(6):2038–47.
6. Roberts Jr JD, Chen TY, Kawai N, et al. Inhaled nitric oxide reverses pulmonary vasoconstriction in the hypoxic and acidotic newborn lamb. Circ Res. 1993;72(2):246–54.
7. Rovira I, Chen TY, Winkler M, Kawai N, Bloch KD, Zapol WM. Effects of inhaled nitric oxide on pulmonary hemodynamics and gas exchange in an ovine model of ARDS. J Appl Physiol. 1994;76(1):345–55.
8. Ichinose F, Adrie C, Hurford WE, Zapol WM. Prolonged pulmonary vasodilator action of inhaled nitric oxide by Zaprinast in awake lambs. J Appl Physiol. 1995;78(4):1288–95.
9. Ivy DD, Ziegler JW, Kinsella JP, Wiggins JW, Abman SH. Hemodynamic effects of dipyridamole and inhaled nitric oxide in pediatric patients with pulmonary hypertension. Chest. 1998;114(1 Suppl):17S.
10. Lepore JJ, Maroo A, Pereira NL, et al. Effect of sildenafil on the acute pulmonary vasodilator response to inhaled nitric oxide in adults with primary pulmonary hypertension. Am J Cardiol. 2002;90(6):677–80.
11. Evgenov OV, Ichinose F, Evgenov NV, et al. Soluble guanylate cyclase activator reverses acute pulmonary hypertension and augments the pulmonary vasodilator response to inhaled nitric oxide in awake lambs. Circulation. 2004;110(15):2253–9.
12. Vermeersch P, Buys E, Pokreisz P, et al. Soluble guanylate cyclase-alpha1 deficiency selectively inhibits the pulmonary vasodilator response to nitric oxide and increases the pulmonary vascular remodeling response to chronic hypoxia. Circulation. 2007;116(8):936–43.
13. Bando M, Ishii Y, Kitamura S, Ohno S. Effects of inhalation of nitroglycerin on hypoxic pulmonary vasoconstriction. Respiration. 1998;65(1):63–70.

14. Gong F, Shiraishi H, Kikuchi Y, et al. Inhalation of nebulized nitroglycerin in dogs with experimental pulmonary hypertension induced by U46619. Pediatr Int. 2000;42(3):255–8.
15. Adrie C, Ichinose F, Holzmann A, Keefer L, Hurford WE, Zapol WM. Pulmonary vasodilation by nitric oxide gas and prodrug aerosols in acute pulmonary hypertension. J Appl Physiol. 1998;84(2):435–41.
16. Palhares DB, Figueiredo CS, Moura AJ. Endotracheal inhalatory sodium nitroprusside in severely hypoxic newborns. J Perinat Med. 1998;26(3):219–24.
17. Goyal P, Kiran U, Chauhan S, Juneja R, Choudhary M. Efficacy of nitroglycerin inhalation in reducing pulmonary arterial hypertension in children with congenital heart disease. Br J Anaesth. 2006;97(2):208–14.
18. Moya MP, Gow AJ, McMahon TJ, et al. *S*-nitrosothiol repletion by an inhaled gas regulates pulmonary function. Proc Natl Acad Sci USA. 2001;98(10):5792–7.
19. Moya MP, Gow AJ, Califf RM, Goldberg RN, Stamler JS. Inhaled ethyl nitrite gas for persistent pulmonary hypertension of the newborn. Lancet. 2002;360(9327):141–3.
20. Lundberg JO, Weitzberg E, Gladwin MT. The nitrate-nitrite-nitric oxide pathway in physiology and therapeutics. Nat Rev Drug Discov. 2008;7(2):156–67.
21. Hunter CJ, Dejam A, Blood AB, et al. Inhaled nebulized nitrite is a hypoxia-sensitive NO-dependent selective pulmonary vasodilator. Nat Med. 2004;10(10):1122–7.
22. Zuckerbraun BS, Shiva S, Ifedigbo E, et al. Nitrite potently inhibits hypoxic and inflammatory pulmonary arterial hypertension and smooth muscle proliferation via xanthine oxidoreductase-dependent nitric oxide generation. Circulation. 2010;121(1):98–109.
23. Ichinose F, Adrie C, Hurford WE, Bloch KD, Zapol WM. Selective pulmonary vasodilation induced by aerosolized zaprinast. Anesthesiology. 1998;88(2):410–6.
24. Ichinose F, Erana-Garcia J, Hromi J, et al. Nebulized sildenafil is a selective pulmonary vasodilator in lambs with acute pulmonary hypertension. Crit Care Med. 2001;29(5):1000–5.
25. Evgenov OV, Kohane DS, Bloch KD, et al. Inhaled agonists of soluble guanylate cyclase induce selective pulmonary vasodilation. Am J Respir Crit Care Med. 2007;176(11):1138–45.
26. Steinhorn RH. Neonatal pulmonary hypertension. Pediatr Crit Care Med. 2010;11(2 Suppl):S79–84.
27. Rao S, Bartle D, Patole S. Current and future therapeutic options for persistent pulmonary hypertension in the newborn. Expert Rev Cardiovasc Ther. 2010;8(6):845–62.
28. Roberts JD, Polaner DM, Lang P, Zapol WM. Inhaled nitric oxide in persistent pulmonary hypertension of the newborn. Lancet. 1992;340(8823):818–9.
29. Kinsella JP, Neish SR, Shaffer E, Abman SH. Low-dose inhalation nitric oxide in persistent pulmonary hypertension of the newborn. Lancet. 1992;340(8823):819–20.
30. Konduri GG, Kim UO. Advances in the diagnosis and management of persistent pulmonary hypertension of the newborn. Pediatr Clin North Am. 2009;56(3):579–600, Table of Contents.
31. Finer NN, Barrington KJ. Nitric oxide for respiratory failure in infants born at or near term. Cochrane Database Syst Rev. 2001;(4):CD000399.
32. The Neonatal Inhaled Nitric Oxide Study Group (NINOS). Inhaled nitric oxide and hypoxic respiratory failure in infants with congenital diaphragmatic hernia. Pediatrics. 1997;99(6):838–45.
33. Roberts Jr JD, Chiche JD, Weimann J, Steudel W, Zapol WM, Bloch KD. Nitric oxide inhalation decreases pulmonary artery remodeling in the injured lungs of rat pups. Circ Res. 2000;87(2):140–5.
34. Tang JR, Markham NE, Lin YJ, et al. Inhaled nitric oxide attenuates pulmonary hypertension and improves lung growth in infant rats after neonatal treatment with a VEGF receptor inhibitor. Am J Physiol Lung Cell Mol Physiol. 2004;287(2):L344–51.
35. McCurnin DC, Pierce RA, Chang LY, et al. Inhaled NO improves early pulmonary function and modifies lung growth and elastin deposition in a baboon model of neonatal chronic lung disease. Am J Physiol Lung Cell Mol Physiol. 2005;288(3):L450–9.
36. Kinsella JP, Walsh WF, Bose CL, et al. Inhaled nitric oxide in premature neonates with severe hypoxaemic respiratory failure: a randomised controlled trial. Lancet. 1999;354(9184):1061–5.

37. Schreiber MD, Gin-Mestan K, Marks JD, Huo D, Lee G, Srisuparp P. Inhaled nitric oxide in premature infants with the respiratory distress syndrome. N Engl J Med. 2003;349(22):2099–107.
38. Van Meurs KP, Wright LL, Ehrenkranz RA, et al. Inhaled nitric oxide for premature infants with severe respiratory failure. N Engl J Med. 2005;353(1):13–22.
39. Kinsella JP, Cutter GR, Walsh WF, et al. Early inhaled nitric oxide therapy in premature newborns with respiratory failure. N Engl J Med. 2006;355(4):354–64.
40. Ballard RA, Truog WE, Cnaan A, et al. Inhaled nitric oxide in preterm infants undergoing mechanical ventilation. N Engl J Med. 2006;355(4):343–53.
41. Mercier JC, Hummler H, Durrmeyer X, et al. Inhaled nitric oxide for prevention of bronchopulmonary dysplasia in premature babies (EUNO): a randomised controlled trial. Lancet. 2010;376(9738):346–54.
42. Rossaint R, Falke KJ, Lopez F, Slama K, Pison U, Zapol WM. Inhaled nitric oxide for the adult respiratory distress syndrome. N Engl J Med. 1993;328(6):399–405.
43. Gerlach H, Rossaint R, Pappert D, Falke KJ. Time-course and dose-response of nitric oxide inhalation for systemic oxygenation and pulmonary hypertension in patients with adult respiratory distress syndrome. Eur J Clin Invest. 1993;23(8):499–502.
44. Adhikari NK, Burns KE, Friedrich JO, Granton JT, Cook DJ, Meade MO. Effect of nitric oxide on oxygenation and mortality in acute lung injury: systematic review and meta-analysis. BMJ. 2007;334(7597):779.
45. Fullerton DA, Jones SD, Jaggers J, Piedalue F, Grover FL, McIntyre Jr RC. Effective control of pulmonary vascular resistance with inhaled nitric oxide after cardiac operation. J Thorac Cardiovasc Surg. 1996;111(4):753–62. discussion 762–3.
46. Beck JR, Mongero LB, Kroslowitz RM, et al. Inhaled nitric oxide improves hemodynamics in patients with acute pulmonary hypertension after high-risk cardiac surgery. Perfusion. 1999;14(1):37–42.
47. Matthews JC, Koelling TM, Pagani FD, Aaronson KD. The right ventricular failure risk score a pre-operative tool for assessing the risk of right ventricular failure in left ventricular assist device candidates. J Am Coll Cardiol. 2008;51(22):2163–72.
48. Drakos SG, Janicki L, Horne BD, et al. Risk factors predictive of right ventricular failure after left ventricular assist device implantation. Am J Cardiol. 2010;105(7):1030–5.
49. Hare JM, Shernan SK, Body SC, Graydon E, Colucci WS, Couper GS. Influence of inhaled nitric oxide on systemic flow and ventricular filling pressure in patients receiving mechanical circulatory assistance. Circulation. 1997;95(9):2250–3.
50. Wagner F, Dandel M, Gunther G, et al. Nitric oxide inhalation in the treatment of right ventricular dysfunction following left ventricular assist device implantation. Circulation. 1997;96(9 Suppl):II-291–6.
51. Argenziano M, Choudhri AF, Moazami N, et al. Randomized, double-blind trial of inhaled nitric oxide in LVAD recipients with pulmonary hypertension. Ann Thorac Surg. 1998;65(2):340–5.
52. Bowers TR, O'Neill WW, Grines C, Pica MC, Safian RD, Goldstein JA. Effect of reperfusion on biventricular function and survival after right ventricular infarction. N Engl J Med. 1998;338(14):933–40.
53. Fujita Y, Nishida O, Sobue K, et al. Nitric oxide inhalation is useful in the management of right ventricular failure caused by myocardial infarction. Crit Care Med. 2002;30(6):1379–81.
54. Fessler MB, Lepore JJ, Thompson BT, Semigran MJ. Right-to-left shunting through a patent foramen ovale in right ventricular infarction: improvement of hypoxemia and hemodynamics with inhaled nitric oxide. J Clin Anesth. 2003;15(5):371–4.
55. Valenti V, Patel AJ, Sciarretta S, Kandil H, Fabrizio B, Ballotta A. Use of inhaled nitric oxide in the treatment of right ventricular myocardial infarction. Am J Emerg Med. 2010 Jul 30. [Epub ahead of print].
56. Inglessis I, Shin JT, Lepore JJ, et al. Hemodynamic effects of inhaled nitric oxide in right ventricular myocardial infarction and cardiogenic shock. J Am Coll Cardiol. 2004;44(4):793–8.
57. McLaughlin VV, Archer SL, Badesch DB, et al. ACCF/AHA 2009 expert consensus document on pulmonary hypertension a report of the American College of Cardiology Foundation Task Force on Expert Consensus Documents and the American Heart Association developed in

collaboration with the American College of Chest Physicians; American Thoracic Society, Inc.; and the Pulmonary Hypertension Association. J Am Coll Cardiol. 2009;53(17):1573–619.
58. Sitbon O, Humbert M, Jagot JL, et al. Inhaled nitric oxide as a screening agent for safely identifying responders to oral calcium-channel blockers in primary pulmonary hypertension. Eur Respir J. 1998;12(2):265–70.
59. Sitbon O, Humbert M, Jais X, et al. Long-term response to calcium channel blockers in idiopathic pulmonary arterial hypertension. Circulation. 2005;111(23):3105–11.
60. Badesch DB, Abman SH, Ahearn GS, et al. Medical therapy for pulmonary arterial hypertension: ACCP evidence-based clinical practice guidelines. Chest. 2004;126(1 Suppl):35S–62.
61. Oliveira EC, Ribeiro AL, Amaral CF. Adenosine for vasoreactivity testing in pulmonary hypertension: a head-to-head comparison with inhaled nitric oxide. Respir Med. 2010;104(4):606–11.
62. Warnes CA, Williams RG, Bashore TM, et al. ACC/AHA 2008 guidelines for the management of adults with congenital heart disease: a report of the American College of Cardiology/American Heart Association Task Force on Practice Guidelines (writing committee to develop guidelines on the management of adults with congenital heart disease). Circulation. 2008;118(23):e714–833.
63. Balzer DT, Kort HW, Day RW, et al. Inhaled nitric oxide as a preoperative test (INOP Test I): the INOP test study group. Circulation. 2002;106(12 Suppl 1):I76–81.
64. Kirklin JK, Naftel DC, Kirklin JW, Blackstone EH, White-Williams C, Bourge RC. Pulmonary vascular resistance and the risk of heart transplantation. J Heart Transplant. 1988;7(5):331–6.
65. Erickson KW, Costanzo-Nordin MR, O'Sullivan EJ, et al. Influence of preoperative transpulmonary gradient on late mortality after orthotopic heart transplantation. J Heart Transplant. 1990;9(5):526–37.
66. Murali S, Kormos RL, Uretsky BF, et al. Preoperative pulmonary hemodynamics and early mortality after orthotopic cardiac transplantation: the Pittsburgh experience. Am Heart J. 1993;126(4):896–904.
67. Chen JM, Levin HR, Michler RE, Prusmack CJ, Rose EA, Aaronson KD. Reevaluating the significance of pulmonary hypertension before cardiac transplantation: determination of optimal thresholds and quantification of the effect of reversibility on perioperative mortality. J Thorac Cardiovasc Surg. 1997;114(4):627–34.
68. Natale ME, Pina IL. Evaluation of pulmonary hypertension in heart transplant candidates. Curr Opin Cardiol. 2003;18(2):136–40.
69. Semigran MJ, Cockrill BA, Kacmarek R, et al. Hemodynamic effects of inhaled nitric oxide in heart failure. J Am Coll Cardiol. 1994;24(4):982–8.
70. Adatia I, Perry S, Landzberg M, Moore P, Thompson JE, Wessel DL. Inhaled nitric oxide and hemodynamic evaluation of patients with pulmonary hypertension before transplantation. J Am Coll Cardiol. 1995;25(7):1656–64.
71. Mahajan A, Shabanie A, Varshney SM, Marijic J, Sopher MJ. Inhaled nitric oxide in the preoperative evaluation of pulmonary hypertension in heart transplant candidates. J Cardiothorac Vasc Anesth. 2007;21(1):51–6.
72. Bocchi EA, Bacal F, Auler Junior JO, Carmone MJ, Bellotti G, Pileggi F. Inhaled nitric oxide leading to pulmonary edema in stable severe heart failure. Am J Cardiol. 1994;74(1):70–2.
73. Loh E, Stamler JS, Hare JM, Loscalzo J, Colucci WS. Cardiovascular effects of inhaled nitric oxide in patients with left ventricular dysfunction. Circulation. 1994;90(6):2780–5.
74. Weinberger B, Laskin DL, Heck DE, Laskin JD. The toxicology of inhaled nitric oxide. Toxicol Sci. 2001;59(1):5–16.
75. Hamon I, Gauthier-Moulinier H, Grelet-Dessioux E, Storme L, Fresson J, Hascoet J. Methaemoglobinaemia risk factors with inhaled nitric oxide therapy in newborn infants. Acta Paediatr. 2010;99(10):1467–73.
76. Hogman M, Frostell C, Arnberg H, Hedenstierna G. Bleeding time prolongation and NO inhalation. Lancet. 1993;341(8861):1664–5.
77. Adrie C, Bloch KD, Moreno PR, et al. Inhaled nitric oxide increases coronary artery patency after thrombolysis. Circulation. 1996;94(8):1919–26.
78. Miller OI, Tang SF, Keech A, Celermajer DS. Rebound pulmonary hypertension on withdrawal from inhaled nitric oxide. Lancet. 1995;346(8966):51–2.

79. Atz AM, Adatia I, Wessel DL. Rebound pulmonary hypertension after inhalation of nitric oxide. Ann Thorac Surg. 1996;62(6):1759–64.
80. Oishi P, Grobe A, Benavidez E, et al. Inhaled nitric oxide induced NOS inhibition and rebound pulmonary hypertension: a role for superoxide and peroxynitrite in the intact lamb. Am J Physiol Lung Cell Mol Physiol. 2006;290(2):L359–66.
81. Ivy DD, Kinsella JP, Ziegler JW, Abman SH. Dipyridamole attenuates rebound pulmonary hypertension after inhaled nitric oxide withdrawal in postoperative congenital heart disease. J Thorac Cardiovasc Surg. 1998;115(4):875–82.
82. Namachivayam P, Theilen U, Butt WW, Cooper SM, Penny DJ, Shekerdemian LS. Sildenafil prevents rebound pulmonary hypertension after withdrawal of nitric oxide in children. Am J Respir Crit Care Med. 2006;174(9):1042–7.
83. Albert J, Norman M, Wallen NH, Frostell C, Hjemdahl P. Inhaled nitric oxide does not influence bleeding time or platelet function in healthy volunteers. Eur J Clin Invest. 1999;29(11):953–9.
84. Gries A, Herr A, Motsch J, et al. Randomized, placebo-controlled, blinded and cross-matched study on the antiplatelet effect of inhaled nitric oxide in healthy volunteers. Thromb Haemost. 2000;83(2):309–15.
85. Beghetti M, Sparling C, Cox PN, Stephens D, Adatia I. Inhaled NO inhibits platelet aggregation and elevates plasma but not intraplatelet cGMP in healthy human volunteers. Am J Physiol Heart Circ Physiol. 2003;285(2):H637–42.
86. Lee JS, Adrie C, Jacob HJ, Roberts Jr JD, Zapol WM, Bloch KD. Chronic inhalation of nitric oxide inhibits neointimal formation after balloon-induced arterial injury. Circ Res. 1996;78(2):337–42.
87. Schmidt U, Han RO, DiSalvo TG, et al. Cessation of platelet-mediated cyclic canine coronary occlusion after thrombolysis by combining nitric oxide inhalation with phosphodiesterase-5 inhibition. J Am Coll Cardiol. 2001;37(7):1981–8.
88. Fox-Robichaud A, Payne D, Hasan SU, et al. Inhaled NO as a viable antiadhesive therapy for ischemia/reperfusion injury of distal microvascular beds. J Clin Invest. 1998;101(11):2497–505.
89. Hataishi R, Rodrigues AC, Neilan TG, et al. Inhaled nitric oxide decreases infarction size and improves left ventricular function in a murine model of myocardial ischemia-reperfusion injury. Am J Physiol Heart Circ Physiol. 2006;291(1):H379–84.
90. Liu X, Huang Y, Pokreisz P, et al. Nitric oxide inhalation improves microvascular flow and decreases infarction size after myocardial ischemia and reperfusion. J Am Coll Cardiol. 2007;50(8):808–17.
91. Mathru M, Huda R, Solanki DR, Hays S, Lang JD. Inhaled nitric oxide attenuates reperfusion inflammatory responses in humans. Anesthesiology. 2007;106(2):275–82.
92. Lang Jr JD, Teng X, Chumley P, et al. Inhaled NO accelerates restoration of liver function in adults following orthotopic liver transplantation. J Clin Invest. 2007;117(9):2583–91.
93. Cannon 3rd RO, Schechter AN, Panza JA, et al. Effects of inhaled nitric oxide on regional blood flow are consistent with intravascular nitric oxide delivery. J Clin Invest. 2001;108(2):279–87.
94. Zuzak KJ, Schaeberle MD, Gladwin MT, Cannon 3rd RO, Levin IW. Noninvasive determination of spatially resolved and time-resolved tissue perfusion in humans during nitric oxide inhibition and inhalation by use of a visible-reflectance hyperspectral imaging technique. Circulation. 2001;104(24):2905–10.
95. Hataishi R, Zapol WM, Bloch KD, Ichinose F. Inhaled nitric oxide does not reduce systemic vascular resistance in mice. Am J Physiol Heart Circ Physiol. 2006;290(5):H1826–9.
96. Minneci PC, Deans KJ, Zhi H, et al. Hemolysis-associated endothelial dysfunction mediated by accelerated NO inactivation by decompartmentalized oxyhemoglobin. J Clin Invest. 2005;115(12):3409–17.
97. Yu B, Raher MJ, Volpato GP, Bloch KD, Ichinose F, Zapol WM. Inhaled nitric oxide enables artificial blood transfusion without hypertension. Circulation. 2008;117(15):1982–90.
98. Yu B, Bloch KD, Zapol WM. Hemoglobin-based red blood cell substitutes and nitric oxide. Trends Cardiovasc Med. 2009;19(3):103–7.

99. Lima B, Forrester MT, Hess DT, Stamler JS. S-nitrosylation in cardiovascular signaling. Circ Res. 2010;106(4):633–46.
100. Duranski MR, Greer JJ, Dejam A, et al. Cytoprotective effects of nitrite during in vivo ischemia-reperfusion of the heart and liver. J Clin Invest. 2005;115(5):1232–40.
101. Gonzalez FM, Shiva S, Vincent PS, et al. Nitrite anion provides potent cytoprotective and antiapoptotic effects as adjunctive therapy to reperfusion for acute myocardial infarction. Circulation. 2008;117(23):2986–94.
102. Nagasaka Y, Fernandez BO, Garcia-Saura MF, et al. Brief periods of nitric oxide inhalation protect against myocardial ischemia-reperfusion injury. Anesthesiology. 2008;109(4):675–82.
103. Bloch KD, Ichinose F, Roberts Jr JD, Zapol WM. Inhaled NO as a therapeutic agent. Cardiovasc Res. 2007;75(2):339–48.

Chapter 13
Pharmacology of Nitrovasodilators

Nathaniel A. Page and Ho-Leung Fung

Key points

- There are two classes of nitrovasodilators used clinically: the organic nitrates (nitroglycerin, isosorbide dinitrate, and isosorbide-5-mononitrate) and the metal nitrosyls (sodium nitroprusside).
- Nitrovasodilators themselves are inactive and must release vasoactive nitric oxide (NO) or its related species to exert their pharmacological effects. The metabolic end-products of nitrovasodilators are nitrite and nitrate, and increased exposure to these ions can be significant especially when nitrovasodilators are used on a chronic basis.
- In addition to their effects on vasotone, nitrovasodilators are also potent anti-platelet agents.
- Nitrovasodilators are used for the management of several cardiovascular pathologies including stable and unstable angina, congestive heart failure, and myocardial infarction.
- Metabolism of organic nitrates can be categorized as mechanism-based (releasing NO) or clearance-based (releasing nitrite). Various enzymes, including mitochondrial aldehyde dehydrogenase and glutathione-*S*-transferase, play key roles in these processes.
- Tolerance to the vasodilatory effects of organic nitrates is mediated via oxidation of multiple protein cysteine groups, producing a myriad of downstream events. A nitrate-free period is used clinically to circumvent this loss of efficacy.
- Organic nitrates can induce endothelial dysfunction through increased oxidative stress and several studies have questioned the long-term safety of these agents.
- Cyanide radical release from the nitroprusside complex and its rapid elimination precludes the use of sodium nitroprusside on a chronic basis.

H.-L. Fung (✉)
Department of Pharmaceutical Sciences, School of Pharmacy and Pharmaceutical Sciences, The University of Buffalo, Buffalo, NY, USA
e-mail: hlfung@buffalo.edu

N.S. Bryan and J. Loscalzo (eds.), *Nitrite and Nitrate in Human Health and Disease*, Nutrition and Health, DOI 10.1007/978-1-60761-616-0_13,

Keywords Nitroglycerin • Nitric oxide • Sodium nitroprusside • Angina • Heart failure • Ald ehyde dehydrogenase

Introduction

Nitrovasodilators are used in the management of various acute and chronic cardiovascular pathologies. Their pharmacologic effects are mediated biochemically via the release of NO in the vasculature independent of endogenous endothelial nitric oxide synthase (eNOS). Once liberated, NO activates soluble guanylyl cyclase (sGC) in the smooth muscle, increases the concentrations of the secondary messenger cyclic GMP (3′,5′ guanosine monophosphate, cGMP), alters calcium flux, and ultimately causes relaxation [1].

In the Unites States, clinical use of nitrovasodilators is restricted to two chemical classes, viz., the organic nitrates (ORN) and metal nitrosyl compounds (see Fig. 13.1 for structures). Nitroglycerin (NTG, also known as glyceryl trinitrate) is the founding member of the ORN family, which also includes isosorbide dinitrate (ISDN) and its clinically administered metabolite isosorbide-5-mononitrate (IS-5-MN). A fourth ORN, pentaerythritol tetranitrate, sees clinical use in Europe and Canada. ORN, as such, have little or no pharmacological activity of their own, and must be

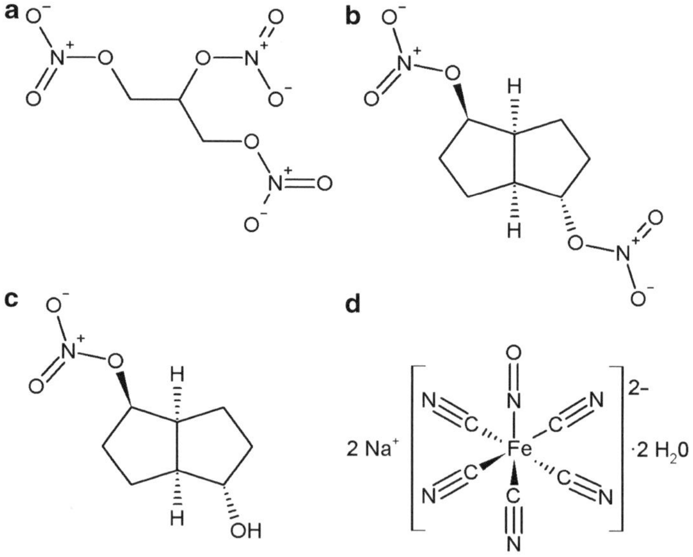

Fig. 13.1 Structures of nitrovasodilators that are clinically used in the United States. (**a**) Nitroglycerin (NTG), (**b**) isosorbide dinitrate (ISDN), (**c**) isosorbide-5-Mononitrate (IS-5-MN), (**d**) sodium nitroprusside (SNP)

enzymatically activated in the vasculature to release NO [2]. The metal nitrosyl group is comprised solely of the agent sodium nitroprusside (SNP). NO liberation from SNP is generally considered spontaneous but endogenous sulfhydryl groups may be involved in this process [3, 4].

Nitrovasodilator administration leads to in vivo production of inorganic nitrite (NO_2^-) and nitrate (NO_3^-) ions, either via ORN denitration or by conversion of the NO produced. However, it has been shown that NO_2^- is not an active intermediate for vascular generation of NO from ORN [5]. Moreover, even at relatively large doses, nitrovasodilators do not typically produce clinically relevant methemoglobinemia [6, 7], the toxicity typically associated with elevated plasma nitrite/nitrate (NO_x^-).

Indications for nitrovasodilator use revolve around correcting ischemia and restoring normal blood flow, reduction of systemic blood pressure, and reducing oxygen demand of the myocardium. Current guidelines from leading organizations support and recommend the use of ORN as mainline therapy for various cardiovascular diseases. For the management of unstable angina, the American College of Cardiology Foundation and the American Heart Association (ACC/AHA) Committee supports the use of sublingual NTG to treat anginal episodes (class I, level of evidence C) [8]. In addition, the Committee recommends the use of a long-acting ORN in conjunction with a beta-adrenergic receptor antagonists (beta-blockers) in patients with stable angina and coronary artery disease in order to improve exercise tolerance and time to onset of angina, or in place of a beta-blocker when its use is contraindicated (class I, level B). The Committee comments that if 24 h anginal relief is necessary, a calcium antagonist may be a better choice due to the need for a daily nitrate-free period to prevent the development of ORN tolerance. It is important to note that in the case of stable angina, ORN therapy is associated only with an improvement in symptoms, and not with improved mortality. These patients should have access to sublingual NTG as needed to abort attacks (class I, level B) [9].

The Committee has also recommended the use of NTG in the management of myocardial infarction. It should be administered sublingually at the onset of an event and then intravenously for the first 24–48 h to manage ischemic pain, control blood pressure, and to manage pulmonary congestion (class I, level C). Continued use past 48 h in the absence of ischemic pain is less well supported (class IIb, level B). Unlike the case of stable angina, ORN use for the management of myocardial infarction is associated with improved outcomes [10].

The rationale for the use of nitrates in congestive heart failure is less clear but may be beneficial in certain cases. Patients receiving angiotensin-converting enzymes (ACE) inhibitors and beta-blockers may benefit from the use of a long-acting ORN in combination with hydralazine if further relief from symptoms is required (class IIa, level A). An ORN/hydralazine combination can also be used in patients who are unable to tolerate an ACE inhibitor or angiotensin receptor blocker (class IIb, level C) [11]. When added to standard therapy, ISDN/hydralazine significantly increased survival and improved quality of life measures in black patients [12]. Intravenous NTG and SNP is indicated for the management of hypertensive crises [13].

Lastly, investigational/off-label uses of ORN include the treatment of erectile dysfunction [14, 15], osteoporosis [16], anal fissures [17, 18], cervical ripening during the induction of labor [19], and the management of esophageal varices due to portal hypertension [20].

Organic Nitrate Metabolism and Nitric Oxide Generation

The metabolism of ORN can be divided into two distinct pathways: clearance-based and mechanism-based [21]. The clearance-based pathway results in reductive denitration of ORN to liberate the very weakly vasoactive NO_2^- ion. In contrast, the mechanism-based pathway results in bioactivation and the release of NO [2]. The enzyme(s) responsible for these distinct metabolic paths of ORN continue to be the subject of much study and debate.

The clearance-based metabolism of ORN takes place extensively in the liver, with extrahepatic metabolism occurring in the kidneys, lungs, blood, and vasculature [22]. Metabolic clearance is thought to be mediated by several enzymes, including various isoforms of glutathione-*S*-transferase and cytochrome p450, but other thiol-containing enzymes also likely contribute to ORN systemic elimination [2, 22–24]. Hepatic and renal glucuronide conjugation of ORN and metabolites form hydrophilic conjugates that can then be excreted renally [22, 25, 26].

The mechanism-based metabolic pathway for ORN has drawn considerable current interest. NO liberation from ORN has been reported to be mediated by numerous enzymes/compounds, including cytosolic and microsomal glutathione-*S*-transferase [23, 27], an unknown cytochrome p450 isoform [24, 28], xanthine oxidase [29], and even free thiols [2]. However, mitochondrial aldehyde dehydrogenase (ALDH2) has recently emerged as the principal enzyme for the bioactivation of NTG [30]. The relevant findings demonstrated that NTG activated sGC in the presence of purified ALDH2 and that nonspecific ALDH2 inhibitors caused impaired relaxation of isolated aortic rings. Follow-up studies using ALDH2 knockout mice and more specific inhibitors of ALDH2 have confirmed the role of this enzyme in NTG bioactivation [31–33]. Clinical studies have also supported the view that ALDH2 is an important enzyme for NTG action [34, 35]. A common ALDH2 polymorph, prevalent in Asian populations, is associated with blunted pharmacological response to NTG [34]. This polymorphism results from a mutation in the cofactor binding region of ALDH2, leading to reduction in its dehydrogenase and ORN reductase activity [36].

While ALDH2 is likely the major enzyme responsible for NTG bioactivation, it appears to play a partial or limited role in the bioactivation of ISDN or IS-5-MN [33, 37]. Purified ALDH2 has the ability to liberated NO from ISDN in vitro [38], but ex vivo studies have demonstrated that its activity is not vital to ISDN-mediated relaxation [33, 37]. The liberation of NO from IS-5-MN has also been shown to be ALDH2-independent both in vitro [38] and ex vivo [37]. Recently, it was demonstrated that purified cytosolic aldehyde dehydrogenase (ALDH1), similar to ALDH2, has the ability to bioactivate NTG [39] and ISDN [38]. However, unlike ALDH2, it can also liberate NO from IS-5-MN [38]. The contribution of this enzyme to the in vivo bioactivation of ISDN and IS-5-MN would require further studies.

NTG denitration produces two separate metabolites, glyceryl 1,2-dinitrate (1,2-GDN), and glyceryl 1,3-dinitrate (1,3-GDN), the relative ratio of which has been used to distinguish the clearance-based vs. mechanism-based metabolic pathways. Production of 1,2-GDN is associated with vascular action and tolerance [21, 40], while that for 1,3-GDN is thought to reflect clearance-based metabolism [41, 42]. Both metabolites possess vasodilator activity, albeit much weaker than that of NTG [43]. Direct measurement of NO accumulation in vivo is impossible due to its extremely short half-life. Thus, as a marker of NO generation, its oxidative end-products (NO_x^-) have been quantified instead, primarily via various fluorometric and spectrophotometric assays [44]. The major limitation of this approach is that it cannot distinguish between NO_x^- generated from NO production from that generated from clearance-based denitration of ORN, or from other sources. Additionally, endogenous cellular or circulating concentrations of these ions are relatively high when compared to those produced by ORN metabolism. Thus, use of NO_x^- to monitor ORN metabolism is often inaccurate.

Investigation of the differential metabolism of ORN to NO_x^- vs. NO can provide insight regarding the clearance-based degradation vs. mechanism-based bioactivation. Incubation of NTG with cultured rat vascular smooth muscle cells demonstrated that nitrite was generated in a 5.7:1 ratio compared to NO [45]. A similar NO_x^-:NO ratio was seen when rat hepatic cytosol was incubated with NTG (5.4:1) and ISDN (3.1:1) [46]. A somewhat higher ratio was demonstrated from the metabolism of NTG by canine coronary artery homogenate (14.5:1) [47] and by nonenzymatic degradation of NTG in the presence of various thiols (10:1) [48]. The exact ratio is expected to vary based on the tissue and species examined, but it is clear that the clearance-based metabolism is the predominate pathway of NO_2^- generation from ORN.

Sodium Nitroprusside Metabolism and Nitric Oxide Generation

Although NO liberation from SNP is generally considered spontaneous [3], several cellular proteins have been shown to facilitate this process [4]. SNP diffuses into erythrocytes where the NO is released via interaction with oxyhemoglobin, which is converted to methoxyhemoglobin. In this process, cyanide radicals, liberated in a 5-to-1 molar ratio vs. NO, reversibly bind to methemoglobin converting it to the nontoxic cyanomethemoglobin, which exists in equilibrium with free cyanide [3]. In the liver, additional cyanide is detoxified via thiosulfate sulfotransferase with thiosulfate as a substrate, converting it to thiocyanate, which is subsequently cleared by the kidneys [49]. A typical adult has the ability to detoxify 50 mg of SNP, with clinical cyanide toxicity occasionally seen once thiosulfate stores are depleted. To prevent cyanide toxicity, thiosulfate may be administered in parallel.

The onset of action of SNP is rapid, but its duration of action is short (about 2 min). SNP is not orally absorbed, so continuous intravenous infusion is needed to maintain its activity [3].

Nitrovasodilators as a Source of Exogenous Nitrite and Nitrate

Clinical doses of nitrovasodilators can result in a large, transient increase in plasma NO_x^- levels. A 0.9 mg dose of NTG given sublingually to migraine headache patients resulted in an increase of approximately 62% in plasma NO_x^- 45 min after administration which returned to baseline within 120 min [50]. In another study, intravenous NTG was administered to patients undergoing hip replacement surgery. The infusion rate (average = 5.13 μg/kg/min) was titrated to result in a reduction in mean arterial pressure of 50–65 mmHg. Plasma nitrite was found to be increased by about 36% during the infusion compared to preadministration levels [51]. The effects of chronic ORN administration on plasma NO_x^- was determined in a population of patients with peripheral vascular disease. Two- and six-week chronic administrations of NTG (as 15 mg/day via transdermal patch) resulted in unchanged plasma NO_x^- at the end of each treatment period [52]. However, animal studies cast doubt on the usefulness of NO_x^- as a measure of NO or NO-donor pharmacodynamics [53].

Pharmacokinetics of Nitrovasodilators

NTG is rapidly eliminated in the body, with a clearance of approximately 50 L/min and a plasma half-life of about 3 min in humans [54]. In animal studies, an arterial–venous concentration was demonstrated, and the systemic clearance was found to be positively related to cardiac output, consistent with extensive extrahepatic metabolism [55]. Thus, the duration of action of NTG is short when given via sublingual or bolus intravenous dosing. The oral bioavailability of NTG itself is poor, requiring much higher doses (up to 40 mg daily in sustained release preparations) [56]. Longer duration of action from organic nitrates can be obtained by dosage form modification of NTG (e.g., transdermal patches or continuous intravenous infusion) or by the use of longer-acting nitrates such as ISDN or IS-5-MN, in either immediate-release or sustained-release oral preparations.

The pharmacokinetic properties of ISDN compare favorably to NTG in terms of longer duration of action and higher oral bioavailability. It has a slower clearance (4 L/min), a longer half-life (approximately 1 h), and a higher oral bioavailability (20%) [25]. ISDN is metabolized predominately to its two mononitrate metabolites, the 5-mononitrate and 2-mononitrate, at a ratio of approximately 4:1 [25]. While its sublingual formulation does not act rapidly enough for the abortion of an angina attack, ISDN is used for prophylaxis in anticipation of an angina-inducing activity.

IS-5-MN, being the least lipophilic ORN in clinical use, has the slowest clearance (115 mL/min) and longest elimination half-life (4.6 h). Because its oral bioavailability is near 100%, this agent has largely become the ORN drug-of-choice for oral prophylaxis.

Cardiovascular Mechanisms of Organic Nitrates

The pharmacological effect of nitrovasodilators is mediated by their vasoactive metabolite NO. The target of NO is the sGC which exists as a heterodimer containing a heme moiety [57]. NO interacts with the ferrous ion forming a ferrous–nitrosyl–heme complex resulting in a conformational change in the sGC, resulting in activation of its catalytic activity. With divalent magnesium as a cofactor, sGC catalyzes guanosine-5′-triphosphate to cGMP which in turn activates various protein kinases such as protein kinase G (PKG). The signaling cascade results in the alteration of several key mediators of the vasotone. Intracellular calcium is sequestered in the sarcoplasmic reticulum via increased uptake by activation of Ca^{2+},Mg^{2+}-ATPase and decreased efflux from calcium channels. In addition, dephosphorylation of the myosin light chain results in decreased sensitivity to calcium [58]. The messenger effect of cGMP is terminated by its hydrolysis by phosphodiesterases (PDEs). The inhibition of PDE in the penis, by drugs like sildenafil for erectile dysfunction, results in potentiation of ORN action, thus leading to possible severe hypotension [59].

ORN possess dose-dependent vasoselectivity which can be exploited to best manage various cardiovascular pathologies. The venous vasculature is significantly more sensitive to the effects of ORN than the arterial side and is preferentially dilated upon administration of low doses of ORN. This results in an increased capacitance in the veins and decreased venous return to the atria, thus reducing cardiac preload and left ventricle end-diastolic pressure [60]. Doses which have substantial effects on venodilation have little effect on arterial vascular resistance and systemic blood pressure. Part of this selectivity is likely due to an increased amount of bioactivating enzymes in the venous smooth muscles cells compared to those in the arterial side [61]. Further venodilation and arterial relaxation occurs with higher doses which decrease cardiac afterload and systemic blood pressure. ORN have no intrinsic effect on cardiac contractility or heart rate, but reflex tachycardia may develop in response to the reduction in blood pressure [60]. Profound hypotension may occur in patients concurrently receiving agents such as beta-blockers or alpha-adrenergic receptor blockers which block the sympathetic tone. SNP, being a more spontaneous donor of NO, has an indiscriminate effect on venous and arterial relaxation [3].

ORN have a complex effect on coronary blood flow, their beneficial effects may not be due to an indiscriminate coronary artery vasodilatation but rather a redistribution of cardiac perfusion [62]. When NTG and ISDN are administered during periods of increased oxygen demand and ischemia, an unequal reduction in coronary blood flow is experienced; flow to well perfused areas is decreased to a greater extent than those that are occluded and experiencing ischemia. While total myocardial flow is reduced due to decreased cardiac output resulting from decreased preload, ischemia is relieved via reduction of myocardial oxygen demand and balancing of the flow based on metabolic demand [63]. In addition, coronary flow to regions supplied by severely occluded arteries is improved by ORN due to an increase in perfusion through collateral vascular beds which circumvent the occlusion [64].

The culmination of these effects is a reduction of myocardial oxygen demand while improving flow to the ischemic regions of the heart. Due to a decrease in preload, ventricular wall tension is reduced which results in a reduction on oxygen demand [60]. Reduction in wall tension also decreases subendocardial resistance allowing for improved perfusion through the cardiac wall [62]. This is further improved upon when higher doses of ORN reduce systemic vascular resistance, which in turn decreases afterload [60]. In severe ventricular dysfunction, the combination of an ORN with hydralazine (an arterial vasodilator) results in decreased preload and afterload, which decreases oxygen demand while improving cardiac output [62]. If hypotension develops, a reduction in coronary perfusion pressure may lead to an increased oxygen debt in the myocardial tissue worsening ischemia.

Effects on Platelet Aggregation

In addition to the vasodilatory effects, ORN and SNP also possess anti-aggregatory effects on platelets. The effects have been demonstrated in isolated platelets [29, 65, 66], animal models [67], and in clinical studies [68]. While there is a lack of evidence supporting ORN use solely for these effects, they contribute beneficially to their therapeutic roles, especially in the management of unstable angina. Like its effects on the vasculature, ORN must be metabolized to NO to be pharmacologically active [29, 65]. Incubation of NTG with whole blood or isolated platelets yield little inhibition, and the addition of a metabolizing system (e.g., vascular tissue [65], isolated vascular cells [66], or a bioactivation enzyme [29]) is needed to produce significant inhibition. The interplay between the vascular tissue and platelets is complex and involves several signaling molecules. Calcitonin gene-related peptide (CGRP), a neuropeptide with potent anti-platelet activity, appears to be one of the more important mediators. ORN induces CGRP release from neurons innervating the vascular tissue, resulting in cGMP and cyclic adenosine monophosphate accumulation, which inhibits platelet function. Blockade of ORN metabolism, vascular CGRP receptor, or the NOS enzyme, all result in reduced anti-aggregatory effect by ORN [65]. However, SNP inhibits platelet aggregation in a NOS and CGRP-independent manner [69].

CGRP is not the exclusive mechanism by which NTG produces its anti-platelet effects. Incubation of isolated platelets with NTG and xanthine oxidase (to bioactivate the ORN) is sufficient to produce inhibition though an unknown mechanism [29]. A possible but uninvestigated mechanism may exist via the modification of the surface platelet ADP receptor, $P2Y_{12}$, the active form of which exists as oligomers stabilized by disulfide bridges [70, 71]. Inhibitors for this receptor (e.g., clopidogrel) exert their pharmacological effect by disrupting these oligomers through reduction of the disulfide bridges [72]. ORN has been shown to react avidly with protein cysteine residues [73], and modification of cysteine linkages in $P2Y_{12}$ may represent a contributory mechanism for the anti-platelet effects of ORN.

Nitrovasodilator Tolerance

The beneficial pharmacological effects of nitrovasodilators are severely limited by the rapid development of tolerance to their vasodilatory effects. In order for ORN to maintain their vasodilatory effects when used clinically, an 8–12 h nitrate-free period needs to be incorporated. This interruption complicates ORN therapy in patients who may require around-the-clock angina protection. Interestingly, in contrast to the vasodilatory effects, tolerance development to the platelet anti-aggregatory effects of ORN does not appear to be significant [67, 68]. The mechanisms of nitrate tolerance have been described as a 130-year-old mystery [2, 74]. Nitrate tolerance exhibits several major features, namely:

1. Sulfhydryl (–SH) involvement: Nitrate-induced vasodilation is facilitated by the presence of free –SH groups. Modifications or depletion of free –SH have been proposed as a mechanism of vascular tolerance [75], widely known classically as the Needleman Sulfhydryl Depletion Hypothesis. However, conflicting findings have been observed based on lack of demonstrated sulfhydryl depletion, as well as incomplete prevention or reversal by exogenous sulfhydryl repletion [76, 77].
2. Multiple consequences of chronic nitrate exposure: Nitrate tolerance is accompanied by a myriad of events, including reduced ORN metabolism to NO and inactivation of ALDH2 [32], neurohormonal counter-regulation, presence of withdrawal rebound, presence of vascular oxidative stress, including increased $O_2^{\bullet-}$ accumulation [78], and extensive alteration of gene expression in the rat aorta [79, 80]. Partial avoidance/reversal of nitrate tolerance can be accomplished by a variety of agents such as N-acetylcysteine [81], vitamin C [82, 83], vitamin E [84], folic acid [85, 86], L-arginine [87], carvedilol [88], hydralazine [89], etc. Although eNOS uncoupling has been proposed as a mechanism of nitrate tolerance [90], eNOS knockout mice also exhibited this phenomenon [91].
3. Other NO donors generally exhibited lack of self-tolerance or cross-tolerance to NTG: Cross-tolerance among ORN has been well demonstrated, both experimentally [92] and clinically [93, 94]. Recently, however, some studies have suggested that pentaerythrityl tetranitrate (PETN) may be less tolerance-inducing than NTG [37, 95]. Cross-tolerance of SNP to NTG was either not observed [96, 97] or required a tenfold higher concentration of NTG, when compared to NTG

self-tolerance [98]. In a dozen studies involving SIN-1 (the active metabolite of molsidomine), neither self-tolerance nor cross-tolerance to NTG was demonstrated (see, e.g., [99–102]). These data indicate that ORN tolerance is not mediated at the level of sGC, since enzyme integrity is required for the action of PETN, SNP, and SIN-1.

The cumulative findings indicate that all these events of ORN tolerance could not have been mediated by one single target protein, e.g., ALDH2. In 2004, we proposed a biochemical mechanism of nitrate tolerance which recognizes the non-selective nature of nitrate-mediated protein cysteine oxidation. We suggested that different cysteine-containing proteins may mediate different events associated with nitrate action (Fig. 13.2). Thus, inactivation of metabolic enzymes, such as ALDH2 or glutathione-*S*-transferase, would result in reduced bioactivation and clearance. Oxidation of signaling proteins such as p21Ras and NFκB would lead to multiple changes in vascular gene expression, which, in turn, may induce withdrawal rebound and vasoconstriction. Finally, oxidation of the cardioprotective protein ALDH2 and remodeling protein matrix metalloproteinase-9 (MMP-9) could be responsible for the increased cardiac events associated with chronic ORN therapy (see later).

Several recent findings lend support to this view. Chronic NTG administration results in increased total protein *S*-glutathionylation in the aortic tissue of a rat model and is correlated with a reduced potency of NTG to vasodilate isolated aortic tissue [73]. In addition, NTG treatment mediated *S*-glutathionylation and increased activity of the GTPase, p21ras, in cultured human vascular endothelial cells [73]. Activated p21ras interacts with its downstream targets, Raf-1 and phosphatidylinositol 3-kinase. Ultimately these targets phosphorylate Akt and extracellular

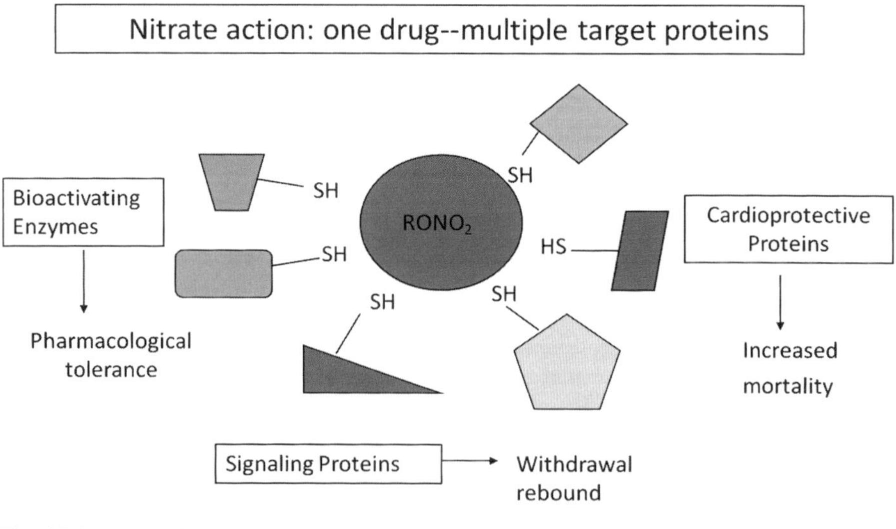

Fig. 13.2 Schematic representation of the interactions between organic nitrates ($RONO_2$) and cellular proteins, resulting in multiple pharmacologic actions

signal-regulated kinase 1/2 (ERK 1/2), and promote increased smooth muscle cell proliferation through gene regulatory pathways [103]. The recent demonstration that NRF2 signaling is upregulated by the spontaneous NO donor spermine NONOate suggests that other signaling pathways are likely to be modulated by ORN as well. Spermine NONOate oxidizes the cysteine residues of NRF2's inhibitor Kelch-like ECH-associated protein 1 (KEAP1). This action mediates the dissociation of the NRF2/KEAP1 complex located in the cytosol, allows NRF2 to translocate to the nucleus, and results in the transcription of a group of proteins whose expression is regulated by antioxidant response elements [104].

Toxicology

Chronic and long-term ORN therapy has been associated with reduced survival when used in patients with coronary artery disease [105, 106], but the underlying mechanisms are not known. It is clear however that ORN therapy leads to increased vascular oxidative stress which in turn can produce endothelial dysfunction. ORN therapy causes an increased production of reactive oxygen species and reactive nitrogen species via several mechanisms including increased activity of NADPH oxidase and uncoupling of both NOS and mitochondrial respiration [41, 107, 108]. The enzymes ALDH2 [109] and xanthine oxidase [73] are also able to generate superoxide when incubated with NTG in vitro and may be another source of ROS in vivo. The induction of endothelial dysfunction in healthy volunteers [35, 110] and animal models [35] by chronic ORN at therapeutic doses has recently been described. This dysfunction was reversed by administration of the antioxidant Vitamin C suggesting that oxidative stress mediates this effect [110]. Superoxide formation was believed to be responsible for tolerance development, but it has been demonstrated, using knockout animal models, that tolerance still occurs despite a reduction in superoxide production, and that NADPH oxidase is not critical in mediating nitrate tolerance [73].

While ALDH2 inactivation in the vasculature may lead to ORN tolerance, its inactivation in the myocardium may have serious effects on cardiac health. It has recently been demonstrated that there is a strong correlation between cardiac ALDH2 activity and infarct size following ischemia/reperfusion in an ex vivo model of myocardial infarction. Preincubation with agents that inhibit ALDH2 (e.g., NTG and cyanamide) resulted in larger infarcts while agents that increased cardiac ALDH2 produced the opposite effect [111]. During periods of ischemia and subsequent reperfusion, toxic and reactive aldehydes (e.g., 4-hydroxy-2-nonenal) are generated via lipid peroxidation and they have the ability to form protein adducts, resulting in cell damage [112]. These aldehydes are detoxified by cardiac ALDH2, and the ability of the myocardium to clear these aldehydes is directly correlated with ALDH2 activity [111].

ORN may also induce toxicity via destabilization of atherosclerotic plaques present in patients [113], via activation of MMP, including MMP-9. MMP-9, like most MMP, is maintained inactive (as proMMP-9) via a "cysteine-switch" which involves a zinc–thiolate linkage at the active site. When the zinc–thiolate bond is broken, e.g., by cysteine oxidation, the switch is turned on and MMP is activated [114]. Using cultured macrophages, it has been demonstrated that NTG incubation increases MMP-9 mRNA, protein levels, and activity [113, 115]. The activation of pro-MMP by NTG was shown to be mediated by cysteine oxidation, and the regulatory effect of NTG on MMP-9 expression is mediated via NFκB [116].

Conclusion

Nitrovasodilators remain a primary therapeutic agent in the pharmacologic management of coronary artery disease, myocardial infarction, and hypertensive crisis. Substantial aspects of their complex pharmacology, metabolism, and toxicology have been uncovered over their past century of use, but much remain to be delineated. Nitrovasodilators have the potential to increase in vivo nitrite and nitrate exposure significantly, especially when given on a chronic basis. The long-term effects of this increased exposure on human health are not clear at present, and would require further studies.

References

1. Ignarro LJ, Napoli C, Loscalzo J. Nitric oxide donors and cardiovascular agents modulating the bioactivity of nitric oxide: an overview. Circ Res. 2002;90(1):21–8.
2. Fung HL. Biochemical mechanism of nitroglycerin action and tolerance: is this old mystery solved? Annu Rev Pharmacol Toxicol. 2004;44:67–85.
3. Friederich JA, Butterworth JF. Sodium nitroprusside: twenty years and counting. Anesth Analg. 1995;81(1):152–62.
4. Kowaluk EA, Seth P, Fung HL. Metabolic activation of sodium nitroprusside to nitric oxide in vascular smooth muscle. J Pharmacol Exp Ther. 1992;262(3):916–22.
5. Kowaluk EA, Chung SJ, Fung HL. Nitrite ion is not an active intermediate in the vascular metabolism of organic nitrates and organic nitrites to nitric oxide. Drug Metab Dispos. 1993;21(5):967–9.
6. Kaplan KJ, Taber M, Teagarden JR, Parker M, Davison R. Association of methemoglobinemia and intravenous nitroglycerin administration. Am J Cardiol. 1985;55(1):181–3.
7. Paris PM, Kaplan RM, Stewart RD, Weiss LD. Methemoglobin levels following sublingual nitroglycerin in human volunteers. Ann Emerg Med. 1986;15(2):171–3.
8. Anderson JL, Adams CD, Antman EM, Bridges CR, Califf KM, Casey DE. ACC/AHA 2007 guidelines for the management of patients with unstable angina/non ST-elevation myocardial infarction: a report of the American College of Cardiology/American Heart Association Task Force on Practice Guidelines (Writing Committee to Revise the 2002 Guidelines for the Management of Patients With Unstable Angina/Non ST-Elevation Myocardial Infarction): developed in collaboration with the American College of Emergency Physicians, the Society for Cardiovascular Angiography and Interventions, and the Society of Thoracic Surgeons: endorsed by the American Association of Cardiovascular and Pulmonary Rehabilitation and the Society for Academic Emergency Medicine. Circulation. 2007;116(7):e148–304.

9. Gibbons RJ, Abrams J, Chatterjee K, et al. ACC/AHA 2002 guideline update for the management of patients with chronic stable angina–summary article: a report of the American College of Cardiology/American Heart Association Task Force on Practice Guidelines (Committee on the Management of Patients With Chronic Stable Angina). Circulation. 2003;107(1):149–58.
10. Antman EM, Anbe DT, Armstrong PW, et al. ACC/AHA guidelines for the management of patients with ST-elevation myocardial infarction: a report of the American College of Cardiology/American Heart Association Task Force on Practice Guidelines (Committee to Revise the 1999 Guidelines for the Management of Patients with Acute Myocardial Infarction). Circulation. 2004;110(9):e82–292.
11. Hunt SA, Abraham WT, Chin MH, et al. ACC/AHA 2005 Guideline Update for the Diagnosis and Management of Chronic Heart Failure in the Adult: a report of the American College of Cardiology/American Heart Association Task Force on Practice Guidelines (Writing Committee to Update the 2001 Guidelines for the Evaluation and Management of Heart Failure): developed in collaboration with the American College of Chest Physicians and the International Society for Heart and Lung Transplantation: endorsed by the Heart Rhythm Society. Circulation. 2005;112(12):e154–235.
12. Taylor AL, Ziesche S, Yancy C, et al. Combination of isosorbide dinitrate and hydralazine in blacks with heart failure. N Engl J Med. 2004;351(20):2049–57.
13. Rhoney D, Peacock WF. Intravenous therapy for hypertensive emergencies, part 1. Am J Health Syst Pharm. 2009;66(15):1343–52.
14. Cavallini G. Minoxidil versus nitroglycerin: a prospective double-blind controlled trial in transcutaneous erection facilitation for organic impotence. J Urol. 1991;146(1):50–3.
15. Meyhoff HH, Rosenkilde P, Bodker A. Non-invasive management of impotence with transcutaneous nitroglycerin. Br J Urol. 1992;69(1):88–90.
16. Wimalawansa SJ. Nitric oxide: novel therapy for osteoporosis. Exp Opin Pharmacother. 2008;9(17):3025–44.
17. Tankova L, Yoncheva K, Kovatchki D, Doytchinova I. Topical anal fissure treatment: placebo-controlled study of mononitrate and trinitrate therapies. International Journal of Colorectal Disease. 2009;24:461–4.
18. Tjandra JJ, Tan JJ, Lim JF, Murray-Green C, Kennedy ML, Lubowski DZ. Rectogesic (glyceryl trinitrate 0.2%) ointment relieves symptoms of haemorrhoids associated with high resting anal canal pressures. Colorectal Dis. 2007;9(5):457–63.
19. Rameez MFM, Goonewardene IMR. Nitric oxide donor isosorbide mononitrate for pre-induction cervical ripening at 41 weeks' gestation: a randomized controlled trial. J Obstet Gynaecol Res. 2007;33(4):452–6.
20. Lo GH, Chen WC, Chan HH, et al. A randomized, controlled trial of banding ligation plus drug therapy versus drug therapy alone in the prevention of esophageal variceal rebleeding. J Gastroenterol Hepatol. 2009;24(6):982–7.
21. Bennett BM, McDonald BJ, Nigam R, Simon WC. Biotransformation of organic nitrates and vascular smooth muscle cell function. Trends Pharmacol Sci. 1994;15(7):245–9.
22. Hashimoto S, Kobayashi A. Clinical pharmacokinetics and pharmacodynamics of glyceryl trinitrate and its metabolites. Clin Pharmacokinet. 2003;42(3):205–21.
23. Ji Y, Bennett BM. Biotransformation of glyceryl trinitrate by rat hepatic microsomal glutathione S-transferase 1. J Pharmacol Exp Ther. 2006;318(3):1050–6.
24. Yuan R, Sumi M, Benet LZ. Investigation of aortic CYP3A bioactivation of nitroglycerin in vivo. J Pharmacol Exp Ther. 1997;281(3):1499–505.
25. Abshagen U. Pharmacokinetics of ISDN, sustained-release ISDN, and IS-5-MN. In: Cohn J, Rittinghausen R, editors. Mononitrates. New York: Springer; 1985. p. 53–66.
26. Raue F, Ritz E, Akpan W, Endele R, Neugebauer G. Kinetics of IS-5-MN and its glucuronide in patients with renal failure. In: Cohn J, Rittinghausen R, editors. Mononitrates. New York: Springer; 1985. p. 78–85.
27. Kenkare SR, Benet LZ. Tolerance to nitroglycerin in rabbit aorta. Investigating the involvement of the mu isozyme of glutathione S-transferases. Biochem Pharmacol. 1996;51(10):1357–63.

28. Minamiyama Y, Imaoka S, Takemura S, Okada S, Inoue M, Funae Y. Escape from tolerance of organic nitrate by induction of cytochrome P450. Free Radic Biol Med. 2001;31(11):1498–508.
29. O'Byrne S, Shirodaria C, Millar T, Stevens C, Blake D, Benjamin N. Inhibition of platelet aggregation with glyceryl trinitrate and xanthine oxidoreductase. J Pharmacol Exp Ther. 2000;292(1):326–30.
30. Chen Z, Zhang J, Stamler JS. Identification of the enzymatic mechanism of nitroglycerin bioactivation. Proc Natl Acad Sci U S A. 2002;99(12):8306–11.
31. DiFabio J, Ji Y, Vasiliou V, Thatcher GR, Bennett BM. Role of mitochondrial aldehyde dehydrogenase in nitrate tolerance. Mol Pharmacol. 2003;64(5):1109–16.
32. Chen Z, Foster MW, Zhang J, et al. An essential role for mitochondrial aldehyde dehydrogenase in nitroglycerin bioactivation. Proc Natl Acad Sci U S A. 2005;102(34):12159–64.
33. Wenzel P, Hink U, Oelze M, et al. Number of nitrate groups determines reactivity and potency of organic nitrates: a proof of concept study in ALDH-2–/– mice. Br J Pharmacol. 2007;150(4):526–33.
34. Mackenzie IS, Maki-Petaja KM, McEniery CM, et al. Aldehyde dehydrogenase 2 plays a role in the bioactivation of nitroglycerin in humans. Arterioscler Thromb Vasc Biol. 2005;25(9):1891–5.
35. Wenzel P, Schulz E, Gori T, et al. Monitoring white blood cell mitochondrial aldehyde dehydrogenase activity: implications for nitrate therapy in humans. J Pharmacol Exp Ther. 2009;330(1):63–71.
36. Larson HN, Zhou J, Chen Z, Stamler JS, Weiner H, Hurley TD. Structural and functional consequences of coenzyme binding to the inactive Asian variant of mitochondrial aldehyde dehydrogenase: roles of residues 475 and 487. J Biol Chem. 2007;282(17):12940–50.
37. Daiber A, Oelze M, Coldewey M, et al. Oxidative stress and mitochondrial aldehyde dehydrogenase activity: a comparison of pentaerythritol tetranitrate with other organic nitrates. Mol Pharmacol. 2004;66(6):1372–82.
38. Page NA, Tsou P-S, Beretta M, Mayer B, Fung H-L. Selective activation of organic nitrates by, and inactivation of, ALDH isoforms. FASEB J. 2009;23(1_MeetingAbstracts):LB374.
39. Beretta M, Gruber K, Kollau A, et al. Bioactivation of nitroglycerin by purified mitochondrial and cytosolic aldehyde dehydrogenases. J Biol Chem. 2008;283(26):17873–80.
40. Fung HL, Poliszczuk R. Nitrosothiol and nitrate tolerance. Z Kardiol. 1986;75 Suppl 3:25–7.
41. Gori T, Parker JD. Nitrate tolerance: a unifying hypothesis. Circulation. 2002;106(19):2510–3.
42. Ji Y, Anderson DJ, Bennett BM. Role of microsomal glutathione transferase 1 in the mechanism-based biotransformation of glyceryl trinitrate in LLC-PK1 cells. Biochem Pharmacol. 2009;77(11):1702–8.
43. Haefeli WE, Gumbleton M, Benet LZ, Hoffman BB, Blaschke TF. Comparison of vasodilatory responses to nitroglycerin and its dinitrate metabolites in human veins. Clin Pharmacol Ther. 1992;52(6):590–6.
44. Ghafourifar P, Parihar MS, Nazarewicz R, et al. Detection assays for determination of mitochondrial nitric oxide synthase activity; advantages and limitations. Methods Enzymol. 2008;440:317–34.
45. Weber AA, Neuhaus T, Seul C, et al. Biotransformation of glyceryl trinitrate by blood platelets as compared to vascular smooth muscle cells. Eur J Pharmacol. 1996;309(2):209–13.
46. Aniya Y, Uehara N, Ishii C, et al. Evaluation of nitric oxide formation from nitrates in pig coronary arteries. Jpn J Pharmacol. 1996;71(2):101–7.
47. Kurz MA, Boyer TD, Whalen R, Peterson TE, Harrison DG. Nitroglycerin metabolism in vascular tissue: role of glutathione S-transferases and relationship between NO. and NO2- formation. Biochem J. 1993;292(Pt 2):545–50.
48. Feelisch M, Noack E, Schroder H. Explanation of the discrepancy between the degree of organic nitrate decomposition, nitrite formation and guanylate cyclase stimulation. Eur Heart J. 1988;9(Suppl A):57–62.
49. Himwich WA, Saunders JP. Enzymatic conversion of cyanide to thiocyanate. Am J Physiol. 1948;153(2):348–54.

50. Costa A, Ravaglia S, Sances G, Antonaci F, Pucci E, Nappi G. Nitric oxide pathway and response to nitroglycerin in cluster headache patients: plasma nitrite and citrulline levels. Cephalalgia. 2003;23(6):407–13.
51. Aoki H, Inoue M, Mizobe T, Harada M, Imai H, Kobayashi A. Platelet function is inhibited by nitric oxide liberation during nitroglycerin-induced hypotension anaesthesia. Br J Anaesth. 1997;79(4):476–81.
52. de Berrazueta JR, Sampedro I, Garcia-Unzueta MT, Llorca J, Bustamante M, Amado JA. Effect of transdermal nitroglycerin on inflammatory mediators in patients with peripheral atherosclerotic vascular disease. Am Heart J. 2003;146(4):746.
53. Janero DR, Bryan NS, Saijo F, et al. Differential nitros(yl)ation of blood and tissue constituents during glyceryl trinitrate biotransformation in vivo. Proc Natl Acad Sci USA. 2004;101(48):16958–63.
54. McNiff EF, Yacobi A, Young-Chang FM, Golden LH, Goldfarb A, Fung HL. Nitroglycerin pharmacokinetics after intravenous infusion in normal subjects. J Pharm Sci. 1981;70(9):1054–8.
55. Fung HL, Blei A, Chong S. Cardiac output is an apparent determinant of nitroglycerin pharmacokinetics in rats. J Pharmacol Exp Ther. 1986;239(3):701–5.
56. Davidov ME, Mroczek WJ. Effect of sustained release nitroglycerin capsules on anginal frequency and exercise capacity: a double-blind evaluation. Angiology. 1977;28(3):181–9.
57. Garthwaite J. New insight into the functioning of nitric oxide-receptive guanylyl cyclase: physiological and pharmacological implications. Mol Cell Biochem. 2010;334(1–2):221–32.
58. Lucas KA, Pitari GM, Kazerounian S, et al. Guanylyl cyclases and signaling by cyclic GMP. Pharmacol Rev. 2000;52(3):375–414.
59. Simonsen U. Interactions between drugs for erectile dysfunction and drugs for cardiovascular disease. Int J Impot Res. 2002;14(3):178–88.
60. Bassenge E, Zanzinger J. Nitrates in different vascular beds, nitrate tolerance, and interactions with endothelial function. Am J Cardiol. 1992;70(8):B23–9.
61. Bauer JA, Fung HL. Arterial versus venous metabolism of nitroglycerin to nitric oxide: a possible explanation of organic nitrate venoselectivity. J Cardiovasc Pharmacol. 1996;28(3):371–4.
62. Winbury WM. Pharmacology of nitrates in relation to antianginal action. In: Lichtlen PR, Engel H-J, Schrey A, Swan HJC, editors. Nitrates III: cardiovascular effects. New York: Springer; 1981. p. 4–11.
63. Engel H-J, Wolf R, Pretschner P, Hundeshagen H, Lichtlen PR. Effects of nitrates on myocardial blood flow during angina: comparison of results obtained by inert gas clearance and 201 thallium imaging. In: Lichtlen PR, Engel H-J, Schrey A, Swan HJC, editors. Nitrates III: cardiovascular effects. New York: Springer; 1981. p. 184–91.
64. Mason DT, Klein RC, Awan NA. Effects of systemic nitroglycerin on perfusion of ischemic myocardium in clinical coronary artery disease. In: Lichtlen PR, Engel H-J, Schrey A, Swan HJC, editors. Nitrates III: cardiovascular effects. New York: Springer; 1981. p. 193–201.
65. Booth BP, Nolan TD, Fung HL. Nitroglycerin-inhibited whole blood aggregation is partially mediated by calcitonin gene-related peptide – a neurogenic mechanism. Br J Pharmacol. 1997;122(3):577–83.
66. Benjamin N, Dutton JA, Ritter JM. Human vascular smooth muscle cells inhibit platelet aggregation when incubated with glyceryl trinitrate: evidence for generation of nitric oxide. Br J Pharmacol. 1991;102(4):847–50.
67. Hebert D, Xiang J-X, Lam JYT. Persistent inhibition of platelets during continuous nitroglycerin therapy despite hemodynamic tolerance. Circulation. 1997;95(5):1308–13.
68. Holmes AS, Chirkov YY, Willoughby SR, Poropat S, Pereira J, Horowitz JD. Preservation of platelet responsiveness to nitroglycerine despite development of vascular nitrate tolerance. Br J Clin Pharmacol. 2005;60(4):355–63.
69. Booth BP, Fung HL. Contribution of vascular tissue to the antiplatelet activity of sodium nitroprusside. J Cardiovasc Pharmacol. 1998;32(1):129–33.

70. Ding Z, Kim S, Dorsam RT, Jin J, Kunapuli SP. Inactivation of the human P2Y12 receptor by thiol reagents requires interaction with both extracellular cysteine residues, Cys17 and Cys270. Blood. 2003;101(10):3908–14.
71. Hollopeter G, Jantzen HM, Vincent D, et al. Identification of the platelet ADP receptor targeted by antithrombotic drugs. Nature. 2001;409(6817):202–7.
72. Savi P, Zachayus JL, Delesque-Touchard N, et al. The active metabolite of Clopidogrel disrupts P2Y12 receptor oligomers and partitions them out of lipid rafts. Proc Natl Acad Sci USA. 2006;103(29):11069–74.
73. Tsou PS, Addanki V, Haas JA, Page NA, Fung HL. Role of glutaredoxin-mediated protein S-glutathionylation in cellular nitroglycerin tolerance. J Pharmacol Exp Ther. 2009;329(2):649–56.
74. Ignarro LJ. After 130 years, the molecular mechanism of action of nitroglycerin is revealed. Proc Natl Acad Sci U S A. 2002;99(12):7816–7.
75. Needleman P, Lang S, Johnson Jr EM. Organic nitrates: relationship between biotransformation and rational angina pectoris therapy. J Pharmacol Exp Ther. 1972;181(3):489–97.
76. Haj-Yehia AI, Benet LZ. In vivo depletion of free thiols does not account for nitroglycerin-induced tolerance: a thiol-nitrate interaction hypothesis as an alternative explanation for nitroglycerin activity and tolerance. J Pharmacol Exp Ther. 1996;278(3):1296–305.
77. Fung HL, Bauer JA. Mechanisms of nitrate tolerance. Cardiovasc Drugs Ther. 1994;8(3):489–99.
78. Munzel T, Sayegh H, Freeman BA, Tarpey MM, Harrison DG. Evidence for enhanced vascular superoxide anion production in nitrate tolerance. A novel mechanism underlying tolerance and cross-tolerance. J Clin Invest. 1995;95(1):187–94.
79. Pi X, Yan C, Kim D, Chen J, Berk BC. Differential expression of genes from nitrate-tolerant rat aorta. J Vasc Res. 2002;39(4):304–10.
80. Wang EQ, Lee WI, Brazeau D, Fung HL. cDNA microarray analysis of vascular gene expression after nitric oxide donor infusions in rats: implications for nitrate tolerance mechanisms. AAPS PharmSci. 2002;4(2):E10.
81. Packer M, Lee WH, Kessler PD, Gottlieb SS, Medina N, Yushak M. Prevention and reversal of nitrate tolerance in patients with congestive heart failure. N Engl J Med. 1987;317(13):799–804.
82. Watanabe H, Kakihana M, Ohtsuka S, Sugishita Y. Randomized, double-blind, placebo-controlled study of ascorbate on the preventive effect of nitrate tolerance in patients with congestive heart failure. Circulation. 1998;97(9):886–91.
83. Bassenge E, Fink N, Skatchkov M, Fink B. Dietary supplement with vitamin C prevents nitrate tolerance. J Clin Invest. 1998;102(1):67–71.
84. Watanabe H, Kakihana M, Ohtsuka S, Sugishita Y. Randomized, double-blind, placebo-controlled study of supplemental vitamin E on attenuation of the development of nitrate tolerance. J Cardiol. 1998;31(3):173–81.
85. Gori T, Burstein JM, Ahmed S, et al. Folic acid prevents nitroglycerin-induced nitric oxide synthase dysfunction and nitrate tolerance: a human in vivo study. Circulation. 2001;104(10):1119–23.
86. Gori T, Saunders L, Ahmed S, Parker JD. Effect of folic acid on nitrate tolerance in healthy volunteers: differences between arterial and venous circulation. J Cardiovasc Pharmacol. 2003;41(2):185–90.
87. Parker JO, Parker JD, Caldwell RW, Farrell B, Kaesemeyer WH. The effect of supplemental L-arginine on tolerance development during continuous transdermal nitroglycerin therapy. J Am Coll Cardiol. 2002;39(7):1199–203.
88. Watanabe H, Kakihana M, Ohtsuka S, Sugishita Y. Preventive effects of carvedilol on nitrate tolerance – a randomized, double-blind, placebo-controlled comparative study between carvedilol and arotinolol. J Am Coll Cardiol. 1998;32(5):1201–6.
89. Bauer JA, Fung HL. Concurrent hydralazine administration prevents nitroglycerin-induced hemodynamic tolerance in experimental heart failure. Circulation. 1991;84(1):35–9.

90. Munzel T, Li H, Mollnau H, et al. Effects of long-term nitroglycerin treatment on endothelial nitric oxide synthase (NOS III) gene expression, NOS III-mediated superoxide production, and vascular NO bioavailability [in process citation]. Circ Res. 2000;86(1):E7–12.
91. Wang EQ, Lee WI, Fung HL. Lack of critical involvement of endothelial nitric oxide synthase in vascular nitrate tolerance in mice. Br J Pharmacol. 2002;135(2):299–302.
92. Needleman P, Johnson Jr EM. Mechanism of tolerance development to organic nitrates. J Pharmacol Exp Ther. 1973;184(3):709–15.
93. Parker JC, Di Carlo FJ, Davidson IW. Comparative vasodilator effects of nitroglycerin, pentaerythritol trinitrate and biometabolites, and other organic nitrates. Eur J Pharmacol. 1975;31(1):29–37.
94. Dalal JJ, Parker JO. Nitrate cross-tolerance: effect of sublingual isosorbide dinitrate and nitroglycerin during sustained nitrate therapy. Am J Cardiol. 1984;54(3):286–8.
95. Dikalov S, Fink B, Skatchkov M, Bassenge E. Comparison of glyceryl trinitrate-induced with pentaerythrityl tetranitrate-induced in vivo formation of superoxide radicals: effect of vitamin C. Free Radic Biol Med. 1999;27(1–2):170–6.
96. Kowaluk EA, Poliszczuk R, Fung HL. Tolerance to relaxation in rat aorta: comparison of an S-nitrosothiol with nitroglycerin. Eur J Pharmacol. 1987;144(3):379–83.
97. Chirkov PYY, Chirkova PLP, Horowitz PJD. Nitroglycerin tolerance at the platelet level in patients with angina pectoris. Am J Cardiol. 1997;80(2):128–31.
98. Zhang LM, Castresana MR, Newman WH. Tolerance to nitroglycerin in vascular smooth muscle cells: recovery and cross-tolerance to sodium nitroprusside. Anesth Analg. 1994;78(6):1053–9.
99. Stewart DJ, Holtz J, Bassenge E. Long-term nitroglycerin treatment: effect on direct and endothelium-mediated large coronary artery dilation in conscious dogs. Circulation. 1987;75(4):847–56.
100. Sutsch G, Kim JH, Bracht C, Kiowski W. Lack of cross-tolerance to short-term linsidomine in forearm resistance vessels and dorsal hand veins in subjects with nitroglycerin tolerance. Clin Pharmacol Ther. 1997;62(5):538–45.
101. Hinz B, Schroder H. Nitrate tolerance is specific for nitric acid esters and its recovery requires an intact protein synthesis. Biochem Biophys Res Commun. 1998;252(1): 232–5.
102. Govantes C, Rodriguez-Martinez MA, Marin J. Vasodilator effect and tolerance induced by the nitrocompound SIN-1 in rabbit femoral artery. Methods Find Exp Clin Pharmacol. 1996;18(6):387–95.
103. Clavreul N, Bachschmid MM, Hou X, et al. S-glutathiolation of p21ras by peroxynitrite mediates endothelial insulin resistance caused by oxidized low-density lipoprotein. Arterioscler Thromb Vasc Biol. 2006;26(11):2454–61.
104. Fourquet S, Guerois R, Biard D, Toledano MB. Activation of NRF2 by nitrosative agents and H2O2 involves KEAP1 disulfide formation. J Biol Chem. 2010;285(11):8463–71.
105. Nakamura Y, Moss AJ, Brown MW, Kinoshita M, Kawai C. Long-term nitrate use may be deleterious in ischemic heart disease: a study using the databases from two large-scale postinfarction studies. Am Heart J. 1999;138(3):577–85.
106. Kanamasa K, Hayashi T, Kimura A, Ikeda A, Ishikawa K. secondary prevention group. Long-term, continuous treatment with both oral and transdermal nitrates increases cardiac events in healed myocardial infarction patients. Angiology. 2002;53(4):399–408.
107. Daiber A, Oelze M, Wenzel P, et al. Nitrate tolerance as a model of vascular dysfunction: roles for mitochondrial aldehyde dehydrogenase and mitochondrial oxidative stress. Pharmacol Rep. 2009;61(1):33–48.
108. Tsou PS, Addanki V, Fung HL. Dissociation between superoxide accumulation and nitroglycerin-induced tolerance. J Pharmacol Exp Ther. 2008;327(1):97–104.
109. Wenzl MV, Beretta M, Gorren AC, et al. Role of the general base Glu-268 in nitroglycerin bioactivation and superoxide formation by aldehyde dehydrogenase-2. J Biol Chem. 2009;284(30):19878–86.

110. Thomas GR, DiFabio JM, Gori T, Parker JD. Once daily therapy with isosorbide-5-mononitrate causes endothelial dysfunction in humans: evidence of a free-radical-mediated mechanism. J Am Coll Cardiol. 2007;49(12):1289–95.
111. Chen CH, Budas GR, Churchill EN, Disatnik MH, Hurley TD, Mochly-Rosen D. Activation of aldehyde dehydrogenase-2 reduces ischemic damage to the heart. Science. 2008;321(5895):1493–5.
112. Esterbauer H, Zollner H, Lang J. Metabolism of the lipid peroxidation product 4-hydroxynonenal by isolated hepatocytes and by liver cytosolic fractions. Biochem J. 1985;228(2):363–73.
113. Death AK, Nakhla S, McGrath KC, et al. Nitroglycerin upregulates matrix metalloproteinase expression by human macrophages. J Am Coll Cardiol. 2002;39(12):1943–50.
114. Springman EB, Angleton EL, Birkedal-Hansen H, Van Wart HE. Multiple modes of activation of latent human fibroblast collagenase: evidence for the role of a Cys73 active-site zinc complex in latency and a "cysteine switch" mechanism for activation. Proc Natl Acad Sci U S A. 1990;87(1):364–8.
115. Krishnatry AS, Brazeau DA, Fung HL. Broad regulation of matrix and adhesion molecules in THP-1 human macrophages by nitroglycerin. Nitric Oxide. 2010;22(1):11–7.
116. Krishnatry AS, Fung SM, Brazeau DA, Soda D, Fung H-L. Nitroglycerin alters matrix remodeling proteins in THP-1 human macrophages and plasma metalloproteinase activity in rats. Nitric Oxide. In Press, Corrected Proof

Chapter 14
Nitrite and Nitrate in Ischemia-Reperfusion Injury

Madhav Lavu, Susheel Gundewar, and David J. Lefer

Key points

- There is a long and rich history of the use of nitrite for medical conditions.
- Nitrite has a number of essential functions in mammalian tissues.
- Nitrite is reduced to nitric oxide in blood and tissues, particularly in hypoxic or ischemic conditions.
- Nitrite protects from ischemia-reperfusion injury due to modulation of mitochondrial respiration.
- Nitrite has been shown in animal models to be beneficial in cardiovascular disease, cerebral, hepatic and renal I-R injury, effective for pulmonary disorders, peripheral artery disease and sickle cell disease.
- Very low levels of nitrite are needed for protection.
- There are now a number of clinical trials for the use of nitrite in a number of disease conditions.
- Nitrite-based therapies may provide a safe and cost-effective strategy for conditions of NO insufficiency.

Keywords Myocardial infarction • Hypoxia • Ischemia • Sickle cell disease • Peripheral artery disease • Cytoprotection • Kidney injury

D.J. Lefer (✉)
Division of Cardiothoracic Surgery, Department of Surgery, Emory University School of Medicine, Emory Clinic Building A, Ste A2223, 1365 Clifton Road NE, Atlanta, GA 30322, USA
e-mail: dlefer@emory.edu

N.S. Bryan and J. Loscalzo (eds.), *Nitrite and Nitrate in Human Health and Disease*, Nutrition and Health, DOI 10.1007/978-1-60761-616-0_14,

Introduction

The therapeutic effects of nitrite and nitric oxide (NO) on the cardiovascular system were initially realized during medieval times in ancient Chinese medicine. In 1900, a Daoist monk discovered thousands of medieval Buddhist manuscripts, paintings, and other documents in a grotto in the ancient Silk Road town of Dunhuang where they were hidden for over 900 years. Anthony Butler, Zhou Wuzong, and John Moffett translated a medical recipe from these manuscripts recently [1]. In this recipe, the patient is instructed to place potassium nitrate under the tongue, and then swallow the saliva to treat symptoms of angina and digital ischemia. The significance of the instructions is that salivary nitrate-reducing bacteria convert the nitrate into nitrite. Therefore, if the patient followed the physician's instructions fully, he or she would, in fact, be consuming nitrite, known to be effective in alleviating pain resulting from angina. In 1859, the English chemist Fredrick Guthrie noted that the vapor from amyl nitrite on inhalation caused an "immediate flushing of the neck, temples and forehead and an acceleration in the action of the heart." These suggestive vasodilatory actions of nitrite inspired Sir Thomas Lauder Brunton to administer amyl nitrite to his patients with angina pectoris, and in what was considered to be the first report on the medical use of nitrite, he described how within a minute after inhalation, a few drops of amyl nitrite caused the anginal pain to disappear.

Close on the heels of organic nitrite, inorganic sodium nitrite was shown to increase coronary flow as well in the 1920s [2]. In comparison to organic amyl nitrite, inorganic sodium nitrite produced a more sustained vasodilatory effect [3] and hence, was considered a better drug.

Although nitrites were used for the treatment of angina and hypertension, they fell out of repute because of concern about multiple side effects. Amyl nitrite was reported to cause blindness, an effect not seen during the last century of its use [4]. Alkyl nitrite administered at very high doses has been associated with increased incidence of Kaposi's Sarcoma probably due to its carcinogenic potential presumably from the formation of N-nitroso compounds [5]. Nitrite therapy at extremely high dosages has the potential for inducing methemoglobinemia (for which reason it is used today as treatment for cyanide poisoning [6]). In addition, amyl nitrite possesses a potential for abuse [7]. These side effects together with increased use of nitroglycerin (NTG) for treatment of angina led to gradual decrease in the use of nitrite in the clinical setting; however, interest in nitrite resurfaced following the discovery of the physiological role of NO in both health and disease states, and with characterization of its metabolism into nitrite and nitrate in mammalian tissues.

NO generated in cells and tissues is oxidized to produce nitrite. Hence, in cells and tissues, inorganic nitrite is an oxidative metabolite of NO. In this regard, nitrite has long been considered to be a physiologically and biochemically defunct byproduct of NO metabolism. Today, however, nitrite is spearheading the field of NO biology with the discovery that it represents a physiologically critical storage form and also a valuable source of NO in blood and tissues. Nitrite storage pools can readily be reduced to NO under certain pathological conditions [8–11].

This intriguing series of discoveries has spawned an entirely new field of research that involves the investigation of the molecular, biochemical and physiological activities of nitrite under a variety of physiological and pathophysiological states. Recent experimental studies clearly demonstrate that nitrite therapy is extremely cytoprotective in a number of animal models of disease. In this chapter, we will highlight the most recent evidence supporting the cytoprotective actions of nitrite in the setting of ischemia-reperfusion (I-R) injury. We will also elucidate its potential role as a therapeutic agent and as a biomarker of cardiovascular health and disease. We then offer comparison between nitrite and classical NO donors in the treatment of ischemic heart disease.

Bioconversion of Inorganic Nitrate and Nitrite to Nitric Oxide

Nitrite in the circulation and tissues is normally derived exclusively from either exogenous dietary sources or endogenous production, following generation of NO by endothelial nitric oxide synthase (eNOS). Exogenously, nitrite is obtained from ingestion of cured meats, green vegetables, and, more importantly, by conversion of dietary nitrate by commensal bacteria in the upper gastrointestinal tract and oral microbiota [12]. Nitrite, when swallowed, is subsequently reduced to NO in the acidic environment in the stomach [13] and additionally generated by gut bacteria with active nitrite reductase enzyme systems [14]. Endogenously, nitrite is formed from NO, which itself is synthesized from the amino acid, L-arginine, primarily by the action of eNOS. Nitrite levels therefore directly correlate with the activity of the eNOS enzyme [15] and have even been proposed as a biomarker of this key enzyme [16]. It is estimated that roughly 50% of the circulating nitrite in blood is derived from nitric oxide production, while the other 50% may be derived from dietary sources and reduction of salivary nitrate [17, 18].

Nitrite in plasma can be reduced to NO under certain conditions by a number of nitrite reductases, including deoxyhemoglobin [19–22], myoglobin [23], neuroglobin, cytoglobin [24], xanthine oxidase [25], mitochondrial enzymes such as cytochrome c [26], and aldehyde dehydrogenase 2 [27, 28], and also by acidic disproportionation [29, 30]. Nitrite, moreover, forms nitrosation products of thiols and nitrosylation products of heme with [31] or without the intermediate formation of NO [32]. Nitrite and these nitrosation products have now evolved to represent principal storage pools of NO in the body rather than be inert metabolites of NO devoid of any biological function [25].

Nitrite has numerous fundamental circulatory functions in mammalian tissues. Nitrite has been shown to mediate vasodilation during hypoxia by both NO-dependent [33] and -independent pathways [34, 35]. Very importantly, in hypoxic or ischemic conditions, nitrite is the primary source of NO, which then leads to vasodilation and cytoprotection [36]. This is because endogenous L-arginine-eNOS-NO pathway requires a constant supply of oxygen to produce NO in tissues, and hypoxic conditions cripple this pathway and its ability to synthesize NO [37–39] (Fig. 14.1). Additionally,

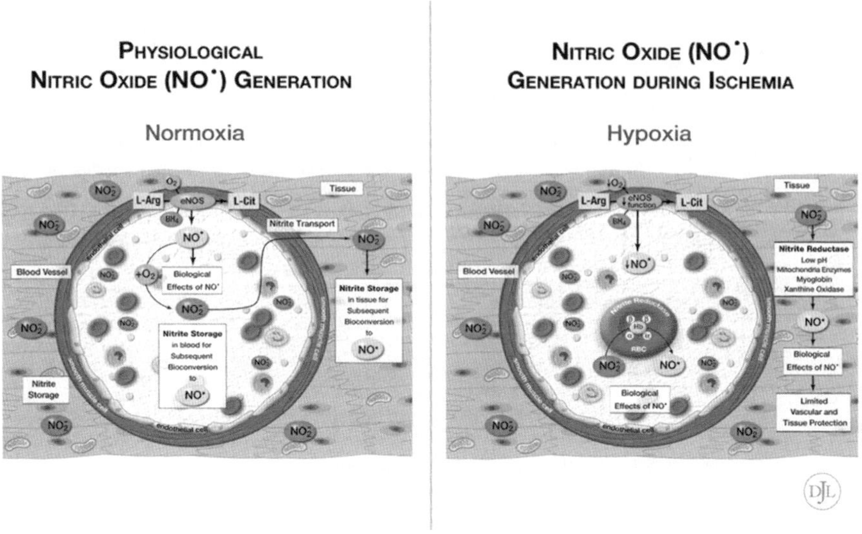

Fig. 14.1 During normoxia, oxygen tension is adequate and endothelial nitric oxide synthase (eNOS) functions normally to produce NO. NO acts on target cells and tissues, and any excess NO constantly replenishes nitrite stores via oxidation reactions with oxygen. Nitrite represents a critical NO storage pool. During hypoxia, eNOS is dysfunctional resulting in a fall in NO synthesis and endogenous nitrite generation. Nitrite stores are very rapidly reduced to NO via the actions of hemoglobin in the blood stream and via the actions of mitochondrial enzymes, myoglobin, and xanthine oxidase in the tissues. Thus, formation of NO from nitrite represents an NO salvage pathway that promotes cytoprotection during hypoxia

xanthine oxidase, a nitrite reductase, is more active under hypoxic and acidic conditions and, hence, is another important source of NO (more so than eNOS) during ischemia [40]. In addition to the proposed role of nitrite in regulating blood flow to organs, nitrite plays a prominent role in defending cells against I-R injury as discussed below.

Mechanisms of Protection by Nitrite and Nitric Oxide in I-R

I-R injury is characterized by numerous detrimental cellular events, such as oxidative damage to proteins and lipids, enzyme release, and inflammatory responses, which ultimately lead to necrosis and apoptosis of tissues [41, 42]. I-R injury propagates mitochondrial dysfunction (leading to reduced ATP synthesis) [43], deranges mitochondrial enzymes leading to formation of reactive oxygen species (ROS) [44, 45], and instigates opening of the mitochondrial permeability transition pore (MPTP) [46]. These changes eventually lead to increased calcium influx into the mitochondria exacerbating mitochondrial and cellular injury [47, 48]. The presence of NO abrogates mitochondrial pathology and leads to substantial protection during I-R. In this regard, NO reversibly inhibits mitochondrial enzymes [49, 50] and prevents formation of ROS [51]. Similarly, nitrite-derived NO reduces mitochondrial ROS generation at the time of reperfusion [52]. NO hinders release of cytochrome c [53] and very potently inhibits apoptosis [54]. Similarly, nitrite inhibits mitochondrial complex 1 following I-R, thereby limiting formation of ROS [55]. Nitrite-derived NO further activates guanylyl cyclase, inhibits cytochrome P450, and modulates HSP 70 and HO-1 activity [56, 57].

Vascular inflammation adversely affects blood vessels and the endothelium in the basal state and exacerbates the severity of I-R injury [58–62]. Plasma nitrite levels correlate well inversely with the severity of endothelial dysfunction [63–66], with decreased plasma nitrite levels observed in persons with known cardiovascular risk factors. Dietary supplementation of nitrite increases levels of the eNOS co-factor, tetrahydrobiopterin, and decreases levels of the inflammatory marker, C-reactive protein, thereby reducing endothelial dysfunction in mice fed a high cholesterol diet [67]. Thus, nitrite therapy to improve NO bioavailability can potentially be used for the numerous conditions associated with endothelial dysfunction (Fig. 14.2). Nitrite therapy also retards the progression of coronary atherosclerosis, hypertension, cardiovascular disease, and potentially attenuates the severity of any I-R injury that may arise in due course.

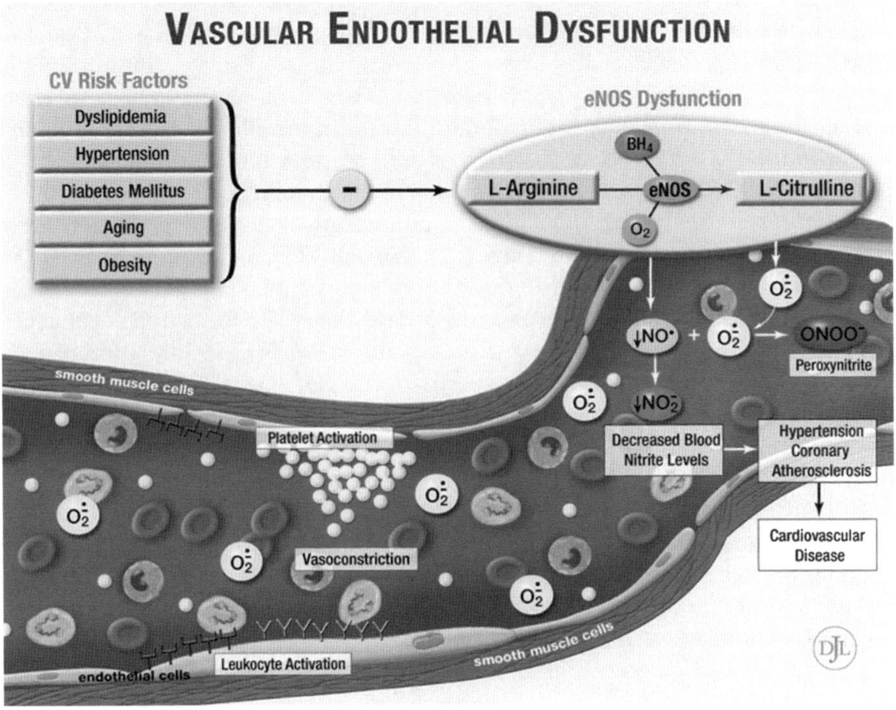

Fig. 14.2 Endothelial dysfunction is the hallmark of numerous cardiovascular risk factors, such as dyslipidemia, hypertension, diabetes, aging, and obesity. In these conditions, the L-arginine-eNOS-NO pathway activity is suppressed owing to dysfunction of eNOS, which leads to decreased nitric oxide and nitrite bioavailability in the blood and tissues. Decreased nitrite level results in abnormal platelet activation, abnormal vasoconstriction, and inability to scavenge oxygen free radicals. In addition, dysfunctional eNOS shifts the above pathway into superoxide generation, leading to heightened oxidative stress in tissues. Ultimately, vascular endothelial dysfunction contributes to the pathology of hypertension, atherosclerosis, acute myocardial infarction, and stroke

Nitrite-Mediated Cytoprotection in Various Organs

Nitrite in Cardiovascular Disease

Endogenous NO derived from NOS plays a vital role in imparting protection during reperfusion injury; its absence leads to increased injury in the heart [17]. Administration of nitrite alleviates reperfusion injury by augmenting NO levels in the coronary circulation and in the myocardium. During ischemia and reperfusion, the decrease in plasma levels of nitrite coincides with increase in levels of S-nitrosated and nitrosylated (heme) proteins [68] suggesting that NO is derived from nitrite during the conditions of ischemia and reperfusion. Even brief elevation in nitrite levels modulates redox status, lowers oxidized glutathione levels, inducing significant cardioprotection in the setting of cardiac injury [69].

Nitrite administered during myocardial I-R reduces infarct size and improves left ventricular (LV) function [70], boosting ejection fraction [71]. In a canine model of myocardial I-R, apart from reducing infarct size and improving LV function [71], nitrite also limits endocardial "no flow" phenomenon [71]. Nitrite administered orally is as beneficial as nitrite given intravenously. Nitrite, administered in the drinking water for 7 days prior to myocardial I-R increases plasma and myocardial nitrite levels, augments S-nitrosothiol levels, and results in decreased infarct size in mice [72]. The cardioprotection observed with nitrite is independent of eNOS-derived NO [68] since eNOS knockout mice exhibit profound protection.

The mechanisms through which nitrite protects myocytes are many and varied. Nitrite protects the heart by activating K_{ATP} channels [73], reducing apoptosis [71], and decreasing the formation of ROS [70]. Of note two enzymes play a key role in nitrite-mediated protection. Xanthine oxidase mediated NO formation is enhanced with nitrite infusions in rats during myocardial I-R injury [74]. Inhibition of xanthine oxidase abolishes the cardioprotective action of nitrite [73]. Myoglobin is another important nitrite reductase in the heart and plays an important role in the course of I-R as well [70].

Nitrite therapy has also been shown to be highly beneficial in the setting of cardiac arrest in animal models. Cardiac arrest and subsequent resuscitation may be viewed as a situation of global reperfusion injury whereby injury occurs in organ systems throughout the body [75]. In a mouse model of asystole for 12 min followed by CPR, a single low dose of nitrite resulted in improved survival, better cardiac function, and improved neurological outcomes [76].

Nitrite Therapy and Cerebral I-R Injury

Cerebral I-R injury is characterized by increased production of ROS, cellular dysfunction, apoptosis, and cellular necrosis. NO donors such as ProliNO reduce ROS and cerebral infarct volume [77]. Similarly, nitrite therapy administered at reperfusion following middle cerebral artery occlusion (MCAO) in rats increases cerebral blood flow, decreases infarct volume, and decreases lipid peroxidation [78]. Nitrite administration within 1.5 h of permanent focal ischemia or within 3 h of transient ischemia to brain also reduces infarct volume and hastens recovery in rats [79]. Nitrite reduces the amount of microhypoxia in rats by minimizing the number of ischemic cells expressing HIF-1alpha [79]. Nitrite co-administered with memantine, a NMDA receptor inhibitor, also leads to reduction in oxidative stress during reperfusion [79]. It may be important to administer nitrite early in the course of reperfusion as delayed treatment does not confer cytoprotection [80].

It has been suggested that dysfunctional eNOS is responsible for the delayed vasospastic response after subarachnoid hemorrhage (SAH) [81]. Nitrite therapy in a primate model of vasospasm (induced by blood clot placement in the cerebral arteries and consequent SAH) resulted in increased nitrite and S-nitrosothiol levels in the blood and cerebrospinal fluid (CSF) with an attributable delay in cerebral vasospasm [82]. Nitrite-based therapy for cerebral vasospasm is currently being tested in humans. There is an ongoing phase 2 clinical trial assessing safety and pharmacokinetics of a 14-day nitrite infusion for prevention of cerebral vasospasm [83]. Hypertension increases risk of stroke in many elderly patients. Following stroke, treatment of labile hypertension is frequently challenging [84]. Nitrite treatment may help control blood pressure in hypertension and improve cerebral blood flow [85] in many of these patients.

Nitrite Therapy and Hepatic I-R Injury

Reperfusion injury in the liver leads to considerable morbidity and mortality. I-R injury commonly occurs with liver transplantation, thermal injury, and shock. During liver I-R, there is an imbalance between vasoconstrictors and vasodilators leading to disruption of microcirculation [86]. There is also a rapid decrease in NO bioavailability [87]. NO protects the liver during I-R injury by inactivation of caspase 3, opening mitochondrial K_{ATP} channels, and inhibition of complex 1 [88]. NO prevents MPTP-dependent necrosis of ischemic hepatocytes after reperfusion through a guanylyl cyclase and cGMP-dependent kinase signaling pathway [89].

In the liver, as with heart and brain, endogenous nitrite is a vital source of NO. During ischemic and hypoxic conditions, endogenous nitrite protects against the development of hepatocellular injury through xanthine oxidase mediated NO release [90]. During ischemia, NO generated from nitrite mediates signaling and formation of nitrosation and nitrosylation protein products in tissues [91]. Endogenous stores of nitrite are undoubtedly consumed to combat the oxidative stress associated with I-R injury. Exogenous administration of nitrite, then, replenishes these depleted endogenous nitrite stores and additionally leads to the formation of N-, S-, and heme-nitrosated or nitrosylated hepatic proteins [68].

As with the heart, nitrite may be administered orally to protect the liver against I-R injury. Regular dietary intake of nitrite and its precursor inorganic nitrate maintains normal levels of nitrite in plasma and deficiency in diet leads to reduced levels in the blood and therefore exacerbated I-R injury [92]. The effect of one time administration of nitrite lasts up to 24 h. Nitrite given orally 24 h prior to I-R inhibits complex 1 in mitochondria, thus reducing the formation of ROS and attenuating both hepatic and myocardial I-R injury [55].

Nitrite Therapy and Renal I-R Injury

NO donors, including those with structures similar to S-nitrosothiols, effectively combat inflammatory responses that occur during I-R injury [93]. NO donor drugs, such as molsidomine [94] and sodium nitroprusside [95], attenuate I-R injury in kidneys; NO derived from these drugs diminishes apoptosis [96], suppresses tubular necrosis [97], and inhibits ET-1 expression [95].

In the kidney, nitrite is an important source of NO during ischemic syndromes [98]. eNOS has also been proposed to be an important nitrite reductase contributing to protection during renal I-R in ischemic conditions [99]. Additionally, xanthine oxidase mediates the reduction of nitrite to NO while the presence of a xanthine oxidase inhibitor prevents NO formation leading to enhanced renal injury [100]. However, nitrite administration must precede renal ischemic events to offer optimal protection because some studies revealed failure of nitrite treatment administered during ischemia [101], but beneficial effects when administered prior to onset of ischemia [102]. Clinically, this pharmacologic "preconditioning" may be applicable only to scenarios such as renal transplantation. Nitrite is also a potential antihypertensive. Administering nitrite orally may replete circulating NO levels, and thus protect the kidney during hypertension [103]. It is possible that the hypotensive effect of a diet rich in fruits and vegetables is due to the release of NO [104, 105]. Nitrite therapy may additionally benefit individuals with renal insufficiency by controlling blood pressure and slowing the progression of hypertension-induced renal injury.

Nitrite Therapy for Pulmonary Disorders

Nitrite therapy is beneficial in several pulmonary disorders. Nitrite given by inhalation in newborn lambs with normoxic pulmonary hypertension has been shown to cause pulmonary vasodilation. This inhalational route has the advantage of limiting the systemic appearance of methemoglobin levels [106] and hence, can be used in neonates. Pulmonary hypoxia leads to depletion of RSNO levels and leads to reflex constriction of the pulmonary arteries [107]. Nitrite attenuates this resultant acute pulmonary vasoconstriction during hypoxia [108]. Nitrite treatment prevents smooth muscle proliferation in pulmonary arteries and attenuates the pathological right ventricular hypertrophy seen in pulmonary hypertension [109]. The beneficial reduction of pulmonary arterial pressure is sustained since it lasts for approximately 1 h after the cessation of nitrite infusion [110]. This sustained pulmonary vasodilation despite plasma nitrite levels returning to baseline value suggests that nitrite has prolonged vasodilatory effects after one time administration. As with other organs, xanthine oxidase reduces nitrite to NO, especially under hypoxic conditions in the lung [111].

Additionally, nitrite treatment conferred benefits during hemolysis and pulmonary embolism. Pulmonary hypertension accompanies hemolysis [112] and nitrite infusion counteracts this elevation of pulmonary pressure [113]. Nitrite treatment has been beneficial for pulmonary thromboembolism where it improved cardiac index and pulmonary and systemic vascular resistance indices [114]. High-altitude pulmonary hypertension is associated with reduced nitrite levels [115]. Replenishing nitrite levels may benefit these patients as well.

Nitrite Therapy for Peripheral Arterial Disease

Plasma nitrite levels correlate with vasodilatory responses in normal individuals [116]. Patients with peripheral arterial disease (PAD) are affected by abnormal vasoreactivity and reduced peripheral vascular perfusion that correlates with deranged nitrite levels [117, 118]. Nitrite-derived NO is known to vasodilate blood vessels during hypoxia and directly stimulate angiogenesis and vasculogenesis [119, 120]. Not surprisingly, NO-based therapies have been shown to induce angiogenesis in the setting of the chronic ischemia that results from PAD [121]. Chronic nitrite therapy leads to endothelial cell proliferation and neovascularization in ischemic limbs [122]. In a clinically relevant model of hind-limb ischemia, low-dose sodium nitrite increased vascular density through endothelial cell proliferation as early as 3 days after ischemia. Therapy with sodium nitrite increased the levels of tissue nitrite and NO metabolites in the ischemic limb. Interestingly, the levels of these metabolites were lower in the nonischemic limb. Thus nitrite therapy leads to an effective and highly tissue-selective angiogenic response in a short period of time after limb ischemia. At present, sodium nitrite is under development for the treatment of PAD in humans as clinical trials have been recently initiated.

Nitrite Therapy for Sickle Sell Disease

Sickle cell disease is the most common genetic disorder in the African American population and is characterized by oxidative stress and endothelial dysfunction [123]. Multiple microvascular vaso-occlusive episodes lead to insufficient perfusion and ultimately to multi-organ injury. Reduced NO bioavailability is now thought to play an important role in pathophysiology of this disease [124]. Hydroxyurea (an NO donor) represents the only FDA approved drug currently for the treatment of this crippling condition. A reduction of nitrite levels has been observed in patients with sickle cell crisis [125]. A human trial testing the ability of nitrite infusions to improve blood flow showed that sickle cell patients did indeed achieve a dose-dependent increase in regional blood flow though to a lesser extent compared to healthy individuals [126]. This suggests that sickle cell patients are more resistant to NO-based therapy than their healthy counterparts. However, nitrite therapy proved to be safe in this study and represents a potential therapeutic drug. Additional safety and blood parameters along with alleviation of pain in sickle cell patients treated with nitrite injections are being currently tested [127]. Results from this latest trial will estimate the efficacy of nitrite therapy for sickle cell disease and establish if further testing is meritorious.

Benefits of Low-Dose Nitrite Therapy

In most of the studies mentioned in this chapter, a low dose of nitrite was employed to attenuate I-R injury or demonstrate other beneficial effects (Table 14.1).

Clinical Trials on Nitrite Therapy

It is surprisingly evident that studies in human subjects on the use of nitrite in health and disease have been scarce, and this void is now being filled with a spate of clinical trials being conducted by the NHLBI in the last few years. Pharmacokinetics

Table 14.1 This table summarizes the concentration and dose of nitrite solutions used in various experiments

References	Nitrite dose
Webb et al. [74]	10–100 μM infusion 15 min prior to I/R
Hendgen-Cotta et al. [70]	0–50 μM 5 min before R
Gonzalez et al. [71]	0.17–0.20 μM/kg/min infusion during last 60 min and last 5 min of ischemia
Cosby et al. [19]	36 μM/mL infusion in healthy individuals
Duranski et al. [68]	48 nmol blood conc. (165 μg/kg i.p.) during hepatic and myocardial I/R
Jung et al. [78]	48 and 480 nmol in 500 μL as infusion at R
Kumar et al. [122]	165 μg/kg via i.p. injection twice daily for 3–7 days after limb ischemia
Stokes et al. [67]	33 or 99 mg/L elemental nitrite in drinking water with high cholesterol rich diet for 3 weeks
Bryan et al. [17]	50 mg/L nitrite in drinking water for 1 week prior to I/R
Bryan et al. [72]	50 mg/L nitrite in drinking water
Dezfulian et al. [76]	0.13 mg/kg iv at the time of CPR

The majority of studies have utilized nitrite in micromolar concentrations of systemic nitrite infusion and NO formation are being evaluated in a phase 1 clinical trial in healthy volunteers aged 21–40. The effects of nitrite on blood pressure and the mechanism by which it increases blood flow are also being studied in this yet unpublished trial [128]. Many other phase 1 clinical trials in healthy volunteers have been recently completed: studying the effects of nitrite on exercise tolerance [129], determining the optimum dose of nitrite for infusion [130], determining whether nitrite can dilate blood vessels through formation of NO [131], confirming the role of nitrite in ischemic preconditioning [132], and whether the amount of nitrites and nitrates in the diet correlates with the amount of NO exhaled [133]. The results of these studies in healthy volunteers will greatly augment the understanding of role of nitrite in physiological regulation of blood pressure, exercise tolerance, ischemic preconditioning, and its pharmacokinetics.

Apart from studies on healthy volunteers, clinical trials of the possible benefits of nitrite in disease have also been started in the last few years. There is an ongoing phase 2 clinical trial on the use of sodium nitrite for acute myocardial infarction. Subjects assigned to sodium nitrite group will receive an initial infusion of 6 nmol/min/kg for 48 h. As a primary outcome measure, the effects of sodium nitrite on myocardial infarct size will be assessed as determined by paired single-photon computed tomography studies with technetium Tc99m sestamibi [134]. The benefit of nitrite therapy prior to or during coronary artery bypass thereby targeting reperfusion injury is being investigated in a phase 2 trial [135]. Another phase 2 trial assessing the role of dietary nitrite in causing NO-induced vasodilation in patients with coronary artery disease has been completed recently, the results of which are awaited [136]. Since nitrite possesses a potential to induce angiogenesis in ischemic limbs, a novel, orally active, extended release formulation of sodium nitrite is being tested in PAD in diabetics in a Phase I clinical trial [137]. The outcome from these studies, both from healthy volunteers and patients with cardiovascular disease, will answer important questions regarding the use of nitrite in combating cardiovascular disease.

Plasma Nitrite Level as a Biomarker of Cardiovascular Disease

Nitrite level in the plasma has been proposed as biomarker of disease risk. Plasma levels of nitrite and of inflammatory biomarkers are elevated in patients with increased cardiovascular risk [138] and in elderly individuals over 80 [139]. Nitrite levels are elevated in individuals with increased lipid levels in the blood [140]; however, plasma nitrite and nitrate levels have been shown to be decreased in patients with hypertension [141]. More data are needed to ascertain the role of plasma nitrite level in predicting disease risk and prognosis. To this end, an observational study evaluating nitrite as a biomarker for cardiovascular disease risk has been completed recently [142]. It would be interesting to know if nitrite would prove to be an effective screening molecule in detecting patients with increased cardiovascular risk.

Implications of Dietary Nitrite and Nitrate Supplementation: Safety and Cost Effectiveness Analysis

Fruits and vegetables have numerous protective components, which include nitrates and nitrites [143]. Nitrate contained in vegetables has excellent bioavailability [144]. Similarly sodium nitrite ingested orally is absorbed well and has low first-pass metabolism in the liver [145]. It is now known that a diet rich in fruits and vegetable reduces the incidence of coronary heart disease [146] and stroke [147]. Fruit and vegetable consumption in women is associated with a decreased incidence of diabetes [148]. A diet rich in fruits, vegetables, and low-fat dairy products along with a low daily intake of total and saturated fats has been shown to reduce blood pressure comparable to a single anti-hypertensive medication [149, 150]. The blood pressure-lowering effects of ingested nitrate interestingly correlate with nitrite levels in plasma but not with nitrate levels, suggesting that nitrite is the final effector of blood pressure reduction [151]. In fact, in a 60-kg adult, high-nitrate DASH diet pattern has been shown to exceed the World Health Organization's Acceptable Daily Intake for nitrate by 550% [104]. Even in normal individuals, diets rich in nitrate have been shown to reduce diastolic blood pressure [152]. Nitrate ingestion has additional benefits such as inhibition of platelet aggregation [153] and reduced oxygen demand. Nitrate utilization by skeletal muscle thereby augments exercise capacity [154–157]. It is highly likely that nitrate in all of these instances is converted to nitrite in the body, which may account for the benefits demonstrated in these studies.

Another source of nitrite is the food processing industry. Sodium nitrite is added during the curing of meats to produce the characteristic dark color and to preserve freshness. There have been concerns of nitrite combining with proteins in meat to form N-nitrosamines, which are carcinogenic. This led to the food industry using antioxidants, such as vitamins C and E, to suppress the formation of N-nitrosamines. Today, governments of many countries regulate nitrate and nitrite levels in beverages and drinking water; however, the evidence is not convincing about causation of cancer by nitrites and nitrates [158]. The risk of development of cancer is outweighed by the health benefits of dietary nitrites and nitrates [104, 105].

Though levels of nitrates and nitrites in dietary sources are being regulated, the safety of chronic pharmacotherapy with organic nitrates has never been questioned. Organic nitrates have been used for the treatment of ischemic heart disease for a very long time due to their perceived beneficial hemodynamic effects. Development of tolerance was viewed merely as decreased sensitivity of patients to the medication eliciting an escalation of dose by care providers. This view has now come under scrutiny since tolerance encompasses much more than purely nonresponsiveness of patients [159].

Conclusions

Nitrite has now emerged as a critically important signaling molecule involved in maintaining perfusion and redox status in tissues, not solely a metabolic product of NO in tissues. It is the principal source of NO in hypoxic conditions and oxidative stress states (Fig. 14.3). Nitrite has several disparate advantages as a therapeutic agent in that it is an extremely stable compound, relatively inexpensive, and can be easily administered as shown by its demonstrable efficacy as an oral, inhalational, and intravenous agent (Fig. 14.4). Nitrite, unlike the current NO donors in clinical use, does not lead to tolerance [160] and, hence, can be used for chronic therapy.

Nitrite offers an immense therapeutic potential for the treatment of the most prevalent diseases and their complications, as they impose huge societal burdens in terms of healthcare resources. Conditions such as ischemic heart disease, stroke, and respiratory disorders represent three of the top five killer diseases in the United States, and thankfully, may be amenable to nitrite therapy. Additionally, nitrite

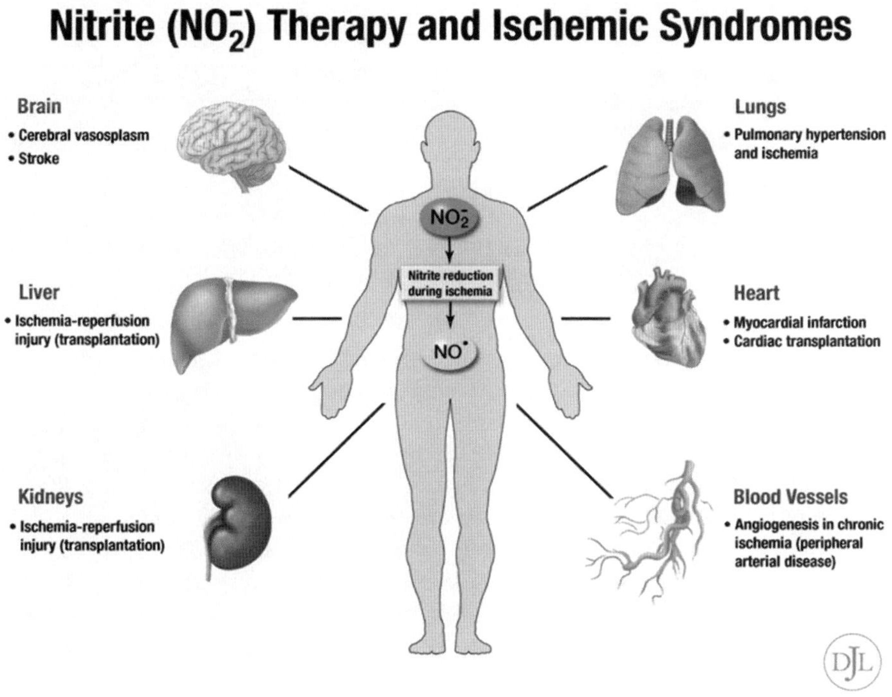

Fig. 14.3 Nitrite is an important source of NO during both acute and chronic ischemic disease states. NO derived from nitrite has been shown to limit both ischemic and reperfusion injury in the brain, lungs, heart, peripheral circulation, kidneys, and liver in a number of preclinical studies. At present, clinical trials are underway to evaluate both the efficacy and safety of nitrite therapy in a variety of ischemic syndromes

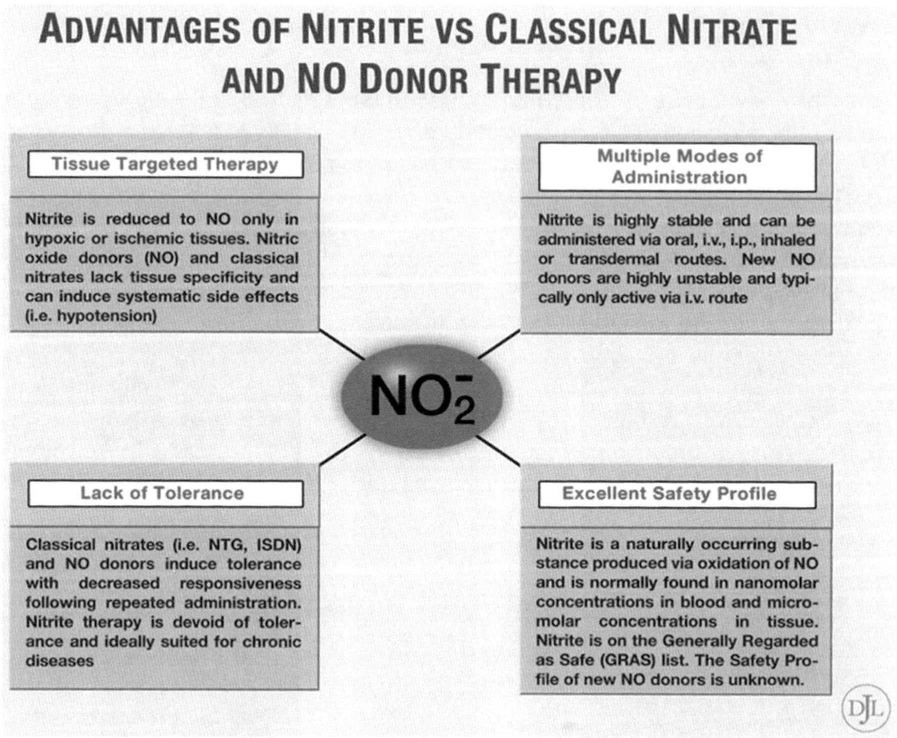

Fig. 14.4 Nitrite has numerous advantages over classical organic nitrates and NO donors. Nitrite can be administered by multiple routes and has proven thus far to be very safe. Specific tissue targeting by nitrite is possible because hypoxic tissues selectively utilize nitrite to generate NO. Repeated administration of nitrite does not induce tolerance unlike classical nitrates

treatment may be beneficial in chronic cardiovascular disease associated with central obesity and diabetes [161]. The future translation of nitrite from bench to bedside will undeniably prove to be an exciting time in the treatment of ischemic states, such as myocardial infarction, stroke, and solid organ transplantation. Further insights gleaned from preclinical investigations will likely provide evidence for the investigation of various forms of sodium nitrite for other clinical conditions.

Despite the overwhelming enthusiasm to develop sodium nitrite therapy for the treatment of cardiovascular disease, it is very important to proceed with caution. There is general agreement that very high levels of NO can promote tissue injury, decrease blood pressure, and reduce cardiac performance. Thus, it is extremely important to consider the timing of administration as well as the dosing when dealing with nitrite or any other agents that augment NO bioavailability. Finally, one must also consider the possible drug interactions between nitrite-based therapies and other agents to avoid further potential toxic effects. When these pharmacokinetic and pharmacodynamic issues have been addressed, the potential clinical benefits of nitrite therapy may be realized.

Disclosure David J. Lefer (D.J.L.) is a participant on two pending United States Patents (patents no. 60/511244 and 61/003150) on the use of sodium nitrite in cardiovascular disease. In addition, D.J.L is also on the scientific advisory board of Theravasc, Inc., a company that is developing sodium nitrite therapy for peripheral arterial disease and other cardiovascular conditions.

References

1. Lundberg JO, Weitzberg E, Gladwin MT. The nitrate-nitrite-nitric oxide pathway in physiology and therapeutics. Nat Rev Drug Discov. 2008;7(2):156–67.
2. Bodo R. The effect of the "heart-tonics" and other drugs upon the heart-tone and coronary circulation. J Physiol. 1928;64(4):365–87.
3. Nossaman VE, Nossaman BD, Kadowitz PJ. Nitrates and nitrites in the treatment of ischemic cardiac disease. Cardiol Rev. 2010;18(4):190–7.
4. Fledelius HC. Irreversible blindness after amyl nitrite inhalation. Acta Ophthalmol Scand. 1999;77(6):719–21.
5. Haverkos HW, Dougherty J. Health hazards of nitrite inhalants. Am J Med. 1988;84(3 Pt 1): 479–82.
6. Gracia R, Shepherd G. Cyanide poisoning and its treatment. Pharmacotherapy. 2004;24(10):1358–65.
7. Wu LT, Schlenger WE, Ringwalt CL. Use of nitrite inhalants ("poppers") among American youth. J Adolesc Health. 2005;37(1):52–60.
8. Dezfulian C, Raat N, Shiva S, Gladwin MT. Role of the anion nitrite in ischemia-reperfusion cytoprotection and therapeutics. Cardiovasc Res. 2007;75(2):327–38.
9. Lefer DJ. Emerging role of nitrite in myocardial protection. Arch Pharm Res. 2009;32(8):1127–38.
10. Jensen FB. The role of nitrite in nitric oxide homeostasis: a comparative perspective. Biochim Biophys Acta. 2009;1787(7):841–8.
11. Manukhina EB, Malyshev IY, Smirin BV, Mashina SY, Saltykova VA, Vanin AF. Production and storage of nitric oxide in adaptation to hypoxia. Nitric Oxide. 1999;3(5):393–401.
12. Lundberg JO, Weitzberg E, Cole JA, Benjamin N. Nitrate, bacteria and human health. Nat Rev Microbiol. 2004;2(7):593–602.
13. Lundberg JO, Weitzberg E. NO generation from inorganic nitrate and nitrite: role in physiology, nutrition and therapeutics. Arch Pharm Res. 2009;32(8):1119–26.
14. Sobko T, Reinders CI, Jansson E, Norin E, Midtvedt T, Lundberg JO. Gastrointestinal bacteria generate nitric oxide from nitrate and nitrite. Nitric Oxide. 2005;13(4):272–8.
15. Kleinbongard P, Dejam A, Lauer T, Rassaf T, Schindler A, Picker O. Plasma nitrite reflects constitutive nitric oxide synthase activity in mammals. Free Radic Biol Med. 2003; 35(7):790–6.
16. Gladwin MT. Haldane, hot dogs, halitosis, and hypoxic vasodilation: the emerging biology of the nitrite anion. J Clin Invest. 2004;113(1):19–21.
17. Bryan NS, Calvert JW, Gundewar S, Lefer DJ. Dietary nitrite restores NO homeostasis and is cardioprotective in endothelial nitric oxide synthase-deficient mice. Free Radic Biol Med. 2008;45(4):468–74.
18. Bryan NS. Cardioprotective actions of nitrite therapy and dietary considerations. Front Biosci. 2009;14:4793–808.
19. Cosby K, Partovi KS, Crawford JH, Patel RP, Reiter CD, Martyr S. Nitrite reduction to nitric oxide by deoxyhemoglobin vasodilates the human circulation. Nat Med. 2003;9(12): 1498–505.

20. Gladwin MT, Schechter AN. NO contest: nitrite versus S-nitroso-hemoglobin. Circ Res. 2004;94(7):851–5.
21. Gladwin MT, Crawford JH, Patel RP. The biochemistry of nitric oxide, nitrite, and hemoglobin: role in blood flow regulation. Free Radic Biol Med. 2004;36(6):707–17.
22. McMahon TJ, Moon RE, Luschinger BP, Carraway MS, Stone AE, Stolp BW. Nitric oxide in the human respiratory cycle. Nat Med. 2002;8(7):711–7.
23. Kim-Shapiro DB, Gladwin MT, Patel RP, Hogg N. The reaction between nitrite and hemoglobin: the role of nitrite in hemoglobin-mediated hypoxic vasodilation. J Inorg Biochem. 2005;99(1):237–46.
24. Petersen MG, Dewilde S, Fago A. Reactions of ferrous neuroglobin and cytoglobin with nitrite under anaerobic conditions. J Inorg Biochem. 2008;102(9):1777–82.
25. Lundberg JO, Weitzberg E. NO generation from nitrite and its role in vascular control. Arterioscler Thromb Vasc Biol. 2005;25(5):915–22.
26. Basu S, Azarova NA, Font MD, King SB, Hogg N, Gladwin MT. Nitrite reductase activity of cytochrome C. J Biol Chem. 2008;283(47):32590–7.
27. Daiber A, Munzel T. Nitrate reductase activity of mitochondrial aldehyde dehydrogenase (ALDH-2) as a redox sensor for cardiovascular oxidative stress. Methods Mol Biol. 2010;594:43–55.
28. Golwala NH, Hodenette C, Murthy SN, Nossaman BD, Kadowitz PJ. Vascular responses to nitrite are mediated by xanthine oxidoreductase and mitochondrial aldehyde dehydrogenase in the rat. Can J Physiol Pharmacol. 2009;87(12):1095–101.
29. Weitzberg E, Lundberg JO. Nonenzymatic nitric oxide production in humans. Nitric Oxide. 1998;2(1):1–7.
30. Zweier JL, Wang P, Samouilov A, Kuppusamy P. Enzyme-independent formation of nitric oxide in biological tissues. Nat Med. 1995;1(8):804–9.
31. Tiravanti E, Samouilov A, Zweier JL. Nitrosyl-heme complexes are formed in the ischemic heart: evidence of nitrite-derived nitric oxide formation, storage, and signaling in post-ischemic tissues. J Biol Chem. 2004;279(12):11065–73.
32. Bryan NS, Fernandez BO, Bauer SM, Garcia-Saura MF, Milsom AB, Rassaf T. Nitrite is a signaling molecule and regulator of gene expression in mammalian tissues. Nat Chem Biol. 2005;1(5):290–7.
33. Lundberg JO, Weitzberg E. NO-synthase independent NO generation in mammals. Biochem Biophys Res Commun. 2010;396(1):39–45.
34. Pinder AG, Pittaway E, Morris K, James PE. Nitrite directly vasodilates hypoxic vasculature via nitric oxide-dependent and -independent pathways. Br J Pharmacol. 2009;157(8):1523–30.
35. Dalsgaard T, Simonsen U, Fago A. Nitrite-dependent vasodilation is facilitated by hypoxia and is independent of known NO-generating nitrite reductase activities. Am J Physiol Heart Circ Physiol. 2007;292(6):H3072–8.
36. Gladwin MT, Raat NJ, Shiva S, Dezfulian C, Hogg N, Kim-Shapiro DB. Nitrite as a vascular endocrine nitric oxide reservoir that contributes to hypoxic signaling, cytoprotection, and vasodilation. Am J Physiol Heart Circ Physiol. 2006;291(5):H2026–35.
37. McQuillan LP, Leung GK, Marsden PA, Kostyk SK, Kourembanas S. Hypoxia inhibits expression of eNOS via transcriptional and posttranscriptional mechanisms. Am J Physiol. 1994;267(5 Pt 2):H1921–7.
38. Fish JE, Yan MS, Matouk CC, St Bernard R, Ho JJ, Gavryushova A, et al. Hypoxic repression of endothelial nitric-oxide synthase transcription is coupled with eviction of promoter histones. J Biol Chem. 2010;285(2):810–26.
39. Tai SC, Robb GB, Marsden PA. Endothelial nitric oxide synthase: a new paradigm for gene regulation in the injured blood vessel. Arterioscler Thromb Vasc Biol. 2004;24(3):405–12.
40. Webb AJ, Milsom AB, Rathod KS, Chu WL, Qureshi S, Lovell MJ. Mechanisms underlying erythrocyte and endothelial nitrite reduction to nitric oxide in hypoxia: role for xanthine oxidoreductase and endothelial nitric oxide synthase. Circ Res. 2008;103(9):957–64.

41. Braunwald E, Kloner RA. Myocardial reperfusion: a double-edged sword? J Clin Invest. 1985;76(5):1713–9.
42. Lefer AM, Lefer DJ. The role of nitric oxide and cell adhesion molecules on the microcirculation in ischaemia-reperfusion. Cardiovasc Res. 1996;32(4):743–51.
43. Sims NR, Muyderman H. Mitochondria, oxidative metabolism and cell death in stroke. Biochim Biophys Acta. 2010;1802(1):80–91.
44. Chen Q, Camara AK, Stowe DF, Hoppel CL, Lesnefsky EJ. Modulation of electron transport protects cardiac mitochondria and decreases myocardial injury during ischemia and reperfusion. Am J Physiol Cell Physiol. 2007;292(1):C137–47.
45. Sack MN. Mitochondrial depolarization and the role of uncoupling proteins in ischemia tolerance. Cardiovasc Res. 2006;72(2):210–9.
46. Kim JS, He L, Qian T, Lemasters JJ. Role of the mitochondrial permeability transition in apoptotic and necrotic death after ischemia/reperfusion injury to hepatocytes. Curr Mol Med. 2003;3(6):527–35.
47. Murphy E, Steenbergen C. Mechanisms underlying acute protection from cardiac ischemia-reperfusion injury. Physiol Rev. 2008;88(2):581–609.
48. Garcia-Rivas GJ, Torre-Amione G. Abnormal mitochondrial function during ischemia reperfusion provides targets for pharmacological therapy. Methodist Debakey Cardiovasc J. 2009;5(3):2–7.
49. Brown GC, Cooper CE. Nanomolar concentrations of nitric oxide reversibly inhibit synaptosomal respiration by competing with oxygen at cytochrome oxidase. FEBS Lett. 1994; 356(2–3):295–8.
50. Cleeter MW, Cooper JM, Darley-Usmar VM, Moncada S, Schapira AH. Reversible inhibition of cytochrome c oxidase, the terminal enzyme of the mitochondrial respiratory chain, by nitric oxide. Implications for neurodegenerative diseases. FEBS Lett. 1994;345(1):50–4.
51. Brookes P, Darley-Usmar VM. Hypothesis: the mitochondrial NO(*) signaling pathway, and the transduction of nitrosative to oxidative cell signals: an alternative function for cytochrome C oxidase. Free Radic Biol Med. 2002;32(4):370–4.
52. Raat NJ, Shiva S, Gladwin MT. Effects of nitrite on modulating ROS generation following ischemia and reperfusion. Adv Drug Deliv Rev. 2009;61(4):339–50.
53. Brookes PS, Salinas EP, Darley-Usmar K, Eiserich JP, Freeman BA, Darley-Usmar VM. Concentration-dependent effects of nitric oxide on mitochondrial permeability transition and cytochrome c release. J Biol Chem. 2000;275(27):20474–9.
54. Kim YM, Kim TH, Seol DW, Talanian RV, Billiar TR. Nitric oxide suppression of apoptosis occurs in association with an inhibition of Bcl-2 cleavage and cytochrome c release. J Biol Chem. 1998;273(47):31437–41.
55. Shiva S, Sack MN, Greer JJ, Duranski M, Ringwood LA, Burwell L. Nitrite augments tolerance to ischemia/reperfusion injury via the modulation of mitochondrial electron transfer. J Exp Med. 2007;204(9):2089–102.
56. Gladwin MT. Nitrite as an intrinsic signaling molecule. Nat Chem Biol. 2005;1(5):245–6.
57. Gladwin MT, Shelhamer JH, Schechter AN, Pease-Fye ME, Waclawiw MA, Panza JA. Role of circulating nitrite and S-nitrosohemoglobin in the regulation of regional blood flow in humans. Proc Natl Acad Sci U S A. 2000;97(21):11482–7.
58. Garcia SC, Pomblum V, Gams E, Langenbach MR, Schipke JD. Independency of myocardial stunning of endothelial stunning? Basic Res Cardiol. 2007;102(4):359–67.
59. Qi XL, Nguyen TL, Andries L, Sys SU, Rouleau JL. Vascular endothelial dysfunction contributes to myocardial depression in ischemia-reperfusion in the rat. Can J Physiol Pharmacol. 1998;76(1):35–45.
60. Granger DN, Rodrigues SF, Yildirim A, Senchenkova EY. Microvascular responses to cardiovascular risk factors. Microcirculation. 2010;17(3):192–205.
61. Carden DL, Granger DN. Pathophysiology of ischaemia-reperfusion injury. J Pathol. 2000;190(3):255–66.
62. Seal JB, Gewertz BL. Vascular dysfunction in ischemia-reperfusion injury. Ann Vasc Surg. 2005;19(4):572–84.

63. Kleinbongard P, Dejam A, Lauer T, Jax T, Kerber S, Gharini P. Plasma nitrite concentrations reflect the degree of endothelial dysfunction in humans. Free Radic Biol Med. 2006;40(2):295–302.
64. Rassaf T, Heiss C, Hendgen-Cotta U, Balzer J, Matern S, Kleinbongard P. Plasma nitrite reserve and endothelial function in the human forearm circulation. Free Radic Biol Med. 2006;41(2):295–301.
65. Rassaf T, Heiss C, Mangold S, Leyendecker T, Kehmeier ES, Kelm M. Vascular formation of nitrite after exercise is abolished in patients with cardiovascular risk factors and coronary artery disease. J Am Coll Cardiol. 2010;55(14):1502–3.
66. Lauer T, Heiss C, Balzer J, Kehmeier E, Mangold S, Leyendecker T. Age-dependent endothelial dysfunction is associated with failure to increase plasma nitrite in response to exercise. Basic Res Cardiol. 2008;103(3):291–7.
67. Stokes KY, Dugas TR, Tang Y, Garg H, Guidry E, Bryan NS. Dietary nitrite prevents hypercholesterolemic microvascular inflammation and reverses endothelial dysfunction. Am J Physiol Heart Circ Physiol. 2009;296(5):H1281–8.
68. Duranski MR, Greer JJ, Dejam A, Jaganmohan S, Hogg N, Langston W. Cytoprotective effects of nitrite during in vivo ischemia-reperfusion of the heart and liver. J Clin Invest. 2005;115(5):1232–40.
69. Perlman DH, Bauer SM, Ashrafian H, Bryan NS, Garcia-Saura MF, Lim CC, et al. Mechanistic insights into nitrite-induced cardioprotection using an integrated metabolomic/proteomic approach. Circ Res. 2009;104(6):796–804.
70. Hendgen-Cotta UB, Merx MW, Shiva S, Schmitz J, Becher S, Klare JP. Nitrite reductase activity of myoglobin regulates respiration and cellular viability in myocardial ischemia-reperfusion injury. Proc Natl Acad Sci U S A. 2008;105(29):10256–61.
71. Gonzalez FM, Shiva S, Vincent PS, Ringwood LA, Hsu LY, Hon YY. Nitrite anion provides potent cytoprotective and antiapoptotic effects as adjunctive therapy to reperfusion for acute myocardial infarction. Circulation. 2008;117(23):2986–94.
72. Bryan NS, Calvert JW, Elrod JW, Gundewar S, Ji SY, Lefer DJ. Dietary nitrite supplementation protects against myocardial ischemia-reperfusion injury. Proc Natl Acad Sci U S A. 2007;104(48):19144–9.
73. Baker JE, Su J, Fu X, Hsu A, Gross GJ, Tweddell JS. Nitrite confers protection against myocardial infarction: role of xanthine oxidoreductase, NADPH oxidase and K(ATP) channels. J Mol Cell Cardiol. 2007;43(4):437–44.
74. Webb A, Bond R, McLean P, Uppal R, Benjamin N, Ahluwalia A. Reduction of nitrite to nitric oxide during ischemia protects against myocardial ischemia-reperfusion damage. Proc Natl Acad Sci U S A. 2004;101(37):13683–8.
75. Binks A, Nolan JP. Post-cardiac arrest syndrome. Minerva Anestesiol. 2010;76(5):362–8.
76. Dezfulian C, Shiva S, Alekseyenko A, Pendyal A, Beiser DG, Munasinghe JP. Nitrite therapy after cardiac arrest reduces reactive oxygen species generation, improves cardiac and neurological function, and enhances survival via reversible inhibition of mitochondrial complex I. Circulation. 2009;120(10):897–905.
77. Pluta RM, Rak R, Wink DA, Woodward JJ, Khaldi A, Oldfield EH. Effects of nitric oxide on reactive oxygen species production and infarction size after brain reperfusion injury. Neurosurgery. 2001;48(4):884–92. discussion 883–92.
78. Jung KH, Chu K, Ko SY, Lee ST, Sinn DI, Park DK. Early intravenous infusion of sodium nitrite protects brain against in vivo ischemia-reperfusion injury. Stroke. 2006;37(11):2744–50.
79. Jung KH, Chu K, Lee ST, Park HK, Kim JH, Kang KM. Augmentation of nitrite therapy in cerebral ischemia by NMDA receptor inhibition. Biochem Biophys Res Commun. 2009;378(3):507–12.
80. Schatlo B, Henning EC, Pluta RM, Latour LL, Golpayegani N, Merrill MJ. Nitrite does not provide additional protection to thrombolysis in a rat model of stroke with delayed reperfusion. J Cereb Blood Flow Metab. 2008;28(3):482–9.
81. Pluta RM. Dysfunction of nitric oxide synthases as a cause and therapeutic target in delayed cerebral vasospasm after SAH. Acta Neurochir Suppl. 2008;104:139–47.

82. Pluta RM, Dejam A, Grimes G, Gladwin MT, Oldfield EH. Nitrite infusions to prevent delayed cerebral vasospasm in a primate model of subarachnoid hemorrhage. JAMA. 2005; 293(12):1477–84.
83. Safety and Pharmacokinetic Evaluation of Nitrite for Prevention of Cerebral Vasospasm. [Clinical Trial].ClinicalTrials.gov Identifier: NCT00873015.
84. Varon J. Diagnosis and management of labile blood pressure during acute cerebrovascular accidents and other hypertensive crises. Am J Emerg Med. 2007;25(8):949–59.
85. Rifkind JM, Nagababu E, Barbiro-Michaely E, Ramasamy S, Pluta RM, Mayevsky A. Nitrite infusion increases cerebral blood flow and decreases mean arterial blood pressure in rats: a role for red cell NO. Nitric Oxide. 2007;16(4):448–56.
86. Teoh NC, Farrell GC. Hepatic ischemia reperfusion injury: pathogenic mechanisms and basis for hepatoprotection. J Gastroenterol Hepatol. 2003;18(8):891–902.
87. Abe Y, Hines IN, Zibari G, Pavlick K, Gray L, Kitagawa Y. Mouse model of liver ischemia and reperfusion injury: method for studying reactive oxygen and nitrogen metabolites in vivo. Free Radic Biol Med. 2009;46(1):1–7.
88. Abe Y, Hines I, Zibari G, Grisham MB. Hepatocellular protection by nitric oxide or nitrite in ischemia and reperfusion injury. Arch Biochem Biophys. 2009;484(2):232–7.
89. Kim JS, Ohshima S, Pediaditakis P, Lemasters JJ. Nitric oxide protects rat hepatocytes against reperfusion injury mediated by the mitochondrial permeability transition. Hepatology. 2004;39(6):1533–43.
90. Lu P, Liu F, Yao Z, Wang CY, Chen DD, Tian Y. Nitrite-derived nitric oxide by xanthine oxidoreductase protects the liver against ischemia-reperfusion injury. Hepatobiliary Pancreat Dis Int. 2005;4(3):350–5.
91. Allen BW, Stamler JS, Piantadosi CA. Hemoglobin, nitric oxide and molecular mechanisms of hypoxic vasodilation. Trends Mol Med. 2009;15(10):452–60.
92. Raat NJ, Noguchi AC, Liu VB, Raghavachari N, Liu D, Xu X. Dietary nitrate and nitrite modulate blood and organ nitrite and the cellular ischemic stress response. Free Radic Biol Med. 2009;47(5):510–7.
93. Garcia-Criado FJ, Rodriguez-Barca P, Garcia-Cenador MB, Rivas-Elena JV, Grande MT, Lopez-Marcos JF, et al. Protective effect of new nitrosothiols on the early inflammatory response to kidney ischemia/reperfusion and transplantation in rats. J Interferon Cytokine Res. 2009;29(8):441–50.
94. Garcia-Criado FJ, Eleno N, Santos-Benito F, Valdunciel JJ, Reverte M, Lozano-Sanchez FS. Protective effect of exogenous nitric oxide on the renal function and inflammatory response in a model of ischemia-reperfusion. Transplantation. 1998;66(8):982–90.
95. Jeong GY, Chung KY, Lee WJ, Kim YS, Sung SH. The effect of a nitric oxide donor on endogenous endothelin-1 expression in renal ischemia/reperfusion injury. Transplant Proc. 2004;36(7):1943–5.
96. Martinez-Mier G, Toledo-Pereyra LH, Bussell S, Gauvin J, Vercruysse G, Arab A. Nitric oxide diminishes apoptosis and p53 gene expression after renal ischemia and reperfusion injury. Transplantation. 2000;70(10):1431–7.
97. Kucuk HF, Kaptanoglu L, Ozalp F, Kurt N, Bingul S, Torlak OA. Role of glyceryl trinitrate, a nitric oxide donor, in the renal ischemia-reperfusion injury of rats. Eur Surg Res. 2006;38(5):431–7.
98. Okamoto M, Tsuchiya K, Kanematsu Y, Izawa Y, Yoshizumi M, Kagawa S. Nitrite-derived nitric oxide formation following ischemia-reperfusion injury in kidney. Am J Physiol Renal Physiol. 2005;288(1):F182–7.
99. Milsom AB, Patel NS, Mazzon E, Tripatara P, Storey A. Mota-Filipe H. Role for endothelial nitric oxide synthase in nitrite-induced protection against renal ischemia-reperfusion injury in mice. Nitric Oxide. 2010;22(2):141–8.
100. Tripatara P, Patel NS, Webb A, Rathod K, Lecomte FM, Mazzon E. Nitrite-derived nitric oxide protects the rat kidney against ischemia/reperfusion injury in vivo: role for xanthine oxidoreductase. J Am Soc Nephrol. 2007;18(2):570–80.
101. Basireddy M, Isbell TS, Teng X, Patel RP, Agarwal A. Effects of sodium nitrite on ischemia-reperfusion injury in the rat kidney. Am J Physiol Renal Physiol. 2006;290(4):F779–86.

102. Nakajima A, Ueda K, Takaoka M, Kurata H, Takayama J, Ohkita M. Effects of pre- and post-ischemic treatments with FK409, a nitric oxide donor, on ischemia/reperfusion-induced renal injury and endothelin-1 production in rats. Biol Pharm Bull. 2006;29(3):577–9.
103. Tsuchiya K, Tomita S, Ishizawa K, Abe S, Ikeda Y, Kihira Y. Dietary nitrite ameliorates renal injury in L-NAME-induced hypertensive rats. Nitric Oxide. 2010;22(2):98–103.
104. Hord NG, Tang Y, Bryan NS. Food sources of nitrates and nitrites: the physiologic context for potential health benefits. Am J Clin Nutr. 2009;90(1):1–10.
105. Milkowski A, Garg HK, Coughlin JR, Bryan NS. Nutritional epidemiology in the context of nitric oxide biology: a risk-benefit evaluation for dietary nitrite and nitrate. Nitric Oxide. 2010;22(2):110–9.
106. Hunter CJ, Dejam A, Blood AB, Shields H, Kim-Shapiro DB, Machado RF. Inhaled nebulized nitrite is a hypoxia-sensitive NO-dependent selective pulmonary vasodilator. Nat Med. 2004;10(10):1122–7.
107. Wu X, Du L, Xu X, Tan L, Li R. Increased nitrosoglutathione reductase activity in hypoxic pulmonary hypertension in mice. J Pharmacol Sci. 2010;113(1):32–40.
108. Egemazarov B, Schermuly RT, Dahal BK, Elliott GT, Hoglen NC, Surber MW. Nebulization of the acidified sodium nitrite formulation attenuates acute hypoxic pulmonary vasoconstriction. Respir Res. 2010;11(1):81.
109. Zuckerbraun BS, Shiva S, Ifedigbo E, Mathier MA, Mollen KP, Rao J. Nitrite potently inhibits hypoxic and inflammatory pulmonary arterial hypertension and smooth muscle proliferation via xanthine oxidoreductase-dependent nitric oxide generation. Circulation. 2010;121(1):98–109.
110. Ingram TE, Pinder AG, Bailey DM, Fraser AG, James PE. Low-dose sodium nitrite vasodilates hypoxic human pulmonary vasculature by a means that is not dependent on a simultaneous elevation in plasma nitrite. Am J Physiol Heart Circ Physiol. 2010;298(2):H331–9.
111. Casey DB, Badejo AM, Jr. Dhaliwal JS, Murthy SN, Hyman AL, Nossaman BD. Pulmonary vasodilator responses to sodium nitrite are mediated by an allopurinol-sensitive mechanism in the rat. Am J Physiol Heart Circ Physiol. 2009;296(2):H524–33.
112. Hsu LL, Champion HC, Campbell-Lee SA, Bivalacqua TJ, Manci EA, Diwan BA. Hemolysis in sickle cell mice causes pulmonary hypertension due to global impairment in nitric oxide bioavailability. Blood. 2007;109(7):3088–98.
113. Minneci PC, Deans KJ, Shiva S, Zhi H, Banks SM, Kern S. Nitrite reductase activity of hemoglobin as a systemic nitric oxide generator mechanism to detoxify plasma hemoglobin produced during hemolysis. Am J Physiol Heart Circ Physiol. 2008;295(2):H743–54.
114. Dias-Junior CA, Gladwin MT, Tanus-Santos JE. Low-dose intravenous nitrite improves hemodynamics in a canine model of acute pulmonary thromboembolism. Free Radic Biol Med. 2006;41(12):1764–70.
115. Berger MM, Dehnert C, Bailey DM, Luks AM, Menold E, Castell C. Transpulmonary plasma ET-1 and nitrite differences in high altitude pulmonary hypertension. High Alt Med Biol Spring. 2009;10(1):17–24.
116. Casey DP, Beck DT, Braith RW. Systemic plasma levels of nitrite/nitrate (NOx) reflect brachial flow-mediated dilation responses in young men and women. Clin Exp Pharmacol Physiol. 2007;34(12):1291–3.
117. Casey DP, Nichols WW, Conti CR, Braith RW. Relationship between endogenous concentrations of vasoactive substances and measures of peripheral vasodilator function in patients with coronary artery disease. Clin Exp Pharmacol Physiol. 2010;37(1):24–8.
118. Allen JD, Miller EM, Schwark E, Robbins JL, Duscha BD, Annex BH. Plasma nitrite response and arterial reactivity differentiate vascular health and performance. Nitric Oxide. 2009;20(4):231–7.
119. Maher AR, Milsom AB, Gunaruwan P, Abozguia K, Ahmed I, Weaver RA. Hypoxic modulation of exogenous nitrite-induced vasodilation in humans. Circulation. 2008;117(5):670–7.

120. Gladwin MT. Evidence mounts that nitrite contributes to hypoxic vasodilation in the human circulation. Circulation. 2008;117(5):594–7.
121. Mendoza MG, Robles HV, Romo E, Rios A, Escalante B. Nitric oxide-dependent neovascularization role in the lower extremity disease. Curr Pharm Des. 2007;13(35):3591–6.
122. Kumar D, Branch BG, Pattillo CB, Hood J, Thoma S, Simpson S. Chronic sodium nitrite therapy augments ischemia-induced angiogenesis and arteriogenesis. Proc Natl Acad Sci U S A. 2008;105(21):7540–5.
123. Wood KC, Granger DN. Sickle cell disease: role of reactive oxygen and nitrogen metabolites. Clin Exp Pharmacol Physiol. 2007;34(9):926–32.
124. Wood KC, Hsu LL, Gladwin MT. Sickle cell disease vasculopathy: a state of nitric oxide resistance. Free Radic Biol Med. 2008;44(8):1506–28.
125. Lopez BL, Barnett J, Ballas SK, Christopher TA, Davis-Moon L, Ma X. Nitric oxide metabolite levels in acute vaso-occlusive sickle-cell crisis. Acad Emerg Med. 1996;3(12):1098–103.
126. Mack AK, McGowan Ii VR, Tremonti CK, Ackah D, Barnett C, Machado RF. Sodium nitrite promotes regional blood flow in patients with sickle cell disease: a phase I/II study. Br J Haematol. 2008;142(6):971–8.
127. Safety and Efficacy of Sodium Nitrite in Sickle Cell Disease. [Clinical Trial].ClinicalTrials.gov Identifier NCT01033227.
128. Evaluation of the Mechanism of NO Formation and Pharmacokinetics of Systemic Nitrite Infusion. [Clinical Trial].ClinicalTrials.gov Identifier: NCT00102271.
129. Evaluation of Systemic Nitrite Infusion and Its Effect on Exercise Physiology and Metabolism. [Clinical Trials].ClinicalTrials.gov Identifier: NCT00105222.
130. Evaluation of the Mechanism of NO Formation and Pharmacokinetics of Long-Term Intravenous Nitrite Infusion in Healthy Volunteers. [Clinical Trial].ClinicalTrials.gov Identifier: NCT00103025.
131. Determination of Nitrite as a Source of Bioactive Nitric Oxide in Human Subjects. [Clinical Trial].ClinicalTrials.gov Identifier: NCT00048477.
132. The Role of Nitrite in Preconditioning Mediated Tolerance to Ischemic Stress. [Clinical Trial].ClinicalTrials.gov Identifier: NCT00250185.
133. The Determination of Variances in Exhaled Nitric Oxide Output in Normal Healthy Male Volunteers Consuming High and Low Nitrate/Nitrite Diets. [Clinical Trial].ClinicalTrials.gov Identifier: NCT00027157.
134. A Safety and Efficacy Evaluation of Sodium Nitrite Injection for the Prevention of Ischemia-Reperfusion Injury Associated With Acute Myocardial Infarction. [Clinical Trial].ClinicalTrials.gov Identifier: NCT00924118.
135. Mechanisms of Nitrite Mediated Cardioprotection in Coronary Artery Bypass Surgery. [Clinical Trial].ClinicalTrials.gov Identifier: NCT01098409.
136. Dietary Nitrate/Nitrite as Sources of Bioactive Nitric Oxide in Patients With Coronary Artery Disease. [Clinical Trial].ClinicalTrials.gov Identifier: NCT00069654.
137. Louisiana pharma startup TheraVasc moves, readies for clinical trials. *Medcity News.* 2009. http://www.medcitynews.com/2009/2006/louisiana-pharma-startup-theravasc-moves-readies-for-clinical-trials/. Accessed on Aug 1, 2010.
138. Caimi G, Montana M, Calandrino V, Caruso M, Carollo C, Catania A. Nitric oxide metabolites (nitrite and nitrate) in young patients with recent acute myocardial infarction. Clin Hemorheol Microcirc. 2008;40(2):157–63.
139. Alusik S, Jedlickova V, Paluch Z, Zecova S. Plasma levels of nitrite/nitrate and inflammation markers in elderly individuals. Bratisl Lek Listy. 2008;109(7):289–92.
140. Choi JW. Enhanced nitric oxide production is closely associated with serum lipid concentrations in adolescents. Clin Chim Acta. 2004;347(1–2):151–6.
141. Sozmen B, Kazaz C, Taskiran D, Alan L, Akyol A, Sozmen E. Plasma antioxidant status and nitrate levels in patients with hypertension and coronary heart disease. Turk J Med Sci. 1998;28(5):525–31.

142. Nitrite as a Marker of Cardiovascular Risk; Development of Novel Biomarkers in Patients With Coronary Artery Disease. [Clinical Trial].ClinicalTrials.gov Identifier: NCT00109018.
143. Lundberg JO, Feelisch M, Bjorne H, Jansson EA, Weitzberg E. Cardioprotective effects of vegetables: is nitrate the answer? Nitric Oxide. 2006;15(4):359–62.
144. van Velzen AG, Sips AJ, Schothorst RC, Lambers AC, Meulenbelt J. The oral bioavailability of nitrate from nitrate-rich vegetables in humans. Toxicol Lett. 2008;181(3):177–81.
145. Hunault CC, van Velzen AG, Sips AJ, Schothorst RC, Meulenbelt J. Bioavailability of sodium nitrite from an aqueous solution in healthy adults. Toxicol Lett. 2009;190(1):48–53.
146. Dauchet L, Amouyel P, Hercberg S, Dallongeville J. Fruit and vegetable consumption and risk of coronary heart disease: a meta-analysis of cohort studies. J Nutr. 2006;136(10):2588–93.
147. Joshipura KJ, Ascherio A, Manson JE, Stampfer MJ, Rimm EB, Speizer FE. Fruit and vegetable intake in relation to risk of ischemic stroke. JAMA. 1999;282(13):1233–9.
148. Bazzano LA, Li TY, Joshipura KJ, Hu FB. Intake of fruit, vegetables, and fruit juices and risk of diabetes in women. Diabetes Care. 2008;31(7):1311–7.
149. Appel LJ, Moore TJ, Obarzanek E, Vollmer WM, Svetkey LP, Sacks FM. A clinical trial of the effects of dietary patterns on blood pressure. DASH Collaborative Research Group. N Engl J Med. 1997;336(16):1117–24.
150. Sacks FM, Svetkey LP, Vollmer WM, Appel LJ, Bray GA, Harsha D. Effects on blood pressure of reduced dietary sodium and the Dietary Approaches to Stop Hypertension (DASH) diet. DASH-Sodium Collaborative Research Group. N Engl J Med. 2001;344(1):3–10.
151. Kapil V, Milsom AB, Okorie M, Maleki-Toyserkani S, Akram F, Rehman F. Inorganic nitrate supplementation lowers blood pressure in humans. Role for nitrite-derived NO. Hypertension. 2010;56:274–81.
152. Sobko T, Marcus C, Govoni M, Kamiya S. Dietary nitrate in Japanese traditional foods lowers diastolic blood pressure in healthy volunteers. Nitric Oxide. 2010;22(2):136–40.
153. Webb AJ, Patel N, Loukogeorgakis S, Okorie M, Aboud Z, Misra S. Acute blood pressure lowering, vasoprotective, and antiplatelet properties of dietary nitrate via bioconversion to nitrite. Hypertension. 2008;51(3):784–90.
154. Larsen FJ, Weitzberg E, Lundberg JO, Ekblom B. Dietary nitrate reduces maximal oxygen consumption while maintaining work performance in maximal exercise. Free Radic Biol Med. 2010;48(2):342–7.
155. Larsen FJ, Weitzberg E, Lundberg JO, Ekblom B. Effects of dietary nitrate on oxygen cost during exercise. Acta Physiol (Oxf). 2007;191(1):59–66.
156. Bailey SJ, Winyard P, Vanhatalo A, Blackwell JR, Dimenna FJ, Wilkerson DP. Dietary nitrate supplementation reduces the O2 cost of low-intensity exercise and enhances tolerance to high-intensity exercise in humans. J Appl Physiol. 2009;107(4):1144–55.
157. Bailey SJ, Fulford J, Vanhatalo A, Winyard PG, Blackwell JR, Dimenna FJ. Dietary nitrate supplementation enhances muscle contractile efficiency during knee-extensor exercise in humans. J Appl Physiol. 2010;109(1):135–48.
158. Katan MB. Nitrate in foods: harmful or healthy? Am J Clin Nutr. 2009;90(1):11–2.
159. Gori T, Parker JD. Nitrate-induced toxicity and preconditioning: a rationale for reconsidering the use of these drugs. J Am Coll Cardiol. 2008;52(4):251–4.
160. Dejam A, Hunter CJ, Tremonti C, Pluta RM, Hon YY, Grimes G. Nitrite infusion in humans and nonhuman primates: endocrine effects, pharmacokinetics, and tolerance formation. Circulation. 2007;116(16):1821–31.
161. Ralt D. Does NO metabolism play a role in the effects of vegetables in health? Nitric oxide formation via the reduction of nitrites and nitrates. Med Hypotheses. 2009;73(5):794–6.

Chapter 15
The Nitrate–Nitrite–Nitric Oxide Pathway in Traditional Herbal Medicine for Heart Disease

Yong-Jian Geng

Key points

- The inorganic anions nitrate and nitrite are important intermediates in the nitrogen cycle.
- The nitrate–nitrite–nitric oxide pathway has been shown to exist in many alternative herbal medicines or dietary supplements.
- Many herbal medicines contain high levels of nitrate and to a less extent nitrite.
- There is an effective system in certain herbal medicines for reducing nitrite to nitric oxide.
- Many herbal medicines for heart disease regulate the activities of NO synthase and nitrate/nitrite reductases, and thus maintain the "Yin-Yang" balance of NO production.
- The ability of herbal medicines to reduce nitrate and nitrite to NO effectively may account for some of their beneficial impacts on cardiovascular disorders.
- Understanding the mechanism of action for herbal medicines will allow for better combinations of herbs for natural disease combat as a key player of alternative Western medicine.

Keywords Herbal medicine • Naturopathy • Complementary and alternative medicine • Natural products

Y.-J. Geng (✉)
Cardiovascular Medicine, The University of Texas School of Medicine at Houston, Houston, TX, USA
e-mail: Yong-Jian.Geng@uth.tmc.edu

N.S. Bryan and J. Loscalzo (eds.), *Nitrite and Nitrate in Human Health and Disease*, Nutrition and Health, DOI 10.1007/978-1-60761-616-0_15,

Introduction

Traditional medicine has been used in some countries or communities for hundreds or thousands of years. Herbs are important part of traditional medicine which has been standing as a major health care system in China and many other countries in Asia and Africa. In some Asian and African countries, 80% of the population depends on traditional medicine for primary health care. Herbal medicine (HM) is considered a complementary or alternative medical system in much of the Western world, mostly due to a lack of understanding of their mechanism of action and/or the active compound(s). However, the use of traditional or alternative medicine is increasing in Western nations, especially in the communities of people with Asian origin. Patients who use alternative or HM in Western countries report that the main reason for using it is due to a "more natural" and potentially safer alternative for the treatment of chronic illness than pharmaceutical drugs or surgery [1]. In many developed countries, including the U.S., as much as 70–80% of the population has used some form of alternative or complementary medicine (e.g., acupuncture or herbal supplements). Herbal treatments are the most popular form of traditional medicine, and are highly lucrative in the international marketplace. Annual revenues in Western Europe reached $5 billion in 2003–2004. In China, sales of products totaled $14 billion in 2005. HM revenue in Brazil was $160 million in 2007, according to WHO Fact Sheet No. 134, 2008.

Traditional HM has its origin in ancient Taoist philosophy, which views a person as an energy system in which the body, mind, and spirit are unified into one when in harmony and disrupted in disease. Traditional medicine practice treats the patient as a whole, not as a part, and it emphasizes a holistic approach that attempts to bring the mind, body, and spirit into harmony. Traditional medicine theory is extremely complex and originated thousands of years ago through meticulous observation of nature, the cosmos, and the human body. There is a growing and sustained interest in alternative medicines fueled by a combination of factors including recognition of benefits, dissatisfaction with and ineffectiveness of traditional Western medicines, an increasing commitment to holistic care, skepticism regarding adverse side-effects of drug therapy, and increasing evidence for the personalized nature of different combinations of herbs for specific disorders [1]. Of the approximately 500 herbs that are in use today, 50 or so are commonly used alone or in combination. Rather than being prescribed individually, single herbs are combined into formulas that are designed to adapt to the specific needs of individual patients. An herbal formula can contain from 3 to 25 herbs. As with diet therapy, each herb has one or more of the four flavors/functions and one of five "temperatures" "氣" (pronounced "chi") (hot, warm, neutral, cool, cold). Herbal formulations work to balance the body from the inside out. Traditional HMs include herbs, herbal materials, herbal preparations, and processed herbal products that contain parts of plants or other plant materials as active ingredients which assist with strengthening the vital energies (chi), blood, and fluids internally. They are typically administered as tablets, tea pills, elixirs, soups, liquid extracts, and teas.

With traditional medicine practices being adopted by new populations in the United States and Europe as well as many other countries around the world, challenges have been emerging recently. To date, most countries have established national policies for traditional medicine. Regulating traditional medicine products, practices, and practitioners is difficult due to variations in definitions and categorizations of traditional medicine therapies. A single herbal product could be defined as either a food, dietary supplement or an HM, depending on the country. This disparity in regulations at the national levels has implications for international access and distribution of products. Scientific evidence from tests done to evaluate the safety and effectiveness of traditional medicine products and practices is limited. Despite the fact that alternative treatments, such as acupuncture, some HMs, and some manual therapies (e.g., massage), have been shown to be effective for specific conditions, no proven underlying mechanism is available. A further study of the products and practices is needed. Many people, including most professionals in the field of biomedicine, believe that because HMs are natural and have been used traditionally they are safe and carry no risk for harm. However, traditional medicines and practices can cause harmful, adverse reactions if the product or therapy is of poor quality, or is taken inappropriately, or in conjunction with other medicines. Increased patient awareness about safe usage is important, as well as more training, collaboration, and communication among providers of traditional and other medicines. Requirements and methods for research and evaluation are complex. It is often difficult to assess the quality of processed herbal products. The safety, effectiveness, and quality of finished HM products depend on the quality of their source materials (which can include hundreds of natural constituents), and how elements are handled through production processes.

There are a number of published reports on the association of traditional medicines and nitric oxide (NO)-related effects [2, 3]. Inorganic nitrite and nitrate have been reported widely as alternative sources of NO production independent of the enzymatic synthesis of NO from L-arginine [4–6]; however, their mechanism of action is far from clear due to the complexity and heterogeneity of the herbal therapies as well as a lack of knowledge of content and ratio or identification of the active compound or compounds. Recent discovery of the nitrate–nitrite–NO pathway in plants and animals [7, 8] offers a new avenue which defines the biological action as well as to assess the safety of certain HMs commonly used for the treatment of cardiovascular disease. By establishing a common molecule of interest acting similarly in many of these concoctions, a purified and concentrated formulation may be developed with even greater efficacy, which may eventually penetrate the pharmaceutical markets in the countries with stricter regulations, such as the United States and European countries. This knowledge is equally important for legal and regulatory reasons, as well as for patient safety related to any contraindications with their concomitant use with other medications. In this chapter, the production of nitrate, nitrite, and NO, and the molecular basis of their biological action will be discussed. Evidence will be presented to support the notion that the levels of nitrite and nitrate as well as the nitrite/nitrate reductase activity ingested as HMs provide a robust and natural system for NO generation that may offer alterative pathway to overcome pathological conditions associated with NO insufficiency and combat cardiovascular disease.

NO Production and Consequences of NO Insufficiency

Previous chapters have established the essential role of NO in the maintenance of cardiovascular health as well as healthy immune or neuronal status. Appropriate levels of NO production are critical in tissue blood perfusion and protection of cardiovascular tissues against ischemia and infarction. Sustained levels of enhanced NO production may result in direct tissue toxicity and contribute to the vascular collapse associated with various pathological conditions. Being a simple molecule, NO is a fundamental player in many different fields of biomedicine, and it was proclaimed the "Molecule of the Year" in 1992 [9]. NO is one of the few gaseous signaling molecules known to play a role in a variety of biological processes. Acting as the "endothelium-derived relaxing factor," or "EDRF" [10], NO is biosynthesized endogenously from L-arginine and oxygen by various nitric oxide synthase (NOS) enzymes. The endothelium of blood vessels uses NO to signal the surrounding smooth muscle to relax, thus resulting in vasodilation and increasing blood flow. Because it has an extremely high reactivity (having a lifetime of a few seconds), NO diffuses freely across membranes over a limited distance. These attributes make NO ideal for a transient paracrine (between adjacent cells) and autocrine (within a single cell) signaling molecule. The production of NO is elevated in populations living at high altitudes, which helps these individuals avoid hypoxia by aiding in pulmonary vascular vasodilation. Nitroglycerin and amyl nitrite serve as vasodilators because they release NO in the body. Sildenafil citrate, popularly known by the trade name *Viagra*, stimulates erections primarily by enhancing signaling through the NO pathway and, thus, dilating the arteries of the penis. NO contributes to vessel homeostasis by inhibiting vascular smooth muscle contraction and growth, platelet aggregation, and leukocyte adhesion to the endothelium. Humans with atherosclerosis, diabetes, or hypertension often show impaired NO pathways. Alternative source of NO production from diets may help compensate for the reduction in NO synthesis catalyzed by NOS [11, 12]. For years, physicians and scientists assumed simply feeding more substrate L-arginine would be sufficient to enhance NO production. It is becoming increasingly clear that this may not be the most effective strategy, especially in patients who are insufficient in vascular NO release due to endothelial dysfunction.

Nitrate and Nitrite in the Nitrogen Cycle

Since HMs are mostly plant-based or extracted from plants, they contain organic nitrogen oxides generated through the environmental nitrogen cycle. Nitrogen constitutes 78% of earth's atmosphere, and serves as a major constituent of all living tissues. As an essential element for life, the nitrogen atom is incorporated into proteins and nuclear acids. In nature, nitrogen molecules (N_2) exist mainly in air. In water and soil, nitrogen can be found predominantly in the form of nitrate and nitrite.

Nitrate and nitrite are naturally occurring anions. Compared to NO, they are chemically stable, and they can be found virtually anywhere in the environment. Nitrite is produced by nitrification or oxidation of ammonia, especially by the action of the nitrifying bacterium, *Nitrosomas*. The nitrites will then be oxidized to nitrates by the bacterium *Nitrobacter*. Nitrate is less toxic than nitrite and is used as a food source by live plants. The process of converting ammonia to nitrate is diagrammed in the nitrogen cycle (Fig. 15.1). Nitrification is most rapid at pH of 7–8 and at temperatures of 25–30°C. Nitrification leads to acidification of water.

The inorganic anions, nitrate (NO_3^-) and nitrite (NO_2^-), were traditionally considered inert end-products of endogenous NO metabolism. However, recent studies show that these supposedly inert anions can be recycled in vivo to form NO, representing an important alternative source of NO to the classical L-arginine–NO-synthase pathway, in particular in hypoxic states. The emerging important biological functions of the nitrate–nitrite–NO pathway implicate the therapeutic potential of many HMs traditionally used for cardiovascular diseases because herbs can provide significant amounts of nitrate and nitrite as well as an inherent ability to reduce nitrite to NO.

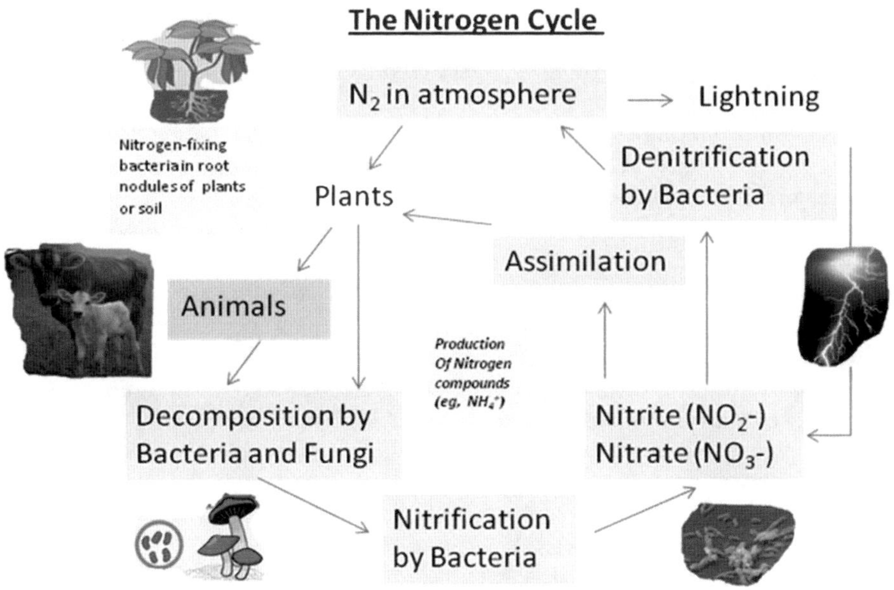

Fig. 15.1 Schematic representation of the nitrogen cycle. The biogeochemical cycle describes the transformations of nitrogen and nitrogen-containing compounds in nature

Medicinal Herbs: An Alternative Source of Exogenous Nitrate and Nitrite

Many HMs for heart diseases are made from the roots and leaves of certain plants that contain abundant nitrate/nitrite or related compounds. In these plants, NO can be produced by at least four routes:

(i) L-arginine-dependent or independent NO-synthesizing enzymes. Several studies [13–16] suggest that as part of the innate defense response, certain plants may express high levels of enzymes that can synthesize NO using L-arginine and other nitro compounds.
(ii) Plasma membrane-bound nitrate/nitrite reductase. Many research groups have demonstrated that plant cells contain nitrate or nitrite reductase [17, 18]. Their reduction may convert them into NO under conditions of low pH or hypoxia.
(iii) Mitochondrial electron transport chain. Mitochondrial compartments are rich in metal enzymes involved actively in the regulation of energy production [19]. The active metabolism in mitochrondria may generate nitro compounds which, in turn, are converted into NO.
(iv) Nonenzymatic reactions. The root of certain plants may contain high levels of unstable nitro compounds, which may release NO spontaneously [20, 21]. Plants exposed to nitro-types of fertilizers typically generate more NO than those grown in the wild.

Similar to that in mammal tissues, plant-derived NO functions as a signaling molecule. It acts mainly against oxidative stress and plays a role in plant–pathogen interactions. Treating cut flowers and other plants with NO has been shown to lengthen the time before wilting. NO can bind to mitochondrial enzymes contain iron-II leading to attenuation of respiration [22]. An important biological reaction of NO is S-nitrosation [23], the conversion of thiol groups, including cysteine residues in proteins, to form *S*-nitrosothiols (RSNOs). S-Nitrosation is a mechanism for dynamic, posttranslational regulation of most or all major classes of protein in both plants and animals, which may also account for some of the NO activity. Inorganic nitrite and nitrate are emerging as key players in NO biology. They are important sources of NO and can be recycled in vivo under specific conditions [4, 5, 24]. There are a number of endogenous systems in mammals capable of reducing nitrite to NO [19]. Dietary nitrite has been shown to protect from tissue injury and restore NO homeostasis in eNOS$^{-/-}$ mice [25–28]. Furthermore, nitrate has also been shown to reduce blood pressure [29, 30], inhibit platelet aggregation [30], and protect the heart from ischemia-reperfusion injury [26]. There are a number of herbal preparations with primary indications for cardiovascular disease. There are increased interests in the role for the nitrate–nitrite–NO system in regulation of cardiovascular function by traditional HM used commonly for cardiovascular disorders, including coronary artery disease (CAD) and heart failure.

Recently, certain HMs or dietary supplements have been reported to generate significant amounts of nitrate, nitrite, and NO bioactivity, which may account for

Table 15.1 The measurements of nitrite, nitrate, nitroso, nitrite reductase activity and the recommended daily dose in several herbs commonly used in Chinese traditional medicine

Latin name	English name	Indication	Nitrite (ng/g)	Nitrate (mg/g)	Nitroso (nmol/g)	NO production (pmol/mg)
Radix Salviae Miltiorrhizae	Danshen root	CAD	330	12,000	120	7
Fructus Trichosanthis	Snakegourd fruit	CAD, acute MI, Hyperlipidemia	260	278	120	46
Bulbus Allii Macrostemi	Longstamen onion bulb	CAD, acute MI, Hyperlipidemia	150	530	842	134
Radix Notoginseng	Sanchi	CAD	210	2,069	73	13
Resina Olibani	Frankincense	Hypertension	980	61	3,210	72
Radix Paeonia Rubra	Red Peony root	CAD	120	37	450	255
Radix Ginseng	Ginseng	Heart failure, CAD	300	243	76	360
Borneolum Syntheticum	Borneol	Increase other herb's function for CAD or brain disease	120	2.99	6	45
Cinnamomum	Borneol	Increase other herb's function for CAD or brain disease	160	2.39	0	875

some of their effects on the heart and vessels [20]. The herbs are commercially available in regular dietary supplement stores. Analysis of nitrite, nitrate, NO-modified proteins, and NO-generating capacity as well as their effects on relaxation of isolated aortic strips have provided convincing evidence that abundant nitrate and nitrate reductase activity are present in certain herbs known to have protective or therapeutic benefits to patients with CAD (Table 15.1), such as the Danshen Root (*radix salviae miltiorrhizae*), Sanchi (*radix notoginseng*), and Hongshen (*radix ginseng*). These herbs, with specific indications for cardiovascular disease, contain high concentrations of nitrite and nitrate, generate NO from nitrite, and relax blood vessels. The therapeutic benefits of HMs are via their ability to generate NO from nitrite providing an alternative source of NO to patients that may be unable to make NO from L-arginine owing to endothelial dysfunction.

In the field of NO biology, the inorganic anions nitrite and nitrate are known NO-donors which act as essential natural regulators of cardiovascular function [5, 31, 32]. Both nitrite and nitrate have been tested in animals and humans, and they show a good effectiveness in attenuating inflammation, reversing endothelial dysfunction, and reducing damage due to ischemia-reperfusion injury [26, 33, 34]. There is an endogenous nitrite reductase activity in animal tissues, such as the liver and aorta, but this inherent biological capacity is low (around 1 pmol/mg protein). The reductase activity in some of these HMs may exceed that detected in the animal tissues [20]. It is estimated that the increased reductase activity may occur by orders of magnitude, almost 1,000 times higher than endogenous production of NO.

This would equate to 300 nmoles/day of NO from a single herbal preparation. The average production of NO in the human body (70 kg) is 1.68 mmol NO/day (based on an NO production rate of 1 μmol/kg/h). By supplying the exogenous nitrate/nitrite and reductase activities, HMs offer an alternative therapeutic strategy to combat or treat heart disease, a condition frequently associated with a deficiency in NO. Maintaining NO homeostasis requires the repletion of nitrite and nitrate through which the ability to generate NO can be restored to compensate for the inability of the endothelium to convert L-arginine to NO in coronary heart disease.

Therapeutic Potential of Herbal Medicine with Active Nitrate–Nitrite–NO Pathway

In traditional Chinese medicine, physicians routinely take the pulse (Fig. 15.2) and examine the rates, rhythms, and strength of arterial pulse. They also inspect the tongue to determine any changes in shape, color, and smell. Based on the patient's symptoms and physical examination, physicians determine whether or not there is a functional abnormality of the cardiovascular system. NO-generating herbs are

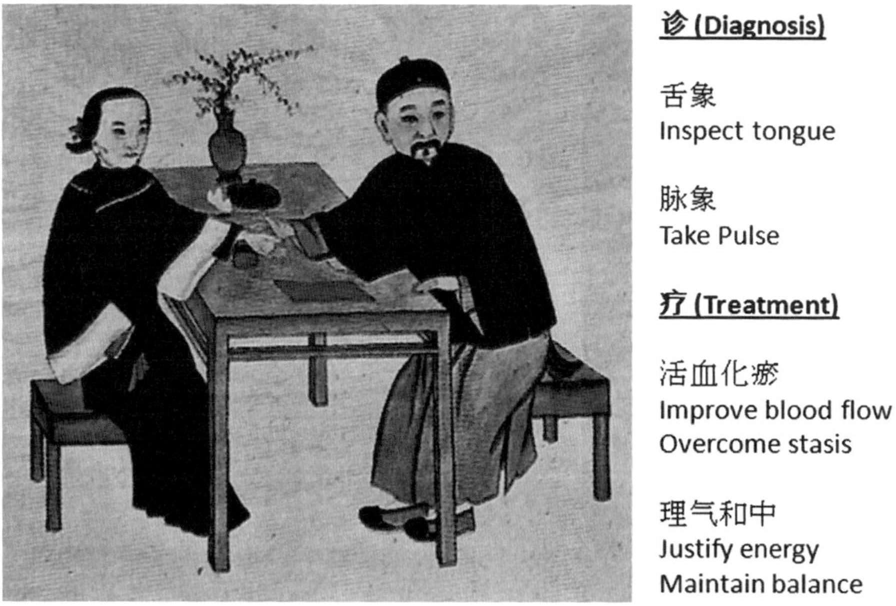

Fig. 15.2 A painting displays an ancient Chinese physician practicing traditional medicine and performing diagnosis and treatment of a heart disorder, such as angina or chest pain. The primary physical examination was to inspect the tongue's shape and color, and measure the pulse rate and strength. The traditional therapy was, in principal, based on improving blood flow, thereby overcoming stasis, and justifying the balance in energy

considered antianginal as they can cause vasodilation, which can help reduce ischemic pain known as angina by improving perfusion and decreasing the cardiac workload as a result of venodilation and a decrease in venous return to the heart per cycle. Nitroglycerin, taken sublingually, is used to prevent or treat acute chest pain. Nitroglycerin reacts with a sulfhydryl group (–SH) to produce NO, which eases the pain by causing vasodilation. Nitrates may be beneficial for the treatment of angina due to reduced myocardial oxygen consumption, either by decreasing preload and afterload or by relaxing directly coronary vessels [32, 35].

Traditional HMs used for thousands of years in Asia and other regions have been proven effective in certain cardiovascular disorders. Some of the HMs have profound NO bioactivity primarily due to the nitrate–nitrite–NO reduction pathway (Fig. 15.3). They contain very large amounts of nitrate/nitrite in the extracts given to patients [20]. The described benefits of these ancient medications may be attributed to their inherent nitrate/nitrite content combined with their robust nitrate/nitrite reductase activity to generate NO independent of the L-arginine–NO pathway [19, 20, 27, 32]. The first use of nitrate for treatment of patients with symptoms that appears to be angina was described in an eighth-century Chinese manuscript uncovered at the Buddhist grotto of Dunhuang. Chinese physicians in traditional medicine have tested the therapeutic effects of Xiao Shi Xiong Huang San (the Nitrum and

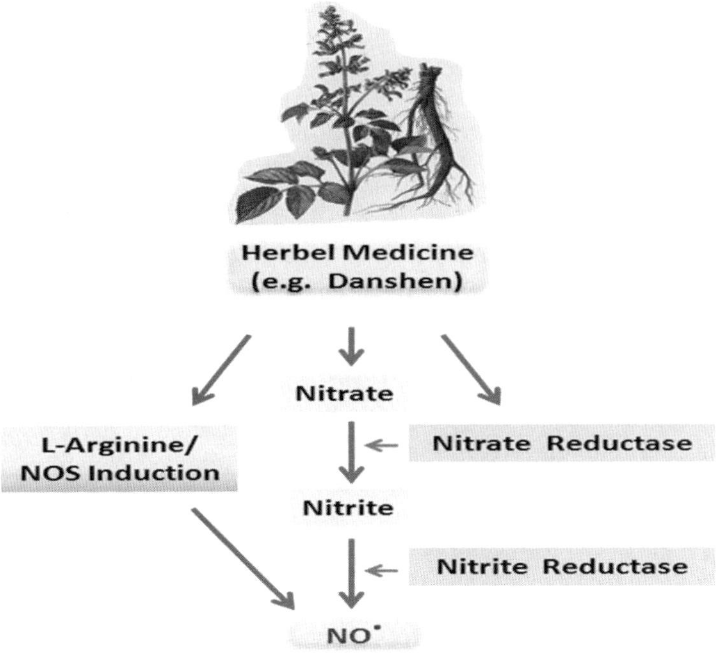

Fig. 15.3 Potential pathways leading to NO production in herbal medicines. Traditional herbal medicines used for cardiovascular disorders, such as Danshen (*Radix salviae miltiorrhizae*), contain large quantities of nitrate and nitrite, and some also express high levels of nitrate reductase activities. They may also enhance eNOS expression and activation

Realgar Powder), one of the Dunhuang prescriptions, on angina pectoris caused by coronary heart disease. The patients were instructed to take Xiao Shi Xiong Huang San, hold it under the tongue for a time, and then swallow the saliva. Compared to nitroglycerin, Xiao Shi Xiong Huang San showed better efficacy and improvement in a clinical trial of 61 patients [36]. The significance of the instructions is that under the tongue, even in a healthy mouth, nitrate-reducing bacteria convert some of the nitrate into nitrite. Therefore, if the patient follows the physician's instructions fully, he or she will be taking nitrite, known to be effective in alleviating pain resulting from angina.

Recent clinical studies have provided new evidence on Danshen, a commonly used HM in China, which may have a similar efficacy to the known NO-donor nitroglycerin [37–39]. The extract of *Salviae Miltiorrhiae*, or Danshen in Chinese, contains large amounts of nitrate [20]. The demonstration of beneficial effects of this herb on ischemic diseases offers an alternative avenue for the management of angina pectoris, myocardial infarction, or stroke [40, 41]. Danshen-related Chinese HMs have been widely used for treatment of coronary heart disease in the East, and a clinical trial is on the way in the United States. Dansen is a routine herbal medication for acute angina pectoris. In addition, it may be effective for dyslipidemia, blood hypervisicosity syndrome, peripheral angiopathy (superficial thrombophlebitis, venous thrombosis, allergic arteriolitis), diabetes mellitus, and cirrhosis, and is also used for altitude sickness. Experimental studies have shown that Danshen dilates coronary arteries, increases coronary blood flow, and scavenges free radicals in ischemic diseases, reducing cellular damage from ischemia and improved heart functions, remarkably similar to known effects of NO and nitrite. However, the nitrate/nitrite reductase activity in Danshen is relatively low. Often, Danshen is mixed with other herbal products. One of them is the extract of cinnamon or borneol. Borneol is consumed widely in China and other Asian countries, particularly in a combined formula for preventing cardiovascular disease. Borneol exerts a concentration-dependent inhibitory effect on venous thrombosis [42]. The antithrombotic activity of borneol contributes to its action in combined formula for preventing cardiovascular diseases. Our recent study has shown that although the natural form of borneol itself contains very little nitrite and nitrate, it displays a potent nitrite/nitrate reductase activity [20]. Another HM made from the root of *Radix ginseng* may also have synergistic effects with Danshen. Ginseng contains modest amounts of nitrate but has stronger reductase activity.

Certain HMs may also exert regulatory effects on expression and activation of endogenous NO-synthesizing enzymes, including eNOS. Recently, we studied whether pretreatment with Tongxinluo (TXL), a mixture of traditional Chinese medicines, can attenuate the no-reflow and ischemia-reperfusion injury in an infarct animal model [43]. TXL is composed of *Radix ginseng*, *Buthus martensi*, *Hirudo*, *Eupolyphaga seu steleophage*, *Scolopendra subspinipes*, *Periostracum cicadae*, *Radix paeoniae rubra*, *Semen ziziphi spinosae*, *Lignum dalbergiae odoriferae*, *Lignum santali albi*, and *Borneolum syntheticum*. To date, TXL has been widely used in China to treat patients with acute coronary syndrome. Pretreatment with TXL at low to high dose for 3 days or with a high loading dose 3 h before ischemia can reduce myocardial no-reflow and infarction in a swine ischemia model.

However, it is not clinically practical to pretreat patients who suffer from acute postinfarct percutaneous coronary intervention with high doses of TXL several hours or days before the operation. Therefore, whether or not low loading dose of TXL just before acute percutaneous coronary intervention can also ameliorate the no-reflow phenomenon and ischemia-reperfusion injury need to be clarified. In a 90-min ischemia and 3-h reperfusion model, miniature pigs were randomly assigned to treatment with TXL (gavaged 1 h prior to ischemia); TXL plus H-89 (protein kinase-A inhibitor intravenously infused before ischemia); or TXL plus N(omega)-nitro-L-arginine (an eNOS inhibitor, intravenously administered prior to ischemia). The results of this study demonstrate that TXL treatment can partially block creatine kinase elevation, improve coronary flow, and reduce infarct size. The effects of TXL may be partially abolished by H-89 or completely reversed by L-NNA. In addition, TXL treatment can enhanced the kinase activity and expression, evidenced by expression of Thr^{198} phosphorated-PKA, Ser^{1179} phosphorated-eNOS (p-eNOS), and Ser^{635} p-eNOS in the ischemic myocardium. Addition of H-89 diminishes the TXL activities. Thus, pretreatment with a single low loading dose of TXL 1 h before ischemia reduces the myocardial no-reflow phenomenon and ischemia-reperfusion injury by up-regulating the phosphorylation of eNOS at Ser^{1179} and Ser^{635}, and this effect is partially mediated by the PKA pathway.

Biosafety of Herbal Medicine Enriching Nitrate and Nitrite

Nitrate and nitrite have long been used for food processing industry for preservation. There are two major health concerns about the safety of food or food supplements that contain abundant inorganic nitrite and nitrate. Previous reports suggest the risk of nitrite uptake for the development of methemoglobinaemia and potential carcinogenic effects of methemoglobinaemia. Compared to nitrite, nitrate is relatively safer, and, in fact, the potential toxicity of the nitrate ion is thought to occur after its bioconversion to nitrite, which is considerably more reactive. Hemoglobin (Hb) in red blood cells plays a key role in preventing the accumulation of nitrite in blood. The formation of methemoglobin is due to oxidation of the oxygen-carrying ferrous ion (Fe^{2+}) of the heme group in the hemoglobin molecule by nitrite to the ferric state (Fe^{3+}). Methemoglobin cannot bind oxygen effectively. In the clinic, significant methemoglobinemia with cyanosis occurs when the levels increase above approximately 5%. In animal studies, intravenous administration of nitrite at doses causing vasodilation increases methemoglobin to very minor levels. This suggests that methemoglobin is not the major obstacle for nitrite administration at these dose ranges. Furthermore, the presence of antioxidants and other organic compounds in herbal formulations may further prevent oxidation of hemoglobin by nitrite and nitrate also contained in the formulations.

As published in 2001 by the US Department of Health and Human Services National Toxicology Program [44], extensive studies in toxicology and carcinogenesis have shown no significant evidence for carcinogenic activity of nitrite, despite dose escalations sufficient to produce profound methemoglobinemia and weight loss in rodents. All the rats and mice were given sodium nitrite in the drinking water for 14-week and 2-year periods. The low dose and short duration of treatment with nitrate/nitrite-containing HMs has a low risk of any carcinogenic effects. Previous reports suggest that the metabolism of dietary nitrate can result in intragastric formation of nitrosamine, a compound that may be carcinogenic. However, after more than 40 years of extensive research, there is still no convincing evidence for a link between nitrate intake and gastric cancer in humans. Furthermore, many HMs contain antioxidants and polyphenols that act as potent inhibitors of nitrosation reactions. The context of nitrite and nitrate delivery along with systems for reduction and inhibition of nitrosation may account for the benefit of HMs. Moreover, it is well known that a diet rich in vegetables is associated with a lower blood pressure and a reduced long-term risk for the development of cardiovascular disease. Recently, a double-blind placebo-controlled crossover evaluation of dietary nitrate supplementation in healthy young volunteers had found a significant reduction in resting blood pressure with a nitrate dose corresponding to the amount found in 150–250 g of green leafy vegetables [29]. Remarkably, the reduction in blood pressure was similar to that described in healthy controls consuming a diet rich in fruits and vegetables in the Dietary Approaches to Stop Hypertension (DASH) trial [8, 45]. Nevertheless, further studies on safety with long-term, high-dose administration of nitrite/nitrate-rich HMs are required.

Summary

Taken together, the nitrate–nitrate–NO pathway plays an important role in the therapeutic effect of many HMs used for cardiovascular disease (Fig. 15.4). The NO-donor-like herbs may serve as an alternative source of nitrate and nitrite

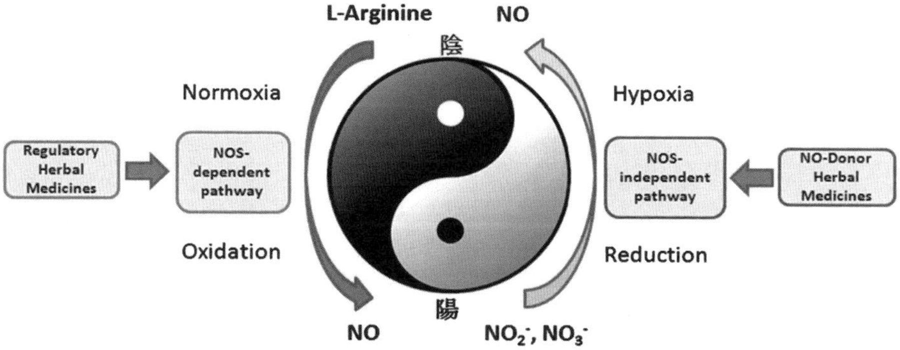

Fig. 15.4 The "Yin-Yang" (陰-陽) balance of NO production and function in association with different types of herbal medicines. As a signaling molecule well established in biology, NO plays a key role in the regulation of the function of the cardiovascular system by herbal medicines. Increasing evidence indicates that herbal medicines can help maintain the "Yin-Yang" balance through regulating the two mechanisms of NO production: oxidation of L-arginine and reduction of nitrate/nitrite

or other NO-donors. They express high activities of nitrate/nitrite reductases that convert the inorganic anions into NO, which, in turn, relaxes blood vessels and prevents thrombosis. Herbs may contain regulatory factors that influence the expression and activity of NOS (Fig. 15.4). The NOS-regulatory herbs may interact with endogenous factors that modulate NOS function. Currently, there is no clear evidence showing that the nitrate/nitrite-rich HMs have any major adverse side-effects, but further investigation is warranted in terms of long-term, high-dose administration of these HMs.

References

1. Davidson P, Hancock K, Leung D, et al. Traditional Chinese medicine and heart disease: what does Western medicine and nursing science know about it? Eur J Cardiovasc Nurs. 2003;2(3):171–81.
2. Gillis CN. Panax ginseng pharmacology: a nitric oxide link? Biochem Pharmacol. 1997;54(1):1–8.
3. Chen X. Cardiovascular protection by ginsenosides and their nitric oxide releasing action. Clin Exp Pharmacol Physiol. 1996;23(8):728–32.
4. Lundberg JO, Weitzberg E. NO generation from inorganic nitrate and nitrite: role in physiology, nutrition and therapeutics. Arch Pharm Res. 2009;32(8):1119–26.
5. Lundberg JO, Weitzberg E. NO-synthase independent NO generation in mammals. Biochem Biophys Res Commun. 2010;396(1):39–45.
6. Lundberg JO, Weitzberg E, Gladwin MT. The nitrate-nitrite-nitric oxide pathway in physiology and therapeutics. Nat Rev Drug Discov. 2008;7(2):156–67.
7. Garg HK, Bryan NS. Dietary sources of nitrite as a modulator of ischemia/reperfusion injury. Kidney Int. 2009;75(11):1140–4.
8. Hord NG, Tang Y, Bryan NS. Food sources of nitrates and nitrites: the physiologic context for potential health benefits. Am J Clin Nutr. 2009;90(1):1–10.
9. Koshland Jr DE. The molecule of the year. Science. 1992;258(5090):1861.
10. Griffith TM, Edwards DH, Davies RL, Harrison TJ, Evans KT. EDRF coordinates the behaviour of vascular resistance vessels. Nature. 1987;329(6138):442–5.
11. Lundberg JO. Cardiovascular prevention by dietary nitrate and nitrite. Am J Physiol Heart Circ Physiol. 2009;296(5):H1221–3.
12. Lundberg JO, Weitzberg E. Nitrite reduction to nitric oxide in the vasculature. Am J Physiol Heart Circ Physiol. 2008;295(2):H477–8.
13. Cueto M, Hernandez-Perera O, Martin R, et al. Presence of nitric oxide synthase activity in roots and nodules of *Lupinus albus*. FEBS Lett. 1996;398(2–3):159–64.
14. Ribeiro Jr EA, Cunha FQ, Tamashiro WM, Martins IS. Growth phase-dependent subcellular localization of nitric oxide synthase in maize cells. FEBS Lett. 1999;445(2–3):283–6.
15. Walker FA, Ribeiro JM, Montfort WR. Novel nitric oxide-liberating heme proteins from the saliva of bloodsucking insects. Met Ions Biol Syst. 1999;36:621–63.
16. Ninnemann H, Maier J. Indications for the occurrence of nitric oxide synthases in fungi and plants and the involvement in photoconidiation of Neurospora crassa. Photochem Photobiol. 1996;64(2):393–8.
17. Morot-Gaudry-Talarmain Y, Rockel P, Moureaux T, et al. Nitrite accumulation and nitric oxide emission in relation to cellular signaling in nitrite reductase antisense tobacco. Planta. 2002;215(5):708–15.
18. Rockel P, Strube F, Rockel A, Wildt J, Kaiser WM. Regulation of nitric oxide (NO) production by plant nitrate reductase in vivo and in vitro. J Exp Bot. 2002;53(366):103–10.

19. Feelisch M, Fernandez BO, Bryan NS, et al. Tissue processing of nitrite in hypoxia: an intricate interplay of nitric oxide-generating and -scavenging systems. J Biol Chem. 2008;283(49):33927–34.
20. Tang Y, Garg H, Geng YJ, Bryan NS. Nitric oxide bioactivity of traditional Chinese medicines used for cardiovascular indications. Free Radic Biol Med. 2009;47(6):835–40.
21. Yamada K, Suzuki E, Nakaki T, Watanabe S, Kanba S. Aconiti tuber increases plasma nitrite and nitrate levels in humans. J Ethnopharmacol. 2005;96(1–2):165–9.
22. Geng YJ, Petersson AS, Wennmalm A, Hansson GK. Cytokine-induced expression of nitric oxide synthase results in nitrosylation of heme and nonheme iron proteins in vascular smooth muscle cells. Exp Cell Res. 1994;214(1):418–28.
23. Janero DR, Bryan NS, Saijo F, et al. Differential nitros(yl)ation of blood and tissue constituents during glyceryl trinitrate biotransformation in vivo. Proc Natl Acad Sci USA. 2004;101(48):16958–63.
24. van Faassen EE, Bahrami S, Feelisch M, et al. Nitrite as regulator of hypoxic signaling in mammalian physiology. Med Res Rev. 2009;29(5):683–741.
25. Bryan NS. Nitrite in nitric oxide biology: cause or consequence? A systems-based review. Free Radic Biol Med. 2006;41(5):691–701.
26. Bryan NS, Calvert JW, Elrod JW, Gundewar S, Ji SY, Lefer DJ. Dietary nitrite supplementation protects against myocardial ischemia-reperfusion injury. Proc Natl Acad Sci USA. 2007;104(48):19144–9.
27. Bryan NS, Calvert JW, Gundewar S, Lefer DJ. Dietary nitrite restores NO homeostasis and is cardioprotective in endothelial nitric oxide synthase-deficient mice. Free Radic Biol Med. 2008;45(4):468–74.
28. Bryan NS, Grisham MB. Methods to detect nitric oxide and its metabolites in biological samples. Free Radic Biol Med. 2007;43(5):645–57.
29. Larsen FJ, Ekblom B, Sahlin K, Lundberg JO, Weitzberg E. Effects of dietary nitrate on blood pressure in healthy volunteers. N Engl J Med. 2006;355(26):2792–3.
30. Webb AJ, Patel N, Loukogeorgakis S, et al. Acute blood pressure lowering, vasoprotective, and antiplatelet properties of dietary nitrate via bioconversion to nitrite. Hypertension. 2008;51(3):784–90.
31. Lundberg JO. Nitric oxide metabolites and cardiovascular disease markers, mediators, or both? J Am Coll Cardiol. 2006;47(3):580–1.
32. Lundberg JO, Govoni M. Inorganic nitrate is a possible source for systemic generation of nitric oxide. Free Radic Biol Med. 2004;37(3):395–400.
33. Stokes KY, Dugas TR, Tang Y, Garg H, Guidry E, Bryan NS. Dietary nitrite prevents hypercholesterolemic microvascular inflammation and reverses endothelial dysfunction. Am J Physiol Heart Circ Physiol. 2009;296(5):H1281–8.
34. Webb A, Bond R, McLean P, Uppal R, Benjamin N, Ahluwalia A. Reduction of nitrite to nitric oxide during ischemia protects against myocardial ischemia-reperfusion damage. Proc Natl Acad Sci USA. 2004;101:13683–8.
35. Milkowski A, Garg HK, Coughlin JR, Bryan NS. Nutritional epidemiology in the context of nitric oxide biology: a risk-benefit evaluation for dietary nitrite and nitrate. Nitric Oxide. 2010;22(2):110–9.
36. Liu X, Ma H, Li C, et al. Sixty-one cases of angina pectoris due to coronary heart disease treated by external use of the paste of nitrum and realgar powder on zhiyang (GV 9). J Tradit Chin Med. 2002;22(4):243–6.
37. Lam FF, Yeung JH, Chan KM, Or PM. Mechanisms of the dilator action of cryptotanshinone on rat coronary artery. Eur J Pharmacol. 2008;578(2–3):253–60.
38. Lam FY, Ng SC, Cheung JH, Yeung JH. Mechanisms of the vasorelaxant effect of Danshen (*Salvia miltiorrhiza*) in rat knee joints. J Ethnopharmacol. 2006;104(3):336–44.
39. Zhou L, Zuo Z, Chow MS. Danshen: an overview of its chemistry, pharmacology, pharmacokinetics, and clinical use. J Clin Pharmacol. 2005;45(12):1345–59.
40. Li MH, Chen JM, Peng Y, Wu Q, Xiao PG. Investigation of Danshen and related medicinal plants in China. J Ethnopharmacol. 2008;120(3):419–26.

41. Li W, Li ZW, Han JP, Li XX, Gao J, Liu CX. Determination and pharmacokinetics of danshensu in rat plasma after oral administration of danshen extract using liquid chromatography/tandem mass spectrometry. Eur J Drug Metab Pharmacokinet. 2008;33(1):9–16.
42. Li YH, Sun XP, Zhang YQ, Wang NS. The antithrombotic effect of borneol related to its anticoagulant property. Am J Chin Med. 2008;36(4):719–27.
43. Li XD, Yang YJ, Geng YJ, et al. Tongxinluo reduces myocardial no-reflow and ischemia-reperfusion injury by stimulating the phosphorylation of eNOS via the PKA pathway. Am J Physiol Heart Circ Physiol. 2010;299:H1255–61.
44. Olden K, Guthrie J. Genomics: implications for toxicology. Mutat Res. 2001;473(1):3–10.
45. Tsuchiya K, Kanematsu Y, Yoshizumi M, et al. Nitrite is an alternative source of NO in vivo. Am J Physiol Heart Circ Physiol. 2005;288(5):H2163–70.

Chapter 16
Nitrite and Nitrate in Cancer

David M. Klurfeld

Key points

- Studies in animal models established the feasibility of sodium nitrite contributing to gastric carcinogenesis primarily via conversion to nitrosamines.
- Most animal studies did not corroborate this assumption.
- Exposure of humans to nitrates is primarily from vegetables.
- Since a fraction of nitrate is reduced to nitrite by oral bacteria, the largest source of nitrite exposure is also from vegetables.
- Meat processed with nitrite has reducing agents added that virtually eliminate the formation of nitrosamines.
- Existing epidemiological studies are not definitive and it is likely that additional studies of the same type will not clarify any putative relationship, due to the inherent weaknesses in epidemiological studies.
- Nitrate and nitrite use has clear benefits for food preservation and only theoretical long-term risk that remains unproven.

Keywords Nitrosamines • Cancer • Epidemiology • Tumor • Carcinogenesis

Introduction

The putative linkage between exposure to nitrites or nitrates and risk of developing cancer derives from studies done in vitro, in animal models or via epidemiological observations. Much of the experimental literature suggesting increased risk is several

D.M. Klurfeld (✉)
United States Department of Agriculture, Agricultural Research Service, Beltsville, MD, USA
e-mail: david.klurfeld@ars.usda.gov

N.S. Bryan and J. Loscalzo (eds.), *Nitrite and Nitrate in Human Health and Disease*, Nutrition and Health, DOI 10.1007/978-1-60761-616-0_16,

decades old, some of it was discounted because of contaminating nitrosamines in drinking water or bedding of test animals, and predates our understanding of the endogenous synthesis and biology of nitrogen-based secondary messengers; it often relied on enormous doses of nitrites/nitrates that have no relevance for normal dietary exposure and usually instilled directly into the stomach of animals. In addition, it was demonstrated that sodium nitrite could be a precursor to carcinogenic *N*-nitroso compounds. Under acidic conditions in the stomach, some nitrites form nitrous acid, which may then nitrosate secondary amines (if present) to form genotoxic and carcinogenic nitrosamines. Such studies only demonstrate plausibility, although it should be recognized that regulatory agencies generally rely on data from animal studies in the absence of reliable exposure/carcinogenicity information in humans. Furthermore, the effects of nitrites on experimental gastric cancer were also demonstrated before it was known that many times as much nitrate, mostly from vegetables whose intake may reduce risk of cancer, is in the food supply as nitrite and that there is significant concentration of nitrate in saliva and conversion of nitrate to nitrite by the oral microflora. Hence, the relationship between exposure to nitrates and nitrites with cancer remains controversial. Informed observers of the environment and cancer generally agree the level of evidence is convincing when there are concordant epidemiological studies showing increased risk of sufficient magnitude supported by animal and cell/molecular studies to demonstrate mechanistic connections. Hill [1] set out nine viewpoints on the information that would contribute to a conclusion of causality from associations. The reader is referred to the original for details of the factors but Hill listed strength of the association as, by far, the most important of them. He also specifically mentioned an example of needing only fair evidence for replacement of a toxic chemical with a safer one that is readily available and economical compared with the requirement of very strong evidence if people were to cease eating sugar and fat in excess.

Animal Studies

Modern standards of testing generally call for lifetime exposure of both sexes from two animal species to a chemical for assessment of carcinogenicity. The dosages tested are often chosen by first determining the maximum tolerated dose and then reducing exposure so that there are no apparent toxic effects. This paradigm has resulted in about half of all chemicals tested being deemed carcinogens and there has been an ongoing debate about whether this experimental scheme is proper since such high doses of many chemicals could result in increased mitosis while lower doses do not [2]; so carcinogenicity in such a situation may be an artifact. Two-year exposure of groups of 50 F344 rats of both sexes to sodium nitrite at 0, 0.125% (125 mg/100 mL = 18.1 mM) or 0.25% (250 mg/100 mL = 36.2 mM) in drinking water or to sodium nitrate in the diet at 0, 2.5% (2.5/100 g = 294 mM) or 5.0% (5/100 g = 588 mM) resulted in two statistically significant cancer differences between treated and control groups – there was a significant decrease in all tumors in females exposed to 0.25% nitrite and all treated groups showed reductions in mononuclear cell leukemias [3], the main cause of death in rats of this strain. These changes were seen even in the presence of a threefold increase of *N*-nitrosodimethylamine in the stomach contents of only male rats fed the higher dose of nitrite. Despite the absence of any carcinogenic effect of either nitrate or nitrite and a significant reduction in two measures of tumorigenesis, the authors concluded by recommending reductions of both compounds in foods as very important because they are precursors of *N*-nitroso compounds. Furthermore, these doses could never be achieved through dietary means. People do not consume sodium nitrate or nitrite directly, so Olsen et al. [4] tested in two generations of rats diets that contained either casein or chopped pork as the sole protein source at 45% by weight of diet; this level is more than double the typical protein content of a rodent diet. Sodium nitrite was added to the pork to make final diet concentrations of 200, 1,000 or 4,000 mg/kg. There was no effect on reproduction or on cancer, except for a non-significant tendency to increased number of tumor-bearing rats fed the highest dose. Again, despite the absence of any statistically significant effect on carcinogenesis, the authors recommended a reduction in the use of nitrite for curing meat.

Human Exposure to Nitrates/Nitrites

The International Agency for Research on Cancer recently evaluated the carcinogenicity of ingested nitrate and nitrites [5]. The reviewers concluded there is inadequate evidence in experimental animals for carcinogenicity of nitrate and limited evidence for nitrite per se, but sufficient evidence for nitrite in combination with amines or amides. In humans, there is inadequate evidence for carcinogenicity of nitrate in food or in drinking water but there is limited evidence for carcinogenicity of nitrite in foods, specifically for stomach cancer. The overall evaluation was that ingested nitrate or nitrite under conditions that result in endogenous nitrosation is probably carcinogenic to humans, placing both compounds in Group 2A (of probable carcinogens) that include well-established carcinogens such as benzo(a) anthracene, 1,2-dimethylhydrazine and *N*-ethyl-*N*-nitrosourea. The IARC report states that the overall evaluation was made despite the observation that none of the human studies linking nitrite to stomach cancer had taken into account potential confounding or effect modification by *Helicobacter pylori*, an important microbial risk factor for stomach cancer.

To assess human exposure to nitrates and nitrites, four primary conditions must be met. First, we need an accurate database of sources of the compounds in food, water and any other relevant sources of exposure. Second, we need accurate methods of determining exposure over time to foods and beverages. Third, we need to know about exposure to preformed carcinogenic nitrosamines, since nitrates and nitrites themselves are not harmful at usual levels of exposure. Fourth, we need to know what proportion of exogenous or endogenous nitrates or nitrites are converted to carcinogenic nitrosamines after ingestion. In addition to exposure factors, we need to know if there are other biological mechanisms by which these compounds affect carcinogenesis.

There are no official government databases that include nitrate or nitrite concentrations in food. The primary reason for this is the marked variability within foods and the fact that drinking water can be a major source of exposure. For example, nitrate content of seven vegetables in Denmark show 20-fold differences within each type with about threefold seasonal variation [6]. From an estimated total intake of 61 mg/day of nitrate and 0.5 mg/day of nitrite, the eight vegetables surveyed contribute 49 and 0.09 mg, respectively. Similar variability was found in vegetables and processed foods likely to contain nitrite (mostly meat and dairy products) in New Zealand [7]. In both countries, lettuce was the food providing the most nitrate. Based on the acceptable daily intake (ADI) values for nitrate and nitrite in New Zealand, the mean adult intake from food and water was 16 and 13% of the ADI. But a maximally exposed adult (a vegetarian consuming well water with high nitrate) was estimated to exceed the nitrate limit by sevenfold. Because of the endogenous conversion of nitrate to nitrite by oral microflora, 10% of the New Zealand population with an average conversion rate of 5% exceeds the ADI for nitrite and among those with a high conversion rate of 20%, half that portion of the population exceeds the ADI. Factors affecting nitrate and nitrite variability of vegetables include variety, soil conditions, growth conditions (including fertilization and water stress), transport and storage conditions. Hord et al. [8] modeled the nitrate and nitrite content of the DASH diet (Dietary Approaches to Stop Hypertension), which stresses high fruit and vegetable intake along with three servings of low-fat dairy and a reduction in meat and eggs, that has been shown in intervention studies to reduce mild hypertension and is one of the dietary patterns recommended in the *Dietary Guidelines for Americans*. Selecting seven low-nitrate fruits and vegetables could yield 174 mg nitrate and 0.41 mg nitrite while choosing high-nitrate plant foods would yield 1,222 mg nitrate plus 0.35 mg nitrite daily, not taking into account the salivary reduction of nitrate to nitrite. Likewise, it has been pointed out that traditional Mediterranean and Japanese diets contain more nitrate than recommended by the World Health Organization [9]. From these data, it should be apparent that dietary patterns recommended for reduction of chronic disease, including various types of cancer, can be notably high in nitrate and/or nitrite content and still afford reduced risk.

Processed meats are the major source of preformed nitrites but given that the conversion of nitrate to nitrite occurs at 5–20% of the total ingested, meats are not the major sources of total exposure to nitrites. Another confounder in assessing exposure is that processing of meats differs geographically. In Europe, total intake of cured meats is lower in Mediterranean countries than in central and northern countries; however, intake of nitrate/nitrite is higher in Spain than in Germany or the United Kingdom because of the differences in the amount of nitrate/nitrite added to meats and storage conditions that may not include refrigeration, or the addition of reducing agents in processing designed to reduce formation of nitrosamines [10]; these aspects call into question the validity of international comparisons of processed meats unless all such parameters are taken into account. In addition, widespread addition of reducing agents to cured meat over the past 35 years has greatly decreased intake of preformed nitrosamines from this source; so much of the older literature is no longer relevant.

When evaluating the potential role of nitrate or nitrite in cancer, it is essential to know what other potential carcinogens are in foods along with the nitrogen ions. This was accomplished by one group through literature searches and included the type of food, cooking method, preservation method, cooking doneness, temperature, time, nitrosamines, heterocyclic amines and polycyclic aromatic hydrocarbons [11]. One limitation of this type of approach is the variation in quality of the analytical methods among published studies. Another limitation in assessing the role of nitrate/nitrite in cancer is factors that associate with higher exposures. In addition to diet, drinking water, occupation and environment, endogenous formation contributes to the overall exposure. We must recognize that nitrite and nitrate are formed endogenously from nitric oxide production and oxidation. Among women in the National Birth Defects Prevention Study, many characteristics distinguished participants according to the level of nitrate/nitrite intake and exposure to nitrosamines that included race/ethnicity, area of residence, household income, education, fat as percent of calories and age at conception [12]; these factors were all associated with adverse birth outcomes and hence it was not possible to clearly separate food components from confounders.

If exposure to nitrate/nitrite is a key factor in cancer, then workers exposed during industrial uses, particularly in years prior to the widespread use of protective breathing devices and improved air handling should prove revealing. In a cohort of 1,327 men working in a fertilizer factory since 1945 and exposed to nitrates for at least 1 year, 537 were considered heavily exposed [13]; this was determined from salivary nitrate and nitrite concentrations and the exposed workers had 212 nM/mL (212 μM) nitrate and 129 nM/mL (129 μM) nitrite in their saliva, which was about double the levels in non-exposed employees of the same company and almost fourfold higher than salivary levels in local residents. Exposure could have been through inhalation, accidental ingestion or dermal contact. Urinary *N*-nitroso compounds were also increased in both the total cohort and heavily exposed workers. Despite this greatly increased chronic exposure to nitrate and its metabolites there was no significant increase in cancers or any other specific diseases, except for laryngeal cancer for which four deaths were observed when 0.9 was expected in the total cohort. Of course, smoking was highly prevalent at that time and that factor was not controlled for. All cause mortality was lower in the exposed workers – 193 deaths observed when expected deaths were 220 in the heavily exposed men and 304 observed versus 368 expected in the total cohort. If industrial exposure that equates to fourfold higher salivary nitrate shows no increase in cancers and about a 15% reduction in total mortality, it is difficult to conclude that there is any risk of serious chronic disease from nitrate at the levels reported.

Unresolved Issues in Epidemiology

There is no doubt that epidemiologic investigation has contributed to our understanding of linkages between lifestyle or environmental factors and chronic diseases. While this has been most useful for cardiovascular disease, the linkages with cancer are less certain, in part, because of our lack of understanding of the causes and various promoters of carcinogenesis that likely interact differently with varying genotypes. The link between cigarette smoking and lung cancer was made first by retrospective case–control studies. This should not be surprising since the relative risk (RR) of lung cancer among smokers is at least tenfold higher than in non-smokers. RR refers to the rate of disease in those exposed to the highest amount of a factor compared with the rate in those exposed to the lowest amount, which can be zero. So, if the absolute lifetime rate of colon cancer is 2 in 100, then a RR of 1.5 reflects a lifetime rate of 3 in 100. Therefore, if nitrate/nitrite carries a theoretical RR of 1.5 for some types of cancer among 100 people who are not exposed, 2 will get cancer and among people who are heavily exposed, 3 will get cancer. But 97 other people will have to avoid exposure to prevent that 1 additional case.

It is rare for a diet–cancer relationship to exceed a RR of 2, and most environmental epidemiologists outside the diet field consider a RR of that magnitude as weak to the point of having little certainty. The only dietary RR exceeding the threshold of 2.0 is the relationship between consumption of moldy cereal grains contaminated with aflatoxin and liver cancer, which carries a RR of 6; this is an important risk factor in parts of Africa and Asia. Substantial controversy exists within the epidemiology community and among other researchers about how capable current epidemiological methods are to accurately determine food intake, how much natural variation occurs in the same foods from different areas or seasons and whether relatively weak RRs really suggest causality. One of the biggest problems facing epidemiologists is the issue of confounding – many lifestyle factors associate with another variable but not all in a causative manner so that only some factors are responsible for the association [14]. These issues will be discussed in detail in the following chapter.

Some epidemiologists working in this area have proposed that the inability of nutritional epidemiology to identify active factors from the diet is not simply a problem of quantitation but an inability to qualitatively identify dietary constituents that are effective [15] because single agents are identified as the active factor in a system in which multiple agents or multiple interacting host factors are responsible for the effect. If this criticism is valid, then the converse is also true – nutritional epidemiology is not capable, with methods currently available, of identifying factors in the diet that increase the risk of cancer and, as a result, larger cohorts will not provide definitive answers because of the inherent collinearities and biases. Willett [16] has argued that prospective studies avoid the biases of recall and selective participation and that diet measurement error would force the RR toward the null. However, it is established that those who volunteer for health research tend to be healthier than the general population. Furthermore, if an error in estimation of nutrient intake is nonrandom, forcing the RR to the null may not occur. Willett's position was written in reaction to the EPIC study's finding that a statistically significant reduction of 4% (RR=0.96) for all cancers was associated with an increment of 200 g/day of fruits and vegetables. Willett concluded from these data that recommending fruits and vegetables to reduce cancer is not justified, even though it is possible that some specific fruit or vegetable may have a greater benefit.

The World Cancer Research Fund (WCRF) in 2007 published what is considered by many a definitive set of systematic reviews and meta-analyses on cancers of various organs and dietary habits [17] even though there is substantial controversy about the conclusions of that report. The only two foods recommended for avoidance were moldy cereals (RR of 6.0 for liver cancer) and processed meats (RR of 1.4 for colorectal cancer). Despite the conclusion about processed meat and colorectal cancer, WCRF noted there was not enough evidence to reach any conclusion about nitrate or nitrite. Some scientists interpreted this report as based on a fragile foundation, given the uncertainties around quantifying the connection and the weak relationships between food and cancer [18].

One of the more provocative and eloquent commentaries on this situation called on cancer epidemiologists to be more circumspect in claiming positive results from the studies of environmental (including diet) and occupational exposures with cancer [19]. The authors pointed out that the more studies conducted on a specific topic, the lower the RR over time, primarily as a result of earlier studies finding statistically significant yet false findings that had arisen by chance leading to other studies attempting to replicate the original observation. A number of commentaries were published in response to that editorial and the same authors were compelled to point out that "committee reports and their conclusions in themselves should not be misconstrued as science; they are consensus documents and opinions with an eye toward closure. In contrast, science is inherently open-ended, provisional and tentative in its findings and conclusions" [20]. If the reader doubts that more research can result in lessening the certainty around most individual epidemiological associations, a mathematical approach has been developed that demonstrates most published research finding are false positives [21], supporting the conclusion that claimed differences may simply be with an accurate measurement of the prevailing bias. This particular author advocates larger studies that offer more statistical power but cautions that large studies are more likely to find a statistically significant difference that is not meaningfully different from the null, and also points out that the totality of the evidence from different types of studies must be considered, again harkening back to the 50-year-old admonition of Hill [1].

One area vital to assessing a relationship of any dietary component and cancer risk is accurate assessment of exposure. Although 24-h diet recalls are often used in small to moderate sized studies, most large prospective cohort studies, considered the strongest design of those currently in use, rely on the food frequency questionnaire (FFQ) that can be computer scored or administered online. Not only does the FFQ have the advantage of lower participant burden and lower cost, it should be able to ascertain foods eaten less often. One inherent problem is that an FFQ is often "validated" by determining correlation for specific nutrients with values derived from one or more 24-h recalls. This would be acceptable if the 24-h recall exhibited both high precision and accuracy; however, most recalls or diaries approximate total energy by plus/minus 25% around the mean, or a range of 50%. Some FFQs are validated against plasma biomarkers such as polyunsaturated fatty acids or carotenoids as markers of intake. However, no FFQ has ever been validated for the numerous components of foods for which they are used to associate with disease endpoints. Just because an FFQ correlates with one or a few markers (usually with modest correlations of up to ~0.5) it does not mean it accurately and precisely captures intake of nutrients calculated from it. Another issue is that subjects who eat more total energy also are exposed to more of all nutrients. One of the responses to this concern has been to adjust for energy intake. The Observing Protein and Energy Nutrition (OPEN) study examined the ability of the FFQ to estimate total energy and protein intake by measuring the attenuation resulting from dietary measurement error [22]. For both energy and protein, the attenuation was substantial enough for the investigators to conclude that the FFQ cannot be recommended. Although multiple 24-h recalls were more accurate, the authors questioned the ability of both these instruments to detect moderate RRs between 1.5 and 2.0 (which is the maximal RR for almost all diet–cancer relationships) even when dietary factors are adjusted for energy intake. These findings led to the assertion by epidemiologists who have made extensive use of the FFQ that we are not likely to learn much more about diet and cancer by continuing to use the standard FFQ [23]. However, it is not clear how an FFQ can be modified to overcome these deficiencies. It is likely the FFQ accurately separates high and low consumers of most foods. But what used to be called semi-quantitative FFQs are now being used to report food intakes to the fraction of a gram and nutrients to the milligram. Also, many studies use a single FFQ to represent intake over the full follow-up period and it is established from studies with FFQs that women change intake of dietary fiber by two or more quintiles over 6 years of follow-up [24]; it is a given that if fiber intake changes to that extent, consumption of other foods must also vary by similar amounts. A key to understanding exposure to a food or nutrient over time is whether the FFQ can capture that and, if so, how to integrate multiple FFQs into a composite exposure index.

Epidemiologic Associations of Nitrates/Nitrites and Cancer

Despite the limitations described in the section above, those charged with deciding about diet and cancer risk feel pressure to use the best available evidence to make a recommendation even though it is not as strong as desired and are rarely definitive. This would be more acceptable if such recommendations carried grades about the strength of evidence but that is not commonly done; in fact, sometimes taking no position may be the correct decision even though this is almost never done. Therefore, a cursory review of the state of the science is presented here, with the assumption that some of the small but statistically significant risks associated with nitrates, nitrites or their food sources may be real, even though that remains speculative.

As mentioned above, the WCRF report [17] stimulated considerable discussion and reaction from the meat industry. Some researchers called upon the industry to reduce or eliminate the use of nitrates/nitrites in processing meat based on the assumption that these additives were proven carcinogens [25]; however, if that conclusion is incorrect, then what is proposed is a major change in the production of foods that have been popular for hundreds of years and whose new formulations are potentially less appealing, less safe and more expensive. So, it is not at all trivial for the food industry to respond to the WCRF report with changes in production.

In 1998, a review of the epidemiological evidence on nitrates, nitrites and *N*-nitroso compounds with regard to cancer came to the conclusion that *N*-nitroso compounds are potent carcinogens in animals but the evidence in humans for all three categories of compounds was inconclusive with regard to cancers of the esophagus, stomach, brain and nasopharynx [26]. The rationale for these specific cancers studied is that most nitrosamines that are carcinogenic (about 90% of those tested) exhibit target-organ specificity. All of the studies summarized in this review were of the case–control variety. This design is inherently susceptible to recall bias – that is, an individual diagnosed with cancer often remembers his or her diet differently from a healthy person when that person is in a research study looking at the link between diet and cancer. Additionally, it is not clear what time period is most important in carcinogenesis other than the obvious requirement that exposure occurs before tumor development. But is diet in the year before diagnosis important, 10 years of continuous exposure required or a specific period of life critical to establishing a relationship? Furthermore, endogenous production of *N*-nitroso compounds may be more important for exposure than exogenous intake. The authors of this review also pointed out, where exposure appears to have a small effect, the amount of uncontrollable confounding inherent in the studies is about as large as the most plausible effect and the absence of actual exposure levels may have resulted in misclassification sufficient to explain negative study results.

A systematic review of the epidemiological evidence on esophageal and gastric cancers with food sources of nitrites and nitrosamines evaluated both case–control and cohort studies [27]. Evidence from case–control studies of gastric cancer was claimed to show elevated risk from high nitrite and nitrosamine intake, but was not considered conclusive. Importantly, no formal meta-analyses were presented in this paper. The three cohort studies summarized showed no elevated risk. Studies of meat, processed meats, preserved fish, smoked foods, preserved vegetables and beer were included in this review although no clear definitions of any food groups were presented and it is safe to assume there was no consistency among the foods, or their nitrate/nitrite contents, making up these food groups across studies. Again, while case–control studies tended to support a correlation between stomach cancer and several of these food categories, the cohort studies neither support such a conclusion nor was there a formal meta-analysis of the data. Another review of various dietary factors and gastric cancer risk concluded that *N*-nitroso compounds such as nitrosamines increase cancer risk but the evidence on nitrate and nitrite are not consistent enough to reach a conclusion [28]. These authors cited a meta-analysis of six prospective studies that found a RR of 1.15 for gastric cancer among the highest consumers of processed meat and made the assumption that this habit increased exposure to *N*-nitroso compounds. However, addition of reducing agents since the 1970s has greatly decreased the formation of these compounds in processed meats and there is little, if any, certainty in judging endogenous production of nitrosamines.

A case–control study in Mexico City examined the intake of nitrates, nitrites and polyphenols in 257 people with stomach cancer and 478 controls via analysis of a FFQ [29]. Cases were distinguished from controls by significantly higher *H. pylori* infection, alcohol use, energy intake, added salt and chili consumption, but lower servings of vegetables. While estimates of nitrate and nitrite consumption by cases were significantly higher, the differences were 6 and 4%, respectively, and there were no differences in these compounds derived from animal sources. Nevertheless, the RR for consumption of nitrate from animal sources was 1.87 (95% confidence interval of 1.19–2.91). Controls ate more vegetables that resulted in significantly greater intake of nitrate, nitrite, cinnamic acids, total lignans, secoisolariciresinol and coumestrol. These increased polyphenols reflect greater consumption of pears, mangoes, legumes, carrots and squash. Since we do not consume nutrients in isolation, except in dietary supplements, perhaps it would be more productive to characterize foods, food groups or dietary patterns associated with reduced risk of cancer as all single nutrient prescriptions for prevention of cancer seem not to hold up over time.

Another study of nitrate intake in relation to stomach cancer was a case–control study in Korea that looked at effect modification by intake of antioxidant vitamins [30]. In Korea, the gastric cancer rate has declined but it still remains the most common form of the disease and accounts for 20% of all cancers. For a perspective, Mexico also has high stomach cancer prevalence while the US has one of the world's lowest; decline in the US paralleled adoption of refrigeration leading to speculation that spoiled foods contributed to the high stomach cancer rate. Korea's experience with stomach cancer differs from most other countries in that the rate in women is almost as high as for men; in most countries, men have twice the prevalence of stomach cancer. One limitation of the Korean study is that only 136 cases and an equal number of controls were studied. *H. pylori* infection was significantly more common among cases than controls. There were nonsignificant trends for shorter usage of refrigerators among cases and a greater history of gastric cancer among first-degree relatives. Cases consumed about the same nitrate as controls but cases ingested significantly less beta-carotene, vitamin C, vitamin E and folate. There were significant correlations between nitrate intake and these plant nutrients, with vitamin C and folate correlations of 0.64 and 0.70 while the correlation for beta-carotene was 0.26. The lower correlation of nitrate with vitamin E (0.11) was not significant. The RR for stomach cancer increased as the ratio of nitrate to these vitamins increased but only the ratio of nitrate:folate remained statistically significant following multivariate adjustment for confounders identified in this study. It should be pointed out that the mean daily intake of nitrate was 534 mg and this is more than 2.4 times the ADI, as a result of the high vegetable intake, particularly Korean radish, cabbage, lettuce and spinach. The authors suggest that the ratio of dietary nitrate to antioxidant vitamins may be a fruitful area of further investigation.

It is enlightening to examine in some detail one of the largest prospective cohort studies on diet and cancer – the NIH-AARP Diet and Health Study, being conducted among more than 567,000 individuals aged 50–71 years at the start who were administered a FFQ at baseline. In the first 6.8 years of follow-up, over

53,000 cancers were diagnosed in 21 different organs or tissues. The mean intake of processed meat by subjects in the highest quintile of consumption was 14 times the average intake in the lowest quintile. The highest quintile had significantly elevated risks of lung (RR = 1.16) and colorectal (RR = 1.20) cancers but significantly reduced risks of leukemia and melanoma, while the other cancers reported did not differ by intake of processed meat [31]. Nitrite and nitrate intakes were not reported in this study. The authors reported the *p* value for the trend of each cancer and those were significant only for the elevated risks as well as for two other cancers for which the 95% confidence interval crossed the null, which traditionally denotes nonsignificant observations. It is compelling to ask if using standard criteria of 95% confidence intervals not crossing unity (RR = 1.0) to determine significance in epidemiological studies that risk of two cancers was elevated, two decreased and 17 unchanged, were the calculated significant differences in trends really meaningful?

Another report from the NIH-AARP cohort described data from 7 years of follow-up in 300,000 from the larger group, with 2,719 colorectal cancers cases diagnosed subsequent to initiation of the study [32]. Subjects were divided by quintile of red meat intake, adjusted for total dietary energy; mean red meat intake in the lowest quintile was 8.9 and 66.5 g/day in the highest quintile per 1,000 kilocalories. However, factors that strongly correlated with red meat intake were male gender, lower education, more smoking, lower physical activity, more calorie intake and less calcium, fiber, fruits and vegetables. Median processed meat intakes in the lowest and highest quintiles were 2.7 and 38.0 g/day/1,000 kilocalories. RR for colorectal cancer was significantly elevated at 1.16 (95% CI, 1.01–1.32). The same RR for nitrate from processed meats was calculated but the risk for nitrite from processed meats (RR = 1.11, 95% CI, 0.97–1.25) was not significant. Failure to calculate nitrite and nitrate intake from other sources makes these data much less certain. Some investigators have suggested other factors in processed meat such as preformed heterocyclic amines or the combination of nitrites and amines make nitrate/nitrite intake from meat dangerous compared with nitrate/nitrite from other foods. But if nitrate or nitrite is designated as a culprit, there is no plausible difference in biology from ingesting larger amounts of nitrogen ions from vegetables, water or meat.

The same large cohort was studied for relation of *N*-nitroso compounds with adult brain cancer [33]. This time 546,000 participants' data were analyzed to identify 585 cases of glioma. No significant trends in risk were found for consumption of red or processed meat, nitrate, or vitamins C or E. Significantly elevated risks were found for total fruit and vegetable consumption (RR = 1.42, 95% CI, 1.08–1.86) and for nitrite from plant sources (RR = 1.59, 95% CI, 1.20–2.10). The authors argued that nitrate from grain products might be harmful because grains do not contain the nitrosation inhibitors found in fruits and vegetables. They admit they did not statistically adjust their results for multiple comparisons, so the findings may be by chance. In addition, this cohort was examined for the risk of prostate cancer in a subset of 175,343 men [34]. Over 10,000 cases of prostate cancer were diagnosed and 1,102 were in advanced disease during 9 years of follow-up. Nitrite and nitrate from meat were not associated with total prostate cancer

(RR = 1.02 for both), but there were significantly elevated risks with advanced prostate cancer (RR = 1.24, 95% CI 1.02–1.51 and 1.31, 95% CI, 1.07–1.61 for nitrite and nitrate, respectively). One would expect if the relationship between a subset of 1,102 cases of prostate cancer and these two compounds were real, then it would be seen in a ninefold larger group of total prostate cancer cases.

The final report from the NIH-AARP cohort to be discussed here reported on total mortality [35]. Red and processed meat intakes were reported to increase the risk of total death during 10 years of follow-up in which there were over 71,000 deaths, with two-thirds of the deaths in men. Men in the quintile of highest red meat intake had a RR of 1.31 (95% CI, 1.27–1.35) and for processed meat the RR was 1.16 (95% CI, 1.12–1.20). The information that put these values in sharpest perspective was the risk of mortality from injuries and sudden deaths. Red meat was associated with an increased RR of 1.26 (95% CI, 1.04–1.54) and white meat associated with a decreased RR of 0.85 (95% CI, 0.70–1.02). There is no plausible link between accidental deaths and greater intake of red meat or a protective effect of poultry/fish consumption unless these different dietary habits reflect other lifestyle decisions that lead to fatal accidents. These data may suggest that RRs in the range reported are not signals, but are noise, even when based on huge numbers compared to most dietary epidemiology studies.

A different cohort study of about 100,000 participants, in contrast to the NIH-AARP study [32], found no association of either red or processed meat consumption with lung cancer in men or women [36]. For processed meat, the RR for lung cancer risk in the highest quintile of consumption was 1.12 (95% CI, 0.83–1.53) in men and 0.98 (95% CI, 0.68–1.41) in women. A wide variety of cancers in other organ sites have been studied in relation to nitrate/nitrite intake and red or processed meat as a proxy for nitrates and nitrites, even though there is no strong correlation between the foods and intake of the compounds in all studies. A meta-analysis of all published studies on red or processed meat and kidney cancer, including 12 case–control studies, three cohorts and the Pooling Project [37] in which 13 cohort studies were evaluated, found no increased risk with high consumption of meat [38]. The RRs ranged from 1.02 to 1.19 and the total findings were interpreted as not supportive of a relationship between intake and cancer risk. Another prospective cohort study in The Netherlands examined a potential relationship between nitrate intake and bladder cancer [39]. In a population of 120,852 middle-aged men and women, followed for 9.3 years, 889 cases of bladder cancer were compared against cohort members without the disease. Nitrate exposures from food, drinking water or total sources were similar and did not elevate risk – RR for nitrate from food was 1.06 (95% CI, 0.81–1.31). Three US cohorts were combined to examine a relationship between nitrates/nitrites and glioma in adults [40]. A total of 237,794 people were followed for up to 24 years and 335 cases of glioma were identified. Risk of glioma was not elevated for highest consumption of nitrate (RR = 1.02, 95% CI, 0.66–1.58) or nitrite (RR = 1.26, 95% CI, 0.89–1.79).

The final epidemiological study to be considered here exemplifies the difficulty in examining compounds found in multiple foods and water with the development of cancer. A case–control study in Iowa, where there are high levels of nitrate in the drinking water, was conducted among 458 cases of non-Hodgkin's lymphoma and 383 controls with records on exposure [41]. There was no elevated risk from high nitrate in drinking water (RR = 1.2, 95% CI 0.6–2.2). Nitrate in diet was associated with decreased risk (RR = 0.54, 95% CI 0.34–0.86) but nitrite was associated with increased risk (RR = 3.1, 95% CI 1.7–5.5); however, this latter increase was attributed to nitrite in breads and cereals.

Conclusions

No consistent relationship appears between nitrate/nitrite exposure and risk of cancer or other adverse health consequences. In fact, experts remain considerably divided on the risk when summarizing and interpreting the same data [42, 43]. A relatively new area of research that challenges traditional biochemical thinking on this topic is that endogenous protein S-nitrosylation may play a substantial role in pathophysiological states, including cancer, and that this phenomenon is mostly independent of exogenous nitrogen ion exposure [44]. S-Nitrosylation is the process in which nitric oxide is post-translationally added to cysteine sulfhydryl groups of proteins. This process represents a form of redox modulation and is becoming increasingly recognized as a regulatory reaction perhaps equally important to phosphorylation. A partial list of proteins related to carcinogenesis shown to undergo S-nitrosylation includes Ras, COX-2, PTEN, NF-κB, p53, Bcl-2, caspases and DNA repair enzymes. Both nitrosylation and denitrosylation of various target proteins contribute to the net biological effect. The activities of nitrite as a signaling molecule and regulator of gene expression, combined with its abundance in vivo and its stability, suggest an important role of nitrite in normal metabolism [45, 46].

As can be seen in the epidemiology section above, some groups of researchers consistently report that nitrate/nitrite increases the risk of cancer. It is not clear why only some scientists find this while others consistently do not. One of the former groups recently reported that men fed 420 g/day of processed meats had a 70-fold increase in fecal excretion of *N*-nitroso compounds compared with those eating a vegetarian diet [47]; however, similar increases in fecal excretion of these compounds were found when subjects consumed a diet supplemented with red meat that contained no nitrite. This strongly suggests that endogenous formation of *N*-nitroso compounds is independent of dietary intake of nitrite. There were five male and 11 female volunteers in this study. Curiously, only the data for the men were published with the rationale that their diets contained the largest amount of meat. The authors of this report proposed that nitrosothiols form in the acidic conditions of the stomach following ingestion of any meat products. However, it is far-fetched that nitrosothiols formed in the stomach would not be absorbed along the length of the gastrointestinal tract. Furthermore, inorganic nitrite under acidic conditions as might occur in the stomach or urinary bladder actually inhibits carcinoma cell growth and this effect can be enhanced by the addition of ascorbic acid [48]. This effect appeared to result from the formation of nitric oxide or other reactive nitrogen oxide species.

The reader is referred to a risk–benefit analysis of dietary nitrate and nitrite that examined the small, uncertain risks of cancer balanced against the certain reduction of food poisoning, including the elimination of botulism and marked reduction in other microbial illnesses from commercially processed meats, and balanced these observations with the potential cardiovascular benefits from the nitrogen ions [49]. It should be clear that scientists and policy makers need to understand the full biological ramifications of exogenous exposure to nitrates/nitrites from all sources, the endogenous synthesis of nitrogen metabolites and their biological consequences in multiple target tissues before any meaningful decisions can be made. The strength of evidence to make such decisions will not be made by counting up the number of publications, or looking at the largest cohort, but by weighing a variety of studies and synthesizing a rational conclusion that does not trade proven and immediate consumer safety for a small, long-term risk that remains theoretical despite decades of research.

References

1. Hill AB. The environment and disease: association or causation? Proc R Soc Med. 1965;58:295–300.
2. Ames BN, Gold LS. Chemical carcinogenesis: too many rodent carcinogens. Proc Natl Acad Sci USA. 1990;87:7772–6.
3. Maekawa A, Ogiu T, Onodera H, Furuta K, Matsuoka C, Ohno Y, et al. Carcinogenicity studies of sodium nitrite and sodium nitrate in F-344 rats. Food Chem Toxicol. 1982;20:25–33.
4. Olsen P, Gry J, Knudsen I, Meyer O, Poulsen E. Animal feeding study with nitrite-treated meat. IARC Sci Publ. 1984;57:667–75.
5. IARC Working Group on the Evaluation of Carcinogenic Risks to Humans. IARC Monogr Eval Carcinog Risks Hum. 2010;94:1–412.
6. Petersen A, Stoltze S. Nitrate and nitrite in vegetables on the Danish market: content and intake. Food Addit Contam. 1999;16:291–9.
7. Thomson BM, Nokes CJ, Cressey PJ. Intake and risk assessment of nitrate and nitrite form New Zealand foods and drinking water. Food Addit Contam. 2007;24:113–21.
8. Hord NG, Tang Y, Bryan NS. Food sources of nitrates and nitrites: the physiologic context for potential health benefits. Am J Clin Nutr. 2009;90:1–10.
9. Lundberg JO, Gladwin MT, Ahluwalia A, Benjamin N, Bryan NS, Butler A, et al. Nitrate and nitrite in biology, nutrition and therapeutics. Nat Chem Biol. 2009;5:865–9.
10. Linseisen J, Rohrmann S, Norat T, Gonzalez CA, Dorronsoro Iraeta M, Morote Gomez P, et al. Dietary intake of different types and characteristics of processed meat which might be associated with cancer risk – results from the 24-hour diet recalls in the European Prospective Investigation into Cancer and Nutrition (EPIC). Public Health Nutr. 2006;9:449–64.
11. Jakszyn P, Agudo A, Ibanez R, Garcia-Closas R, Pera G, Amiano P, et al. Development of a food database of nitrosamines, heterocyclic amines, and polycyclic aromatic hydrocarbons. J Nutr. 2004;134:2011–4.
12. Griesenbeck JS, Brender JD, Sharkey JR, Steck MD, Huber JC Jr, Rene AA, et al. Maternal characteristics associated with the dietary intake of nitrates, nitrites, and nitrosamines in women of child-bearing age: a cross-sectional study. Environ Health. 2010;9:10–6.

13. Forman D, Al-Dabbagh S, Knight T, Doll R. Nitrate exposure and the carcinogenic process. Ann NY Acad Sci. 1988;534:597–603.
14. van den Brandt P, Voorrips L, Hertz-Picciotto I, Shuker D, Boeing H, Speijers G, et al. The contribution of epidemiology. Food Chem Toxicol. 2002;40:387–424.
15. Meyskens FL, Szabo E. Diet and cancer: the disconnect between epidemiology and randomized clinical trials. Cancer Epidemiol Biomarkers Prev. 2005;14:1366–9.
16. Willett WC. Fruit, vegetables, and cancer prevention: turmoil in the produce section. J Natl Cancer Inst. 2010;102:1–2.
17. World Cancer Research Fund/American Institute for Cancer Research. Food, nutrition, physical activity, and the prevention of cancer: a global perspective. Washington, DC: AICR; 2007.
18. Boyle P, Boffetta P, Autier P. Diet, nutrition and cancer: public, media and scientific confusion. Ann Oncol. 2008;19:1665–7.
19. Boffetta P, McLaughlin JK, La Vecchia C, Tarone RE, Lipworth L, BLot WJ. False-positive results in cancer epidemiology: a plea for epistemological modesty. J Natl Cancer Inst. 2008;100:988–95.
20. Boffetta P, McLaughlin JK, La Vecchia C. A further plea for adherence to the principles underlying science in general and the epidemiologic enterprise in particular. Int J Epidemiol. 2009;38:678–9.
21. Ioannidis JPA. Why most published research findings are false. PLoS Med. 2005;2:0676–701.
22. Schatzkin A, Kipnis V, Carroll RJ, Midthune D, Subar AF, Bingham S, et al. A comparison of a food frequency questionnaire with a 24-hour recall for use in an epidemiological cohort study: results from the biomarker-based Observing Protein and energy Nutrition (OPEN) study. Int J Epidemiol. 2003;32:1954–62.
23. Kristal AR, Peters U, Potter JD. Is it time to abandon the food frequency questionnaire? Cancer Epidemiol Biomarkers Prev. 2005;14:2826–8.
24. Fuchs CS, Giovannucci EL, Colditz GA, Hunter DJ, Stampfer MJ, Rosner B, et al. Dietary fiber and the risk of colorectal cancer and adenoma in women. N Engl J Med. 1999;340:169–76.
25. Demeyer D, Honikel K, De Smet S. The World Cancer Research Fund report 2007: a challenge for the meat processing industry. Meat Sci. 2008;80:953–9.
26. Eichholzer M, Gutzwiller F. Dietary nitrates, nitrites, and N-nitroso compounds and cancer risk: a review of the epidemiologic evidence. Nutr Rev. 1998;56:95–105.
27. Jakszyn P, Gonzalez CA. Nitrosamine and related food intake and gastric and oesophageal cancer risk: a systematic review of the epidemiological evidence. World J Gastroenterol. 2006;12:4296–303.
28. Liu C, Russell RM. Nutrition and gastric cancer risk: an update. Nutr Rev. 2008;66:237–49.
29. Hernandez-Ramirez RU, Galvan-Portillo MV, Ward MH, Agudo A, Gonzalez CA, et al. Dietary intake of polyphenols, nitrate and nitrite and gastric cancer risk in Mexico City. Int J Cancer. 2009;125:1424–30.
30. Kim HJ, Lee SS, Choi BY, Kim MK. Nitrate intake relative to antioxidant vitamin intake affects gastric cancer risk: a case-control study in Korea. Nutr Cancer. 2007;59:185–91.
31. Cross AJ, Leitzmann MF, Gail MH, Hollenbeck AR, Schatzkin A, Sinha R. A prospective study of red and processed meat intake in relation to cancer risk. PLoS Med. 2007; 4:e325.
32. Cross AJ, Ferrucci LM, Risch A, Graubard BI, Ward MH, Park Y, et al. A large prospective study of meat consumption and colorectal cancer risk: an investigation of potential mechanisms underlying this association. Cancer Res. 2010;70:2406–14.
33. Dubrow R, Darefsky AS, Park Y, Mayne ST, Moore SC, Kilfoy B, et al. Dietary components related to *N*-nitroso compound formation: a prospective study of adult glioma. Cancer Epidemiol Biomark Prev. 2010;19:1709–22.
34. Sinha R, Park Y, Graubard BI, Leitzmann MF, Hollenbeck A, Schatzkin A, et al. Meat and meat-related compounds and risk of prostate cancer in a large prospective cohort study in the United State. Am J Epidemiol. 2009;170:1165–77.

35. Sinha R, Cross AJ, Graubard BI. Meat intake and mortality. A prospective study of over half a million people. Arch Intern Med. 2009;169:562–71.
36. Tasevska N, Cross AJ, Dodd KW, Ziegler RG, Caporaso NE, Sinha R. No effect of meat, meat cooking preferences, meat mutagens or heme iron on lung cancer risk in the Prostate, Lung, Colorectal, and Ovarian Cancer Screening Trial. Int J Cancer. 2011;128:402–11.
37. Lee JE, Spiegelman D, Hunter DJ, Albanes D, Bernstein L, van den Brandt PA, et al. Fat, protein, and meat consumption and renal cell cancer risk: a pooled analysis of 13 prospective studies. J Natl Cancer Inst. 2008;100:1695–706.
38. Alexander DD, Cushing CA. Quantitative assessment of red meat or processed meat consumption and kidney cancer. Cancer Detect Prev. 2009;32:340–51.
39. Zeegers MP, Selen RFM, Kleinjans JCS, Goldbohm RA, van den Brandt PA. Nitrate intake does not influence bladder cancer risk: The Netherlands Cohort Study. Environ Health Perspect. 2006;114:1527–31.
40. Michaud DS, Holick CN, Batchelor TT, Giovannucci E, Hunter DJ. Prospective study of meat intake and dietary nitrates, nitrites, and nitrosamines and risk of adult glioma. Am J Clin Nutr. 2009;90:570–7.
41. Ward MH, Cerhan JR, Colt JS, Hartge P. Risk of non-Hodgkin lymphoma and nitrate and nitrite from drinking water and diet. Epidemiology. 2006;17:375–82.
42. Powlson DS, Addiscott TM, Benjamin N, Cassman KG, de Kok TM, van Grinsven H, et al. When does nitrate become a risk for humans? J Environ Qual. 2008;37:291–5.
43. Ward MH, deKok TM, Levallois P, Brender J, Gulis G, Nolan BT, et al. Workgroup report: drinking-water nitrate and health – recent findings and research needs. Environ Health Perspect. 2005;113:1607–14.
44. Foster MW, Hess DT, Stamler JS. Protein S-nitrosylation in health and disease: a current perspective. Trends Mol Med. 2009;15:391–404.
45. Bryan NS, Fernandex BO, Bauer SM, Garcia-Saura MF, Milson AB, Rassaf T, et al. Nitrite is a signaling molecule and regulator of gene expression in mammalian tissues. Nat Chem Biol. 2005;1:290–7.
46. Angelo M, Singel DJ, Stamler JS. An S-nitrothiol (SNO) synthase function of hemoglobin that utilizes nitrite as a substrate. Proc Natl Acad USA. 2006;103:8366–71.
47. Joosen AM, Kuhnle GG, Aspinall SM, Barrow TM, Lecommandeur E, Azqueta A. Effect of processed and red meat on endogenous nitrosation and DNA damage. Carcinogenesis. 2009;30:1402–7.
48. Morcos E, Carlsson S, Weitzberg E, Wiklund NP, Lundberg JO. Inhibition of cancer cell replication by inorganic nitrite. Nutr Cancer. 2010;62:501–4.
49. Milkowski A, Garg HK, Coughlin JR, Bryan NS. Nutritional epidemiology in the context of nitric oxide biology: a risk-benefit evaluation for dietary nitrite and nitrate. Nitric Oxide. 2010;22:110–9.

Chapter 17
Looking Forward

Nathan S. Bryan and Joseph Loscalzo

Key points

- The Nitrate-Nitrite-Nitric Oxide pathway may be the redundant system for maintenance of NO homeostasis in physiology.
- Utilizing nitrite and nitrate as a means to restore and replete NO homeostasis may provide a first line of defense for conditions associated with NO insufficiency.
- Identification of foods with NO activity may allow for rationale design of diets or meal plans for people with cardiovascular disease.
- More studies are needed in order to firmly establish the risk benefit of dietary nitrite and nitrate in specific patient populations.
- Developing safe and effective strategies to restore and replete NO homeostasis will have a profound impact on public health and patient care.

Keywords Drug development • Nutritional interventions • Dietary guidelines • Risk benefit evaluation

Eating a well-balanced, nutritious diet and performing moderate exercise comprise the ideal model for routine good health and disease prevention. Probably not coincidently, these two activities are known to affect nitric oxide production positively, which, one can argue, is a key molecular mediator of the salubrious effect of both interventions. What is becoming more apparent is that NO derived from dietary sources of nitrite and nitrate comprises another key pathway in NO biology. Notwithstanding the essential nature of nitrite and nitrate in the environmental nitrogen cycle, historical use of nitrates and nitrites as medicinal agents, and the fact that these anions are produced naturally in the body from the oxidation of nitric oxide, public perception remains that these are harmful substances in our food and water supply [1].

N.S. Bryan (✉)
Brown Foundation Institute of Molecular Medicine,
The University of Texas, Health Science Center at Houston,
1825 Pressler Street 530C, Houston, TX 77030, USA
e-mail: Nathan.Bryan@uth.tmc.edu

N.S. Bryan and J. Loscalzo (eds.), *Nitrite and Nitrate in Human Health and Disease*, Nutrition and Health, DOI 10.1007/978-1-60761-616-0_17,

There exists a number of nitrogen-containing molecules that are essential and fundamental for all life. Nitrogen is the largest single constituent of the earth's atmosphere; it is created by fusion processes in stars, and is estimated to be the seventh most abundant chemical element by mass in the universe. Nitric oxide, nitrite, and nitrate in the environment and in the body make up part of the overall biological nitrogen cycle along with proteins and nucleic acids, and are important intermediates or mediators of a number of biological actions. The realm of nitrogen-based chemistry is historical and complex. Just as Alfred Nobel found great irony in the fact that he was prescribed nitroglycerin for his angina later in life, which was the very substance that he had patented for safe delivery of dynamite, the current view of the biological activity of nitrite and nitrate shares a similar historical irony. The discovery that a toxic gas (NO) was produced by the inner lining of our blood vessels was shocking and revolutionary. This is not to say that NO is not poisonous or dangerous because its chemical properties define it as a poison or toxin; however, in the right context and in an ideal environment, it is without doubt the most important molecule produced in blood vessels. The same argument can be made for nitrite and nitrate. Their inherent chemistry makes them toxic, but in the right context and the right environment, they can have extremely beneficial actions in the human body. The illustration in Fig. 17.1 shows the label of NO, sodium nitrite, and sodium nitrate as poisonous and toxic. Yet, we appreciate and understand the essential functions that NO has in the body. Becoming more apparent are the same beneficial effects of nitrite and nitrate *in the proper context*. As an analogous example, water is absolutely essential to all life on earth and is completely safe and innocuous; however, if one were to drink a large excess of water within a few minutes, the cells in the body would swell and burst owing to the hypotonic effects. In the words of Paracelsus, "the dose makes the poison." With respect to nitric oxide and nitrite, one could add the duration, location, context of exposure, and method of delivery [2] as defining the poison.

Despite NO being recognized by the scientific and medical community as one of the most important molecules produced within the body and being named

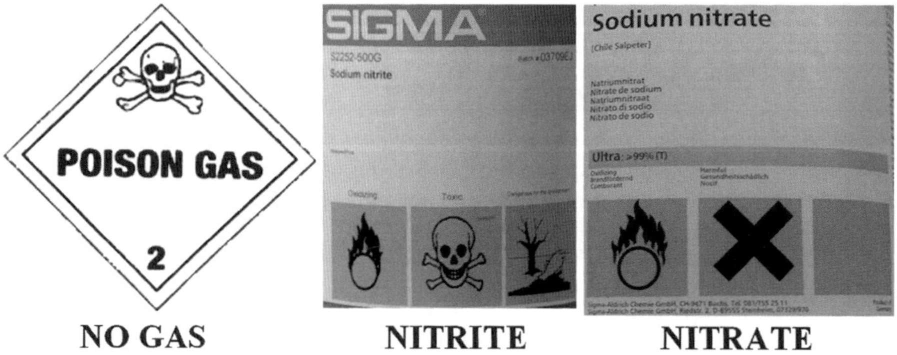

Fig. 17.1 The labels from a cylinder of nitric oxide gas, and bottles of sodium nitrite and sodium nitrate

"Molecule of the Year" by *Science* in 1992 and a Nobel Prize in Physiology or Medicine awarded for its discovery in 1998, there are currently only three products in the market directly related to NO: (1) organic nitrates, such as nitroglycerin for the treatment of acute angina (these have been used for many decades long before the discovery of NO); (2) inhaled NO therapy for neonates for treatment of pulmonary hypertension due to underdeveloped lungs; and (3) phosphodiesterase inhibitors, such as sildenafil (Viagra®), which do not directly affect NO production but act through affecting the downstream second messenger of NO, cyclic guanosine monophosphate (cGMP). There are a number of other NO-based therapies in development, including technologies designated to affect posttranslational protein modifications through S-nitrosation. The method of delivery of NO is of utmost importance. We know that delivery of NO through controlled and enzymatic metabolism of organic nitrates is safe and effective treatment for angina, but still not without some adverse effects as presented in Chap. 13. The safe delivery of NO gas through inhalation therapy is now also in practice. Inhaled NO is currently approved by the US Food and Drug Administration for hypoxemia in term and near-term infants with pulmonary hypertension. Although more clinical trails are underway in other disease states, currently, this is the only approved use of inhaled NO.

We expect the same considerations for the safe and effective delivery of nitrite and/or nitrate. Understanding their underlying chemistry and metabolism is intrinsic to developing strategies for delivery. The risk–benefit spectrum from nitrite and nitrate may very well depend upon the specific metabolism and the presence of other components that may be concomitantly ingested or available at the time of administration or ingestion. The stepwise reduction of nitrate to nitrite and NO may account for these benefits, while pathways leading to nitrosation of low-molecular-weight amines or amides may account for the health risks of nitrite and nitrate exposure. Understanding and affecting those pathways will certainly help in mitigating the risks. The early concerns about nitric oxide reactivity and the propensity to undergo unwanted nitrosation reactions also are invariably related to nitrite and nitrate biochemistry. We know from Chap. 4 that these reactions can occur, but there are also very effective inhibitors of nitrosation reactions including vitamins C and E, as well as polyphenols, that are present in many foods that contain nitrite and nitrate, particularly vegetables.

The first pathway to be discovered for the endogenous production of NO was that involving L-arginine. For years scientists and physicians have investigated L-arginine supplementation as a means to enhance NO production. This strategy has been shown to work effectively in young healthy individuals with functional endothelium or in older patients with high levels of asymmetric dimethyl L-arginine (ADMA) where the supplemental L-arginine can outcompete this natural inhibitor of NO production. Patients with endothelial dysfunction, however, by definition, are unable to convert L-arginine to NO and, therefore, this strategy has failed in clinical trials. In fact, concerning L-arginine therapy in acute myocardial infarction, the Vascular Interaction with Age in Myocardial Infarction (VINTAGE MI) randomized clinical trial published in the *JAMA* in 2006, concluded that L-arginine, when added to standard postinfarction therapies, did not improve vascular stiffness measurements or ejection fraction and was associated with higher postinfarction mortality [3]. L-Arginine should, therefore, not be recommended following acute myocardial infarction (MI). However, there are also a number of studies showing benefit to patients taking L-arginine and just as many showing no benefit, no harm. Collectively, the literature suggests that strategies to enhance NO production through L-arginine supplementation are equivocal at best.

NOS and Nitrite: A Concert in NO Homeostasis

The nitrate–nitrite–nitric oxide pathway may be a redundant system for overcoming the body's inability to make NO from L-arginine. The emerging literature suggests this, and the information from Chap. 3 presents a strong case for such a hypothesis. It appears we have at least two systems for affecting NO production/homeostasis. The first is through the classical L-arginine–NO pathway. As we know from Chap. 11, this is a complex and complicated pathway, and if any of the co-factors become limiting, then NO production from NOS shuts down, and in many cases NOS then produces superoxide instead. The enzymatic production of NO is, indeed, a very complex and coordinated effort that normally proceeds very efficiently. However, in disease characterized by oxidative stress where essential NOS cofactors become oxidized, NOS uncoupling, or conditions of hypoxia where oxygen is limiting, this process can no longer maintain NO production. Therefore, one can argue saliently that there has to be an alternate route for NO production. It is highly unlikely that Nature devised such a sophisticated mechanism of NO production as a sole source of a critical molecule.

This alternate route involves the provision of nitrate and nitrite reductively recycled to NO. Nitrite reduction to NO can occur in a much simpler mechanism than nitrate. The 1-electron reduction of nitrite can occur by ferrous heme proteins (or any redox active metal) through the following reaction:

$$NO_2^- + Fe^{(II)} + H^+ \leftrightarrow NO + Fe^{(III)} + OH^-$$

This is the same biologically active NO as that produced by NOS, with nitrite rather than L-arginine as the precursor, and is a relatively inefficient process [4]. Therefore, for this reaction to occur, the tissues or biological compartment must have a sufficient pool of stored nitrite. Since plasma nitrite is a direct measure of NOS activity [5], a compromised NOS system can also affect downstream nitrite production and metabolism, which can perhaps exacerbate any condition associated with decreased NO bioavailability. Replenishing nitrate and nitrite through dietary means may then act as a protective measure to compensate for insufficient NOS activity under conditions of hypoxia or in a number of conditions characterized by NO insufficiency. It is very likely that exogenous nitrite contributes to whole-body NO production and homeostasis. Considerable published support for this theory derives from the following facts: NO produced from nitrite in the upper intestine is up to 10,000 times the concentrations that occur in tissues from enzymatic synthesis [6]; nitrite can act as a circulating NO donor [7]; and nitrite can itself perform many actions previously attributable to NO [8] without the intermediacy of NO [9].

There are known strategies that will affect both pathways positively. Exercise has been shown to enhance endothelial production of L-arginine. Since the burden of exposure of nitrite and nitrate comes from the diet, eating a diet rich in green leafy vegetables or other NO_x-rich foods can fuel the second pathway. We believe that both systems complement one another. When we are young and healthy, the endothelial production of NO through L-arginine is efficient and sufficient to produce NO; however, as we age we lose our ability to synthesize endothelial-derived NO. Taddei et al. [10] have shown that there is a gradual decline in endothelial function due to aging with greater than 50% loss in endothelial function in the oldest age group tested as measured by forearm blood flow assays. Egashira et al. [11] reported more dramatic findings in the coronary circulation of aging adults whereby there was a loss of 75% of endothelium-derived nitric oxide in 70–80 year old patients compared to young, healthy 20 year olds. Vita and colleagues [12] demonstrated that increasing age was one predictor of abnormal endothelium-dependent vasodilation in atherosclerotic human epicardial coronary arteries. Gerhard et al. [13] concluded from their 1996 study that age was the most significant predictor of endothelium-dependent vasodilator responses by multiple stepwise regression analysis. Collectively, these important findings illustrate that endothelium-dependent vasodilation in resistance vessels declines progressively with increasing age. This abnormality is present in healthy adults who have no other cardiovascular risk factors, such as diabetes, hypertension, or hypercholesterolemia. Most of these studies found that impairment of endothelium-dependent vasodilation was clearly evident by the fourth decade. In contrast, endothelium-independent vasodilation does not change significantly with aging, demonstrating that the responsiveness to NO did not change, but only the ability to generate it did. These observations enable us to conclude that reduced availability of endothelium-derived nitric oxide occurs as we age, and to speculate that this abnormality may create an environment that is conducive to atherogenesis and other vascular disorders. It is that early event, the inability to produce sufficient NO under the right preclinical conditions, that enhances the risk for a number of diseases that plague the older population. If true, then there exists an opportunity to intervene early during this process, implement strategies to restore NO homeostasis, and, perhaps, delay or prevent the onset and progression of certain diseases. This gradual loss of NO activity with age can be sped up or slowed down based on individual lifestyle and diet. This idea is illustrated in the hypothetical graphical representation in Fig. 17.2. Adopting healthy habits such as a good diet and exercise can prolong the precipitous drop in NO production with age. To the contrary, a poor diet along with physical inactivity can accelerate the process and lead to a faster decline in NO production at a younger age. Therapeutic strategies directed at improving endothelial function or providing an alternative source of NO should be the primary focus because they may reduce the incidence of atherosclerosis or other diseases that occur with aging, even perhaps Alzheimer's Disease.

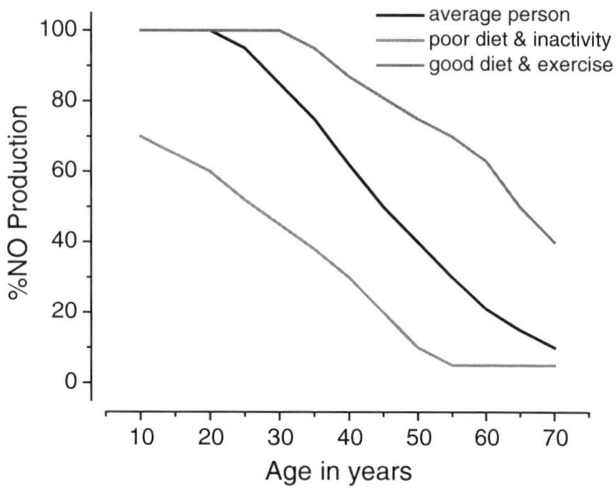

Fig. 17.2 Hypothetical representation of NO production based on diet and lifestyle

Exercise training has been shown, in many animal and human studies, to augment endothelial, NO-dependent vasodilatation in both large and small vessels (reviewed in [14, 15]); however, the response to exercise is diminished in patients with age-dependent endothelial dysfunction [16, 17]. In fact, plasma nitrite has been shown to predict exercise capacity [18] and further demonstrates that endothelial production of NO declines with age and increasing risk factor burden [19]. More recent data reveal that dietary supplementation with nitrate reduces oxygen costs of low-intensity exercise and enhances tolerance to high-intensity exercise [20–22]. This effect is due to an increase in plasma nitrite. These data support the notion that one can compensate through dietary nitrite and nitrate for the endothelial production of NO during exercise. This dietary pathway may also extend beyond exercise. Dietary supplementation of nitrite and nitrate in animals has been shown to reverse endothelial dysfunction, suppress microvascular inflammation, and reduce levels of C-reactive protein in mice subjected to a high cholesterol diet [23] and to protect from ischemia reperfusion injury [24–26]. This proof-of-concept has now been extended to humans supplemented with dietary sources of nitrate. Dietary nitrate has also been shown to reduce blood pressure [27–29], inhibit platelet aggregation [29], and restore endothelial function [29] in humans. What is clearly emerging is that there are two pathways for NO production, one through endothelial production via the L-arginine pathway and one through dietary sources of nitrite, nitrate, and anti-oxidants. This overview is illustrated in Fig. 17.3. The L-arginine pathway becomes dysfunctional with age, and we, therefore, need a back up system to compensate. Eating a diet rich in NO potential, i.e., sufficient nitrite and nitrate along with antioxidants to facilitate reduction to NO, can appear to overcome an insufficiency in endothelium-derived NO. This dietary pathway does not appear to be affected by age. However, overuse of antibiotics or antiseptic mouthwashes can affect this pathway by eliminating the commensal bacteria that are essential for the first step of nitrate reduction to nitrite. Furthermore, use of proton pump inhibitors

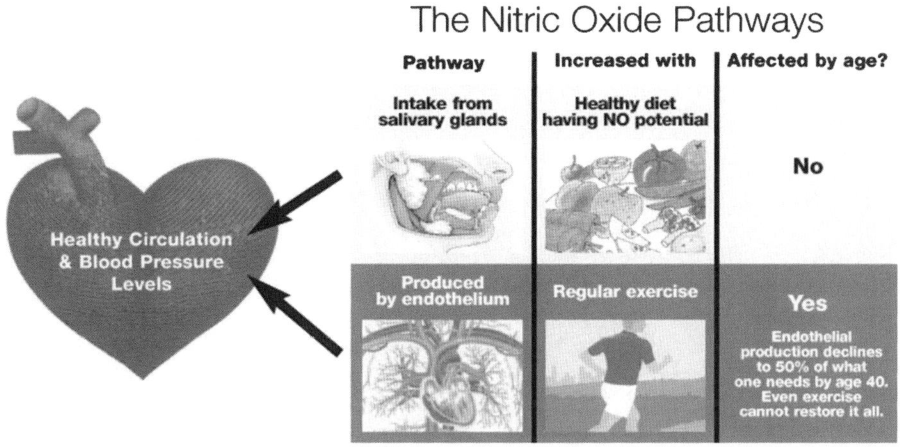

Fig. 17.3 The two pathways for NO production; one from the reduction of the nitrate and nitrite from the foods we eat and the other from the oxidation of L-arginine

can decrease the acid secretion in the stomach, thereby affecting the acidic disproportionation of nitrite to NO. This dietary pathway is reliant upon recognizing foods that are rich in NO potential. The inherent NO bioactivity of certain foods or diets is a delicate balance between nitrite and nitrate content as well as antioxidant capacity to facilitate reduction to NO and to inhibit any unwanted nitrosation reactions. Antioxidant status can be generically estimated by the Oxygen Radical Absorbance Capacity (ORAC) score. An antioxidant's strength reflects its ability to eliminate oxygen free radicals to prevent scavenging of nitric oxide as well as provide the reductive capacity to activate dietary nitrate and nitrite to NO. Foods with a high ORAC score are thought to protect cells and their components from oxidative damage. The ORAC score combined with the inherent nitrite and nitrate content of certain foods may provide a novel scoring system for NO potential or NO index. The Bryan lab has created a nitric oxide index below by applying an algorithm considering the nitrite/nitrate concentrations of foods as well as their reported ORAC values. We define the NO index below:

$$[(\text{Nitrite} + \text{nitrate}(\text{mg/100g}) \times \text{ORAC}(\mu\text{mol/100g}))] / 1,000 = \text{Nitric Oxide Index}$$

Similar to the glycemic index for diabetics, the nitric oxide index may be a useful tool for people with vascular disease or any conditions associated with NO insufficiency. A list of the NO index of select foods is included in Table 17.1. We suspect the context of a Nitric Oxide Index whereby nitrite and nitrate along with antioxidants will help define the health benefits of certain foods or diets.

Although dietary nitrite and nitrate have been shown to replenish blood and tissue stores of NO, it is still not clear how this pathway is regulated or controlled. The production of NO through the NOS enzymes is a precise, spatially, and temporally regulated, controlled event that generates NO in a local environment upon need or specific stimulus. Gladwin and colleagues have proposed that nitrite reduction

Table 17.1 Nitric Oxide Index of select foods

	Nitrite + nitrate (mg/100 g)	ORAC (μmol/100 g)	Nitric Oxide Index
High			
Rocket	612	2,373	1,452
Chicory	625	1,500	938
Wild radish	465	1,750	814
Bok choy	310	2,500	775
Beets	174	3,632	632
Spinach	388	1,515	588
Chinese cabbage	161	3,100	499
Lettuce	268	1,447	388
Cabbage	125	2,496	312
Mustard greens	116	1,946	226
Cauliflower, raw	202	829	167
Parsely	115	1,301	150
Kohlrabi	177	769	136
Carrot	190	666	127
Broccoli	39.5	3,083	122
Medium			
Cole slaw	55.9	1,500	84
Asparagus	50	1,644	82
Celery	160	497	80
Watercress	33	2,200	73
Artichoke	9.6	6,552	63
Eggplant	42	933	39
Strawberry	9.4	3,577	34
Potato	20	1,300	26
Garlic	3.4	5,708	19
Tomato	39.2	367	14
Vegetable soup	20.9	500	10
Cereal	4.9	2,000	10
Melons	68	142	10
Low			
String beans	30	300	9
Cured, dried sausage (uncooked)	78.8	100	8
Figs	2	3,383	7
Prunes	1	5,770	6
Sweet potato, raw, uncooked	5.4	902	5
Blackberries	1	5,347	5
Raspberries	1	4,882	5
Raisins	1.2	3,037	4
Banana	4.5	879	4
Cherries	1	3,365	3
Cucumber	14	214	3
Onions	3.2	1,034	3
Bean sprouts	1	1,510	2

(continued)

Table 17.1 (continued)

	Nitrite + nitrate (mg/100 g)	ORAC (μmol/100 g)	Nitric Oxide Index
Hot dog	9	100	1
Kiwi	1	882	1
Cucumber	9.3	100	1
Bacon	5.5	100	1
Apricot	1	1,115	1
Papaya	1	500	1
Chickpeas	1	847	1
Bacon, nitrite-free	3	100	0
French fries	2	150	0
Juices	mg/100 ml		
Beet root	279	1,727	482
Vegetables juice	20	548	11
Carrot	6.8	195	1
Pomegranate	1.3	2,681	3
Cranberry	1		0
Red wine	1	3,000	3
Green tea	0.02	1,253	0
Acai	0.06	1,767	0

to NO is allosterically linked to oxygen saturation of hemoglobin [30, 31], thereby providing a sensing mechanism for production of NO from nitrite during hypoxic vasodilation. Myoglobin has also been implicated in serving such a function in the heart [32]. More work is needed to determine if nitrite alone is sufficient and necessary to elicit specific cellular events as has been well described for NO. Although it may be argued that nitrite is simply a storage pool that can be reduced to NO under appropriate conditions, we consider this an unlikely role for nitrite in normal physiology, i.e., the normoxic and neutral pH conditions under which nitrite is stable. Alternatively, nitrite acts as an important molecule in its own right, but has regulatory effects on the NO pathway. It may be that both hypoxic reduction of nitrite to NO and normoxic metabolism of nitrite represent an advantageous oxygen-sensing system, which is a vestige of denitrifying microorganisms that existed long before the advent of aerobic respiration and the emergence of an NO synthase system [33]. The fact that both systems still exist today highlight the importance of nitrite in all cellular processes throughout the entire physiological oxygen gradient [8]. More research is needed in order to completely delineate the precise and regulated cell-signaling aspects of nitrite, both from endogenous sources as well as from dietary sources.

The role of diet in the prevention and control of morbidity and premature mortality due to non-communicable diseases has been well established by vast population-based epidemiological studies carried out during the last decade [34]. Nothing affects our health more than what we choose to eat. Nitric oxide is essential for maintaining normal blood pressure, preventing adhesion of blood cells to the endothelium, and preventing platelet aggregation; it may, therefore, be argued that this single abnormality, the inability to generate NO, puts us at risk for diseases that plague us later in life, such as atherosclerosis, myocardial infarction, stroke, and peripheral vascular disease. Therefore, developing strategies and new technologies designed to restore NO availability is essential for inhibiting the progression of certain common chronic diseases. The provision of dietary nitrate and nitrite may allow for such a strategy. The information presented throughout this text illustrates that the beneficial effects of nitrite and nitrate are seen at doses and levels that are easily achievable by eating certain foods enriched in nitrite and nitrate, particularly green leafy vegetables. The fact that exposure to nitrite and nitrate from certain foods or diets exceeds the World Health Organizations limit for acceptable daily intake (ADI) calls into question the current regulatory limits.

Hord and Bryan have shown that people following the Dietary Approaches to Stop Hypertension (DASH) diet exceed the ADI for nitrate by greater than 500% [35], and infants consuming human breast milk and some formulas can also exceed the regulatory limits [36]. The DASH diet was developed by the US National Institutes of Health to lower blood pressure without medication. The DASH diet is based on the research studies and has been proven to lower blood pressure, reduce cholesterol, and improve insulin sensitivity [37]. The DASH diet provides more than just the traditional low-salt or low-sodium diet plans to help lower blood pressure. It is based on an eating plan proven to lower blood pressure, a plan rich in fruits, vegetables, and low-fat or nonfat dairy. DASH diet is recommended by The National Heart, Lung, and Blood Institute (one of the National Institutes of Health of the US Department of Health and Human Services), The American Heart Association, The Dietary Guidelines for Americans, and US guidelines for treatment of high blood pressure, and the DASH diet formed the basis for the USDA MyPyramid. So how can such a diet exceed regulatory limits for the molecule that may be responsible for its blood-pressure-lowering effects? The same argument can be made for breast milk [36]. Human breast milk is recommended to serve as the exclusive food for the first 6 months of life and to continue, along with safe, nutritious complementary foods, for up to 2 years [38, 39]. Breast milk is nature's most perfect food. In fact, the US Centers for Disease Controls in 2010 acknowledged, "Breast milk is widely acknowledged as the most complete form of nutrition for infants, with a range of benefits for infants' health, growth, immunity, and development." Breast milk is a unique nutritional source for infants that cannot adequately be replaced by any other food, including infant formula. It remains superior to infant formula from the perspective of the overall health of both mother and child. Human milk is known to confer significant nutritional and immunological benefits for the infant [40–42], and yet, human breast milk and colostrum is enriched in nitrite and nitrate [36, 43–45]. It may be time to reconsider new regulatory guidelines for dietary sources of nitrite and nitrate given the last decade of research showing remarkable health benefits at levels that do not pose any significant risk.

We hope this work on nitrite and nitrate in human health and disease will provide the reader with a comprehensive body of knowledge on these molecules.

The last 20 years of research has drastically altered the landscape of how we think about nitrite and nitrate. The discovery of the nitric oxide pathway revealed that these two anions are naturally produced within our bodies and are not simply synthetic food additives. Prior to this discovery, much of the scientific research focused on the toxic properties due to exposure in both industrial settings and from cured and processed foods. Surprisingly, the context for nitrite and nitrate in human health and disease has not been adequately addressed since. We are pleased to be able to communicate this information for the Nutrition Series. Nutrition can play a key and cost-effective role in decreasing the risks of different chronic diseases. Identifying the key components of nutrition that may be responsible for these effects will help in refining dietary guidelines and designing optimal preventive nutrition regimens for specific disease. Future clinical studies will determine if nitrite and nitrate can provide a nutritional approach to prevention and/or treatment of specific diseases and if such positive effects will outweigh any negative health effects traditionally attributed to these anions. The ground is now fertile for such studies.

References

1. L'Hirondel JL. Nitrate and man: toxic, harmless or beneficial? Wallingford, UK: CABI; 2001.
2. Wagemakers JJ, Prynne CJ, Stephen AM, Wadsworth ME. Consumption of red or processed meat does not predict risk factors for coronary heart disease; results from a cohort of British adults in 1989 and 1999. Eur J Clin Nutr. 2007;63:303–11.
3. Schulman SP, Becker LC, Kass DA, Champion HC, Terrin ML, Forman S, et al. L-arginine therapy in acute myocardial infarction: the vascular interaction with age in myocardial infarction (VINTAGE MI) randomized clinical trial. JAMA. 2006;295(1):58–64.
4. Feelisch M, Fernandez BO, Bryan NS, Garcia-Saura MF, Bauer S, Whitlock DR, et al. Tissue processing of nitrite in hypoxia: an intricate interplay of nitric oxide-generating and -scavenging systems. J Biol Chem. 2008;283(49):33927–34.
5. Kleinbongard P, Dejam A, Lauer T, Rassaf T, Schindler A, Picker O, et al. Plasma nitrite reflects constitutive nitric oxide synthase activity in mammals. Free Radic Biol Med. 2003; 35(7):790–6.
6. McKnight GM, Smith LM, Drummond RS, Duncan CW, Golden M, Benjamin N. Chemical synthesis of nitric oxide in the stomach from dietary nitrate in humans. Gut. 1997;40(2):211–4.
7. Dejam A, Hunter CJ, Schechter AN, Gladwin MT. Emerging role of nitrite in human biology. Blood Cells Mol Dis. 2004;32(3):423–9.
8. Gladwin MT, Schechter AN, Kim-Shapiro DB, Patel RP, Hogg N, Shiva S, et al. The emerging biology of the nitrite anion. Nat Chem Biol. 2005;1(6):308–14.
9. Bryan NS, Fernandez BO, Bauer SM, Garcia-Saura MF, Milsom AB, Rassaf T, et al. Nitrite is a signaling molecule and regulator of gene expression in mammalian tissues. Nat Chem Biol. 2005;1(5):290–7.
10. Taddei S, Virdis A, Ghiadoni L, Salvetti G, Bernini G, Magagna A, et al. Age-related reduction of NO availability and oxidative stress in humans. Hypertension. 2001;38(2):274–9.
11. Egashira K, Inou T, Hirooka Y, Kai H, Sugimachi M, Suzuki S, et al. Effects of age on endothelium-dependent vasodilation of resistance coronary artery by acetylcholine in humans. Circulation. 1993;88(1):77–81.

12. Vita JA, Treasure CB, Nabel EG, McLenachan JM, Fish RD, Yeung AC, et al. Coronary vasomotor response to acetylcholine relates to risk factors for coronary artery disease. Circulation. 1990;81(2):491–7.
13. Gerhard M, Roddy MA, Creager SJ, Creager MA. Aging progressively impairs endothelium-dependent vasodilation in forearm resistance vessels of humans. Hypertension. 1996;27(4):849–53.
14. Green DJ, Maiorana A, O'Driscoll G, Taylor R. Effect of exercise training on endothelium-derived nitric oxide function in humans. J Physiol. 2004;561(Pt 1):1–25.
15. Ignarro LJ, Balestrieri ML, Napoli C. Nutrition, physical activity, and cardiovascular disease: an update. Cardiovasc Res. 2007;73(2):326–40.
16. Lauer T, Heiss C, Balzer J, Kehmeier E, Mangold S, Leyendecker T, et al. Age-dependent endothelial dysfunction is associated with failure to increase plasma nitrite in response to exercise. Basic Res Cardiol. 2008;103(3):291–7.
17. Allen JD, Miller EM, Schwark E, Robbins JL, Duscha BD, Annex BH. Plasma nitrite response and arterial reactivity differentiate vascular health and performance. Nitric Oxide. 2009;20(4):231–7.
18. Rassaf T, Lauer T, Heiss C, Balzer J, Mangold S, Leyendecker T, et al. Nitric oxide synthase-derived plasma nitrite predicts exercise capacity. Br J Sports Med. 2007;41(10):669–73; discussion 673.
19. Kleinbongard P, Dejam A, Lauer T, Jax T, Kerber S, Gharini P, et al. Plasma nitrite concentrations reflect the degree of endothelial dysfunction in humans. Free Radic Biol Med. 2006;40(2):295–302.
20. Larsen FJ, Weitzberg E, Lundberg JO, Ekblom B. Effects of dietary nitrate on oxygen cost during exercise. Acta Physiol (Oxf). 2007;191(1):59–66.
21. Bailey SJ, Winyard P, Vanhatalo A, Blackwell JR, Dimenna FJ, Wilkerson DP, et al. Dietary nitrate supplementation reduces the O_2 cost of low-intensity exercise and enhances tolerance to high-intensity exercise in humans. J Appl Physiol. 2009;107(4):1144–55.
22. Larsen FJ, Weitzberg E, Lundberg JO, Ekblom B. Dietary nitrate reduces maximal oxygen consumption while maintaining work performance in maximal exercise. Free Radic Biol Med. 2010;48(2):342–7.
23. Stokes KY, Dugas TR, Tang Y, Garg H, Guidry E, Bryan NS. Dietary nitrite prevents hypercholesterolemic microvascular inflammation and reverses endothelial dysfunction. Am J Physiol Heart Circ Physiol. 2009;296(5):H1281–8.
24. Bryan NS, Calvert JW, Elrod JW, Gundewar S, Ji SY, Lefer DJ. Dietary nitrite supplementation protects against myocardial ischemia-reperfusion injury. Proc Natl Acad Sci USA. 2007;104(48):19144–9.
25. Bryan NS, Calvert JW, Gundewar S, Lefer DJ. Dietary nitrite restores NO homeostasis and is cardioprotective in endothelial nitric oxide synthase-deficient mice. Free Radic Biol Med. 2008;45(4):468–74.
26. Shiva S, Sack MN, Greer JJ, Duranski MR, Ringwood LA, Burwell L, et al. Nitrite augments tolerance to ischemia/reperfusion injury via the modulation of mitochondrial electron transfer. J Exp Med. 2007;204(9):2089–102.
27. Larsen FJ, Ekblom B, Sahlin K, Lundberg JO, Weitzberg E. Effects of dietary nitrate on blood pressure in healthy volunteers. N Engl J Med. 2006;355(26):2792–3.
28. Kapil V, Milsom AB, Okorie M, Maleki-Toyserkani S, Akram F, Rehman F, et al. Inorganic nitrate supplementation lowers blood pressure in humans: role for nitrite-derived NO. Hypertension. 2010;56(2):274–81.
29. Webb AJ, Patel N, Loukogeorgakis S, Okorie M, Aboud Z, Misra S, et al. Acute blood pressure lowering, vasoprotective, and antiplatelet properties of dietary nitrate via bioconversion to nitrite. Hypertension. 2008;51(3):784–90.
30. Cosby K, Partovi KS, Crawford JH, Patel RK, Reiter CD, Martyr S, et al. Nitrite reduction to nitric oxide by deoxyhemoglobin vasodilates the human circulation. Nat Med. 2003;9:1498–505.

31. Huang Z, Shiva S, Kim-Shapiro DB, Patel RP, Ringwood LA, Irby CE, et al. Enzymatic function of hemoglobin as a nitrite reductase that produces NO under allosteric control. J Clin Invest. 2005;115(8):2099–107.
32. Hendgen-Cotta UB, Merx MW, Shiva S, Schmitz J, Becher S, Klare JP, et al. Nitrite reductase activity of myoglobin regulates respiration and cellular viability in myocardial ischemia-reperfusion injury. Proc Natl Acad Sci USA. 2008;105(29):10256–61.
33. Feelisch M, Martin JF. The early role of nitric oxide in evolution. Trends Ecol Evol. 1995;10(12):496–9.
34. World Health Organization. Report on diet, nutrition and the prevention of chronic diseases. Geneva: World Health Organization; 2003.
35. Hord NG, Tang Y, Bryan NS. Food sources of nitrates and nitrites: the physiologic context for potential health benefits. Am J Clin Nutr. 2009;90(1):1–10.
36. Hord NG, Ghannam JS, Garg HK, Berens PD, Bryan NS. Nitrate and nitrite content of human, formula, bovine and soy milks: implications for dietary nitrite and nitrate recommendations. Breastfeed Med. 2010 Oct 19. [Epub ahead of print].
37. Appel LJ, Moore TJ, Obarzanek E, Vollmer WM, Svetkey LP, Sacks FM, et al. A clinical trial of the effects of dietary patterns on blood pressure. DASH Collaborative Research Group. N Engl J Med. 1997;336(16):1117–24.
38. Heinig MJ. The American Academy of Pediatrics recommendations on breastfeeding and the use of human milk. J Hum Lact. 1998;14(1):2–3.
39. Gartner LM, Morton J, Lawrence RA, Naylor AJ, O'Hare D, Schanler RJ, et al. Breastfeeding and the use of human milk. Pediatrics. 2005;115(2):496–506.
40. Hoddinott P, Tappin D, Wright C. Breast feeding. BMJ. 2008;336(7649):881–7.
41. James DC, Lessen R. Position of the American Dietetic Association: promoting and supporting breastfeeding. J Am Diet Assoc. 2009;109(11):1926–42.
42. Ip S, Chung M, Raman G, Chew P, Magula N, DeVine D, et al. Breastfeeding and maternal and infant health outcomes in developed countries. Evidence report/technology assessment. Rockville, MD: Tufts-New England Medical Center Evidence-based Practice center, under Contract NO. 290-02-0022; 2007.
43. Iizuka T, Sasaki M, Oishi K, Uemura S, Koike M, Shinozaki M. Non-enzymatic nitric oxide generation in the stomachs of breastfed neonates. Acta Paediatr. 1999;88(10):1053–5.
44. Ohta N, Tsukahara H, Ohshima Y, Nishii M, Ogawa Y, Sekine K, et al. Nitric oxide metabolites and adrenomedullin in human breast milk. Early Hum Dev. 2004;78(1):61–5.
45. Cekmen MB, Balat A, Balat O, Aksoy F, Yurekli M, Erbagci AB, et al. Decreased adrenomedullin and total nitrite levels in breast milk of preeclamptic women. Clin Biochem. 2004;37(2):146–8.

Index

N.S. Bryan and J. Loscalzo (eds.), *Nitrite and Nitrate in Human Health and Disease*, Nutrition and Health, DOI 10.1007/978-1-60761-616-0,

Series Editor

Dr. Adrianne Bendich has recently retired as Director of Medical Affairs at GlaxoSmithKline (GSK) Consumer Healthcare where she was responsible for leading the innovation and medical programs in support of many well-known brands including TUMS and Os-Cal. Dr. Bendich had primary responsibility for GSK's support for the Women's Health Initiative (WHI) intervention study. Prior to joining GSK, Dr. Bendich was at Roche Vitamins Inc. and was involved with the groundbreaking clinical studies showing that folic acid-containing multivitamins significantly reduced major classes of birth defects. Dr. Bendich has co-authored over 100 major clinical research studies in the area of preventive nutrition. Dr Bendich is recognized as a leading authority on antioxidants, nutrition and immunity and pregnancy outcomes, vitamin safety and the cost-effectiveness of vitamin/mineral supplementation.

Dr. Bendich, who is now President of Consultants in Consumer Healthcare LLC, is the editor of ten books including **"Preventive Nutrition: The Comprehensive Guide For Health Professionals, Fourth Edition"** co-edited with Dr. Richard Deckelbaum, and is **Series Editor of "Nutrition and Health"** for Springer/

Humana Press (www.springer.com/series/7659). The Series contains 40 published volumes - major new editions in 2010 -2011 include **Vitamin D, Second Edition** edited by Dr. Michael Holick; **"Dietary Components and Immune Function"** edited by Dr. Ronald Ross Watson, Dr. Sherma Zibadi and Dr. Victor R. Preedy; **"Bioactive Compounds and Cancer"** edited by Dr. John A. Milner and Dr. Donato F. Romagnolo; **"Modern Dietary Fat Intakes in Disease Promotion"** edited by Dr. Fabien DeMeester, Dr. Sherma Zibadi, and Dr. Ronald Ross Watson; **"Iron Deficiency and Overload"** edited by Dr. Shlomo Yehuda and Dr. David Mostofsky; **"Nutrition Guide for Physicians"** edited by Dr. Edward Wilson, Dr. George A. Bray, Dr. Norman Temple and Dr. Mary Struble; **"Nutrition and Metabolism"** edited by Dr. Christos Mantzoros and **"Fluid and Electrolytes in Pediatrics"** edited by Leonard Feld and Dr. Frederick Kaskel. Recent volumes include: **"Handbook of Drug-Nutrient Interactions"** edited by Dr. Joseph Boullata and Dr. Vincent Armenti; **"Probiotics in Pediatric Medicine"** edited by Dr. Sonia Michail and Dr. Philip Sherman; **"Handbook of Nutrition and Pregnancy"** edited by Dr. Carol Lammi-Keefe, Dr. Sarah Couch and Dr. Elliot Philipson; **"Nutrition and Rheumatic Disease"** edited by Dr. Laura Coleman; **"Nutrition and Kidney Disease"** edited by Dr. Laura Byham-Grey, Dr. Jerrilynn Burrowes and Dr. Glenn Chertow; **"Nutrition and Health in Developing Countries"** edited by Dr. Richard Semba and Dr. Martin Bloem; **"Calcium in Human Health"** edited by Dr. Robert Heaney and Dr. Connie Weaver and **"Nutrition and Bone Health"** edited by Dr. Michael Holick and Dr. Bess Dawson-Hughes.

Dr. Bendich served as Associate Editor for "Nutrition" the International Journal; served on the Editorial Board of the Journal of Women's Health and Gender-based Medicine, and was a member of the Board of Directors of the American College of Nutrition.

Dr. Bendich was the recipient of the Roche Research Award, is a *Tribute to Women and Industry* Awardee and was a recipient of the Burroughs Wellcome Visiting Professorship in Basic Medical Sciences, 2000-2001. In 2008, Dr. Bendich was given the Council for Responsible Nutrition (CRN) Apple Award in recognition of her many contributions to the scientific understanding of dietary supplements. Dr Bendich holds academic appointments as Adjunct Professor in the Department of Preventive Medicine and Community Health at UMDNJ and has an adjunct appointment at the Institute of Nutrition, Columbia University P&S, and is an Adjunct Research Professor, Rutgers University, Newark Campus. She is listed in Who's Who in American Women.

About the Editors

Nathan S. Bryan, Ph.D.
Assistant Professor of Molecular Medicine
The University of Texas Health Science Center at Houston Houston, Texas

Dr. Nathan S. Bryan is an Assistant Professor of Molecular Medicine within the Brown Foundation Institute of Molecular Medicine, part of the School of Medicine at the University of Texas Health Science Center at Houston. He is also on faculty within the Department of Integrative Biology and Pharmacology and Graduate School of Biomedical Sciences at the UT Houston Medical School. Dr. Bryan earned his undergraduate Bachelor of Science degree in Biochemistry from the University of Texas at Austin and his doctoral degree from Louisiana State University School of Medicine in Shreveport where he was the recipient of the Dean's Award for Excellence in Research. He pursued his post-doctoral training as a Kirschstein Fellow at Boston University School of Medicine in the Whitaker Cardiovascular Institute. Dr. Bryan joined the Institute of Molecular Medicine, University of Texas Health Science Center in Houston, in June 2006 with his primary appointment within the Texas Therapeutics Institute. In 2007, he was recognized as one of the University's Most Accomplished Young Investigators. He is an active member of the Nitric Oxide Society, Society for Free Radical Biology and Medicine and the American Heart Association.

Dr. Bryan's research is dedicated to providing a better understanding of the interactions of nitric oxide and related metabolites with their different biological targets at the molecular and cellular level and the significance of these reactions for physiology and pathophysiology. Attempts are made to identify what particular changes in NO-related signaling pathways and reaction products occur in disease states such as endothelial dysfunction, ischemia/reperfusion, tissue/cardiac protection, diabetes, atherosclerosis and inflammation with the aim of testing their amenability as biomarkers for diagnosis and/or treatment of specific disease. Current research is directed to understand the interactions of exogenous dietary nitrite/nitrate (NOx) on the endogenous NO/cGMP pathway and how perturbations in each system affect cardiovascular health. Dr. Bryan and colleagues recently discovered that nitrite is a biologically active molecule which was previously thought to be an inert breakdown product of NO production. These findings have unveiled many beneficial effects of nitrite in the treatment and prevention of human disease. These discoveries may provide the basis for new preventive or therapeutic strategies in diseases associated with NO insufficiency and new guidelines for optimal health. Dr. Bryan has published a number of highly cited papers and authored or edited 4 books.

Joseph Loscalzo, M.D., Ph.D.
Hersey Professor of the Theory and Practice of Medicine,
Harvard Medical School,
Chairman, Department of Medicine and Physician-in-Chief,
Brigham and Women's Hospital, Boston, Massachusetts

Dr. Joseph Loscalzo is Hersey Professor of the Theory and Practice of Medicine at Harvard Medical School, and Chairman of the Department of Medicine, and Physician-in-Chief at Brigham and Women's Hospital. Dr. Loscalzo received his A.B. degree, summa cum laude, his Ph.D. in biochemistry, and his M.D. from the University of Pennsylvania. His clinical training was completed at Brigham and Women's Hospital and Harvard Medical School, where he served as Resident and Chief Resident in medicine and Fellow in cardiovascular medicine.

After completing his training, Dr. Loscalzo joined the Harvard faculty and staff at Brigham and Women's Hospital in 1984. He rose to the rank of Associate Professor of Medicine, Chief of Cardiology at the West Roxbury Veterans Administration Medical Center, and Director of the Center for Research in Thrombolysis at Brigham and Women's Hospital. He joined the faculty of Boston University in 1994, first as Chief of Cardiology and, in 1997, Wade Professor and Chair of Medicine, Professor of Biochemistry, and Director of the Whitaker Cardiovascular Institute. He returned to Harvard and Brigham and Women's Hospital and Harvard Medical School in 2005.

Dr. Loscalzo is recognized as an outstanding cardiovascular scientist, clinician, and teacher. He has received many awards, including the Clinician-Scientist Award, the Distinguished Scientist Award, the Research Achievement Award, and the Paul Dudley White Award from the American Heart Association; a Research Career Development Award, a Specialized Center of Research in Ischemic Heart Disease Award, and a MERIT Award from the National Institutes of Health; the George W. Thorn Award for Excellence in Teaching at Brigham and Women's Hospital, and Educator of the Year Award in Clinical Medicine from Boston University; the

Glaxo Cardiovascular Research Award, and the Outstanding Investigator Prize from the International Society for Heart Research; and election to the American Society for Clinical Investigation, the Association of American Physicians, and the Institute of Medicine of the National Academy of Sciences. He has served on several NIH study sections and editorial boards, and has chaired the Gordon Conference on Thrombolysis. He served as an associate editor of the New England Journal of Medicine for nine years, Chair of the Cardiovascular Board of the American Board of Internal Medicine, Chair of the Research Committee of the American Heart Association, Chair of the Scientific Board of the Stanley J. Sarnoff Society of Fellows for Research in the Cardiovascular Sciences, and Chair of the Board of Scientific Counselors of the National Heart, Lung, and Blood Institute of the National Institutes of Health. He is currently Editor-in-Chief of Circulation, a senior editor of Harrison's Principles of Internal Medicine, a recent member of the Advisory Council of the National Heart, Lung, and Blood Institute, and a recent member of the Council of Councils of the National Institutes of Health.

Dr. Loscalzo has been a visiting professor at many institutions, holds two honorary degrees, has authored or co-authored more than 600 scientific publications, has authored or edited 27 books, and holds 31 patents for his work in the field of nitric oxide. He is also the recipient of many grants from the NIH and industry for his work in the areas of vascular biology, thrombosis, and atherosclerosis over the past twenty-five years.